Journal of Applied Logics - IfCoLog
Journal of Logics and their Applications

Volume 8, Number 2

March 2021

Disclaimer

Statements of fact and opinion in the articles in Journal of Applied Logics - IfCoLog Journal of Logics and their Applications (JALs-FLAP) are those of the respective authors and contributors and not of the JALs-FLAP. Neither College Publications nor the JALs-FLAP make any representation, express or implied, in respect of the accuracy of the material in this journal and cannot accept any legal responsibility or liability for any errors or omissions that may be made. The reader should make his/her own evaluation as to the appropriateness or otherwise of any experimental technique described.

ISBN 978-1-84890-360-9
ISSN (E) 2631-9829
ISSN (P) 2631-9810

College Publications
Scientific Director: Dov Gabbay
Managing Director: Jane Spurr

http://www.collegepublications.co.uk

EDITORIAL BOARD

SCOPE AND SUBMISSIONS

This journal considers submission in all areas of pure and applied logic, including:

pure logical systems
proof theory
constructive logic
categorical logic
modal and temporal logic
model theory
recursion theory
type theory
nominal theory
nonclassical logics
nonmonotonic logic
numerical and uncertainty reasoning
logic and AI
foundations of logic programming
belief change/revision
systems of knowledge and belief
logics and semantics of programming
specification and verification
agent theory
databases

dynamic logic
quantum logic
algebraic logic
logic and cognition
probabilistic logic
logic and networks
neuro-logical systems
complexity
argumentation theory
logic and computation
logic and language
logic engineering
knowledge-based systems
automated reasoning
knowledge representation
logic in hardware and VLSI
natural language
concurrent computation
planning

This journal will also consider papers on the application of logic in other subject areas: philosophy, cognitive science, physics etc. provided they have some formal content.

Submissions should be sent to Jane Spurr (jane@janespurr.net) as a pdf file, preferably compiled in LaTeX using the IFCoLog class file.

CONTENTS

ARTICLES

vii

Some Remarks on Assertion and Proof

Massimiliano Carrara

*Department of Philosophy, Sociology, Education and Applied Psychology
(FISPPA) – Padua University*
massimiliano.carrara@unipd.it

Daniele Chiffi

DAStU, Politecnico di Milano
daniele.chiffi@polimi.it

Ciro De Florio

Faculty of Economics, Università Cattolica, Milan
ciro.deflorio@unicatt.it

Abstract

In our *introduction* we make some remarks on the main topics of this issue:
assertion and proof. We briefly describe how each of the papers in the present
publication has contributed from either different or complementary perspectives
to the logical reflection on assertion and proof, while also specifying the relation
between them.

1 Introductory remarks

It may sound like a philosophical *cliché*, but one could not stress enough the importance of the notions of assertion and proof for logic, philosophy of logic and philosophy of language.

Although these two notions have been undergoing development since the second half of the 19th century in a relatively independent way within research programs in logic and linguistics alike, the conceptual relationships between them are undeniable. In this short contribution to the Special Issue (**Assertion and Proof**) we will illustrate some of the (possible) links between proof and assertion.

With "assertion" we denote *prima facie* at least two rather different entities; the first is a kind of act, i.e. an illocutionary act, namely the act of asserting something;

the other entity is the outcome of the same act, that is, the asserted thing. We will see that such *prima facie* duality of assertion is reflected on its logical treatment.

Consider a proposition. One and the same proposition can occur both asserted and unasserted in different contexts. In particular, Geach [10, p. 449] pointed out that "[a] thought may have just the same content whether you assent to its truth or not; a proposition may occur in a discourse now asserted, now unasserted, and yet be recognizably the same proposition". A standard example is the justification of *modus ponens:* assuming (1) $\alpha \to \beta$ and (2) α, one infers (3) β. In (2), α is usually considered asserted, while in (1), α is not asserted, because only the whole conditional $\alpha \to \beta$ is asserted (on this, see Russell 1903). This means that the very same proposition may be used in both its asserted and unasserted forms[1].

Moreover, the same line of thought could be, somehow, extended: within any argumentative structure, the same propositions may be the subject of various illo-cutionary acts. For instance, one could assume that φ, then hypothesize that ψ, conjecture that θ, and so on.

In the light of previous considerations, one can also ask how to provide the expressive resources in order to describe the formal features of assertions. In a letter to Frege, Peano wrote that the various positions in which a formula can be placed determines whether it occurs asserted or unasserted in some truth-functional context. Using Peano's words, "the several positions that a proposition can have in a formula completely determine what is asserted of it" [12, p. 191]. Frege observed that this is because "the principal relation sign invariably carries assertoric force" [8, p. 148] without any specific sign for assertion being present in the notation. This means that in Peano's notation it is impossible to show whether a complex formula occurs asserted or unasserted.

In this regard, Frege introduced an *ad hoc* sign of assertion as a notational requirement of the logical language. He indicated it with "⊢", which stands for the sign of assertion. (Nowadays, the sign "⊢", has acquired the name "turnstile" and expresses the concept of derivability or provability.)

Once given the expressive resources in order to describe assertions, it becomes crucial to provide some criteria concerning the logical behaviour of assertions. From this point of view, it is normally assumed that assertions, intended as acts, cannot be iterated and cannot be connected by truth-functional connectives [15]. Reichenbach

---

[1]Bell [1] comments Russell's views on *modus ponens* in the following way: "Now this would imply that either all inferences having the *modus ponens* form (to take but one example) are invalid, or, at least, that all those with either a true antecedent or a true consequent in the conditional premiss are invalid. This is, of course, quite unacceptable" (pp. 87-88). On the analysis of Russell's views on *modus ponens* and Bell's untoward consequences, see the justification of *modus ponens* in pragmatic logic [3].

[15, p. 346] claims this on the basis of the fact that the term "assertion" is used in three different ways. Namely, "it denotes, first, the act of asserting; second, the result of this act, i.e., an expression of the form '⊢p'; third, a statement which is asserted, i.e. a statement 'p' occurring within an expression '⊢p'". Regarding the result of an assertion, Reichenbach claims that "since assertive expressions are not propositions, they cannot be combined by propositional operators" [15, §57, p. 337]. The assertion sign works, according to him, in its "pragmatic capacity", since it cannot be, for instance, negated with a propositional connective. And if this is so, then inferences can be understood as processes that allow us to derive justified asserted conclusions once the asserted premises are also justified. This means that there can be no nested or iterated occurrences of the assertion sign, because the truth-functional connectives only operate on propositions and never on assertions. Furthermore, inferences operate only on assertions and never on propositions.

The structure of an elementary assertion is thus composed by a unique assertion sign prefixed to the asserted content. This is, for instance, the case with Frege's [8] *Begriffsschrift*. A similar restriction is presented in Reichenbach's treatise, where the distinction between assertions and (propositional) content is developed.

Of course, the logic of assertions must engage with the possibility of connecting elementary assertions in order to construct more complex ones. At play here is in fact the intertwining between the logical form of asserted contents and the logic of assertions. It is plausible to claim that asserting a conjunction $(\varphi \wedge \psi)$ is equivalent to juxtaposing two elementary assertions, respectively of φ and ψ.

Things are not always so easy. Asserting a disjunction $(\varphi \vee \psi)$ does not seem to be equivalent to asserting φ or asserting ψ. And the same holds for implication. The logic of assertions is, thus, more complex and, somehow, different from the logic dealing with the asserted content.

This discrepancy becomes more evident if we look at Dalla Pozza's system [5]. Within it, complex assertions may be logically combined by an application of intuitionistic-like connectives. This means that complex asserted formulas may be expressed through connectives that explicate intuitionistic meanings of logical constants, without a truth-functional behaviour. Moreover, intuitionistic connectives can indicate the (pragmatic) justification-conditions for (acts of) assertions.

On the basis of this short overview on some fundamental features of the logic of assertion, we are able to cast some light on the link between the concept of assertion and the concept of proof.

As we said, when assertions are intended as acts, they are neither true nor false. Therefore, it is quite natural to employ a non-truth-functional kind of semantics in order to construe the formula of a logic for assertion. A natural candidate is the concept of *proof*: an assertion is justified (or unjustified) depending on the existence (or not) of a proof of the asserted content. The reference to the concept of proof emphasizes the constructive feature of the logic of assertions.

By way of example, it is interesting to notice the justification clause pertaining to the assertion according to Dalla Pozza and Garola's [5] approach: in that system, the implication between two assertions ($\vdash \varphi \supset \vdash \psi$) is justified if and only if we have a proof which transforms any proof of φ into a proof of ψ at our disposal. It is, thus, clear why the outer logic of assertion can be different from the inner logic of asserted contents. We can, for instance, be justified in asserting every instance of excluded middle ($\varphi \lor \neg\varphi$) without being justified either to assert φ or to assert $\neg\varphi$. It is sufficient to assume that φ describes a proposition for which we do not have conclusive proofs.

We said that an assertion is justified if and only if we have a proof at disposal. But what is a *proof*? Here below we propose some introductory remarks. Let us start by observing that this is a central notion of the *proof-theorethic approach to logic*. According to this, a consequence is identified with *deducibility*: an argument $\langle \Gamma \therefore \alpha \rangle$ is valid if and only if there exists a *proof* or a derivation of α from Γ, each of whose steps is intuitively sound. The account is formal, insofar as logical consequence is identified with derivability in a system of rules of a certain form. *Proof-theoretic semantics* is standardly taken as an alternative to *truth-condition semantics*. In a nutshell, *proof-theoretic semantics* is based on the assumption that the meanings of the logical constants are assigned in terms of *proof* and of their inferential role rather than in terms of *truth*.

Proof-theoretic accounts of consequence are sometimes quickly dismissed. Field, for example, writes that "proof-theoretic definitions proceed in terms of some definite proof procedure", and observes that "it seems pretty arbitrary which proof procedure one picks" and "it isn't very satisfying to rest one's definitions of fundamental metalogical concepts on such highly arbitrary choices" [7, p. 2]. Etchemendy similarly observes that "the intuitive notion of consequence cannot be captured by any *single* deductive system" [6, p. 2], since the notion of consequence is neither tied to any particular language, nor to any particular deductive system.

In order to understand what a proof is one can first specify a notion of *formal proof* or of *informal proof*. The formalisation of the idea of *proof* as a *given set of sequences of symbols* underlies the meta-mathematical research pioneered by Hilbert and Bernays and subsequently developed by Gödel, Gentzen and others. Boolos [2] famously explored the intensional representation of formal proof through systems of modal logic in which ($\Box$) is interpreted as "it is provable that" in a formal sense. A characteristic axiom (called the "Gödel-Löb axiom") of the notion of formal provability is $\Box(\Box\varphi \rightarrow \varphi) \rightarrow \Box\varphi$. If this axiom is added to the modal system K4, we obtain the modal logic G. Such a system was formalized by Solovay [17]. It is complete with respect to transitive and conversely well-founded frames. In this system the reflection principle $\Box\varphi \to \varphi$ does not hold.

Things are different in case of *informal proofs* (intended, for instance, as good mathematical arguments) which are usually assumed to justify truth, thus accepting the reflection principle.

The notion of *informal* or *naïve proof* received some attention when Gödel, in the *Gibbs lecture* [11], asserted his famous *dichotomy* concerning the nature of the human mind. Then, Priest characterised a naïve proof as a process of deductive argumentation by which one establishes certain mathematical claims to be true. So, supposing there is a mathematical assertion whose truth or falsity is to be established, one can look for a proof or a refutation to justify it or not. The informal deductive arguments from basic statements are, according to Priest, "naïve proofs" [14, 40]. It is interesting to observe that Priest [13], in his "The logic of paradox," developed a controversial argument, grounded in the notion of naïve proof, showing some possible connections between Gödel's first incompleteness theorem and the presence of *dialetheias* (viz., sentences that are both true and false) in the standard model of arithmetic. This last point has been criticized especially regarding the notion of naïve proof itself.

2   The contents of the issue

Each of the papers in this special issue contributes from different and complementary perspectives to the logical reflection on assertion and proof as well as on the relations between these two concepts.

Barés Gómez and Fontaine in *Defeasibility and non-monotonicity in dialogues* show how to introduce the notions of defeasibility and non-monotonicity in dialogical logic, and discuss them in a framework of adaptive dialogical logics.

Bellucci, Chiffi and Pietarinen in *Beta assertive graphs: Proofs of assertions with quantification* introduce and investigate quantification in the diagrammatic system of assertive graphs.

Carrara and Strollo in **DLEAC** *and the rejection paradox* develop a Dialetheic Logic with exclusive assumptions and conclusions, both understood as speech acts. A new paradox – the *rejectability paradox* – is (first informally, then formally) introduced. Its derivation is possible in an extension of **DLEAC** contanining the *rejectability* predicate.

Chiffi in *Asserting boo! and horray! Pragmatic logic for assertion and moral attitudes* proposes a pragmatic logic for expressivist moral attitudes in order to deal with the logical problems of expressivism such as the Frege-Geach problem, the negation problem, etc. The second part of the paper makes some analytic comparisons with other classical logical systems for expressivist sentences.

D'Agostino, Larese and Modgil in *Towards depth-bounded natural deduction for classical first-order logic* present a new proof-theory for classical first-order logic that allows for a natural characterization of a notion of inferential depth. Unlike natural deduction, in this framework the rules fixing the meaning of the logical operators are symmetrical with respect to assent and dissent and do not involve the discharge of formulas.

De Florio in *Reflections on logics for assertion and denial* discusses and refines the justification conditions for assertion and denial in an extension of Dalla Pozza's pragmatic logic.

Fait and Primiero in *HTLC: hyperintensional typed lambda calculus* introduce a new logical system termed "HTLC". The system extends the typed lambda-calculus with hyperintensions and rules to govern them. This allows us to reason with expressions for extensional, intensional and hyperintensional entities.

Francez in *Bilateralism based on corrective denial* presents a new variant of bilateralism based on a strong notion of denial, called "corrective denial". In this framework, a ground for denial is an incompatible atomic alternative to the denied formula.

Jespersen in *Two tales of the turnstile* criticizes, from a hyperintensional perspective, the view held by act-theoretic 'internalists' who invert the Frege-Geach point by making force integral to content.

Kürbis in *Normalisation for bilateral classical logic with some philosophical remarks* presents two bilateral connectives, comparable to Prior's *tonk*, for which, unlike the case of *tonk*, there are reduction steps for the removal of maximal formulas, arising from introducing and eliminating formulas with those connectives as main operators.

Lemanski in *Extended syllogistics in calculus CL* addresses the question regarding to what extent a syllogistic representation in *CL* (Lange's *Cubus Logicus*) diagrams can be seen as a form of extended syllogistics. The author shows that the ontology of CL enables numerically exact assertions and inferences.

Morato in *Assertions of counterfactuals and epistemic irresponsibility* discusses the so-called "reverse Sobel sequences", problematic for the variably strict semantics for counterfactuals. Morato shows, in particular, some limitations of the "principle of epistemic irresponsibility", which is assumed to ground the pragmatic view on this type of counterfactual sequences.

Finally, Schang in *A general semantics for logics of affirmation and negation* proposes some semantic considerations on the notions of affirmation and negation that may help us understand the possible translations among different logical systems.

Acknowledgements

The papers collected in the present issue have been presented at the international conference "Assertion and Proof", held in Lecce (Italy), 12-14 September 2019, https://sites.google.com/view/assertionproof. A *book of abstracts* of the conference has been published and it is available at this link: `https://sites.google.com/view/assertionproof/book-of-abstracts?authuser=0`. We would like to thank Caterina Annese for her extremely valid help in organizing the conference.

References

[1] Bell, D. (1979). *Frege's Theory of Judgment.* New York: Oxford University Press.

[2] Boolos, G (1995). *The Logic of Provability.* Cambridge: Cambridge University Press.

[3] Chiffi, D. & Di Giorgio, A. (2017). Assertions and Conditionals: A Historical and Pragmatic Stance. *Studia Humana,* 6(1): 25-38.

[4] Dalla Pozza, C. (1991). Un'interpretazione pragmatica della logica proposizionale intuizionistica. In G. Usberti (Ed.), *Problemi fondazionali nella teoria del significato.* Firenze: Leo S. Olschki.

[5] Dalla Pozza, C., Garola, C. (1995). A Pragmatic Interpretation of Intuitionistic Propositional Logic. *Erkenntnis,* 43(1): 81-109.

[6] Etchemendy, J. (1990). *The Concept of Logical Consequence.* CSLI Stanford.

[7] Field, H. (1991). Metalogic and Modality. *Philosophical Studies,* 62(1), 1-22.

[8] Frege, G. (1879), *Begriffsschrift, eine der arithmetischen nachgebildete Formelsprache des reinen Denkens.* Halle: L. Nebert. Translated as *Begriffsschrift, a Formula Language, Modeled upon that of Arithmetic, for Pure Thought.* In *From Frege to Gödel,* edited by Jean van Heijenoort. Cambridge, MA: Harvard University Press, 1967.

[9] Frege, G. (1991). *Collected Papers on Mathematics, Logic, and Philosophy,* edited by B. McGuinness. Oxford: Blackwell.

[10] Geach, P.T. (1965). Assertion. *The Philosophical Review,* 74(4): 449-465.

[11] Gödel, K. (1951). Some Basic Theorems in the Foundations of Mathematics and Their Implications. In S. Feferman et. al., editors, *Collected works,* volume III, pp. 304-323. Oxford: Oxford University Press.

[12] Peano, G. (1958). *Opere scelte.* Volume II. *Logica matematica. Interlingua ed algebra della grammatica.* Roma: Edizioni Cremonese.

[13] Priest, G. (1979). The Logic of Paradox. *Journal of Philosophical Logic,* 8:219–241.

[14] Priest, G. (2006). *In Contradiction.* Oxford: Oxford University Press. Expanded edition (first published in 1987 Dordrecht: Kluwer).

[15] Reichenbach, H. (1947). *Elements of Symbolic Logic*. London: Macmillan.

[16] Russell B. (1903). *The Principles of Mathematics*. Cambridge: Cambridge University Press.

[17] Solovay, R.M. (1976). Provability Interpretations of Modal Logic. *Israel Journal of Mathematics*, 25: 287-304.

Received 16 December 2020

Defeasibility and Non-Monotonicity in Dialogues

Cristina Barés Gómez
Universidad de Sevilla
crisbares@gmail.com

Matthieu Fontaine
Universidad de Sevilla
fontaine.matthieu@gmail.com

Abstract

Although dialogical logic was originally defined to model deductive reasoning, in particular intuitionistic logic, it may be useful to model other kinds of inferences. Dialogical logic should include the possibility of involving some kind of defeasibility, whether it be at the play level or at the level of the strategies. Whereas the former only involves the application of rules of interactions, the latter is concerned with the notion of validity. Is it possible to introduce defeasibility in Dialogical Logic? According to Dutilh Novaes [7], monotonicity and non-defeasibility are consequences of a strategic requisite inherent to dialogical games. But, according to Rahman et al. [18], this position relies on a confusion between the play and the strategy levels. Actually, the rules of dialogical games do not involve any strategic component. As a consequence, there is room for defeasibility and non-monotonicity in dialogues. We finally discuss this possibility in the context of recent developments of adaptive dialogical logics, in particular *IAD* of Beirlaen and Fontaine [3], and put forward a distinction between a notion of defeasible move defined at the play level and a notion of dialogical non-monotonicity defined at the strategy level.

We are particularly thankful to Caterina Annese (Zei, Lecce), Daniele Chiffi (Politecnico di Milano), Massimiliano Carrara (University of Padova), and all the participants of the "Workshop: Assertion and Proof" (Lecce, Italy, September 2019) for their helpful comments. We also thank Angel Nepomuceno, Olga Pombo and Shahid Rahman for their support and their advice. Matthieu Fontaine acknowledges financial support of the postdoctoral program of the FCT-Portugal (Grant Number SFRH/BPD/116494/2016) and the financial support of FCT CFCUL UID/FIL/00678/2019. Cristina Barés Gómez acknowledges the support of VPPI-US (Contrato de acceso al Sistema Espanol de Ciencia, Tecnología e Innovación para el desarrollo del programa propio I+D+i de la Universidad de Sevilla).

1 Introduction

As stressed by Pollock [15, p. 481], a "common misconception about reasoning is that reasoning is deducing, and in good reasoning the conclusions follow logically from the premises." Indeed, it is widely recognized that a good deal of human reasoning and argumentative practices lies beyond deductive reasoning. We often infer conclusions defeasibly; that is, by being prepared to revise them in the light of new information. In the context of his naturalized logic, Woods [21] explains that a theory of inference must take into account scant resources strategies, linked to limited memory, computational skills, time, access to information, and so on. This brings us to the field of cognitive economy, from the perspective of which mathematical rigor of deductive reasoning may not be the best standard of inference[1]: Less costly forms of reasoning are sometimes more apt and efficient than research of accurate results. Depending on the rational agents and their targets, it is often a more efficient strategy to defeasibly infer conclusions, on which we can act despite a lack of precision, and to pursue strategies of management of errors, than to refrain from inferring and acting at all. If human beings learn through errors and corrections processes, as it is sometimes (if not usually) the case, then a theory of inference must also be a theory of defeasible reasoning.

Although there is no consensus with respect to its definition, we can say that an argument is defeasible if its premises provide support for the conclusion, even though it is possible for the premises to remain true and the conclusion to be revised in the light of new information. Formally, we may approach defeasibility through logics that do not respect the monotony property; i.e. if $\Sigma \vdash \varphi$, then $\Sigma \cup \Sigma' \vdash \varphi$ (where $\Sigma \subseteq \Sigma'$). In this paper, we discuss the possibility of approaching such notions in the context of dialogical logic by taking advantage of the distinction between the play level (the level of interaction) and the strategy level (the level of inference).

Dialogical games were initially defined for intuitionistic logic (see Lorenzen and Lorenz [12]. Studies in dialogues took a pluralist turn with Rahman and his collaborators.[2] Excepted from rare exceptions, most of them focus on deductive logic (classical, intuitionistic, free logics, modal logics, and so on). If we wish to conceive dialogical games as a framework for actual human reasoning, they should also include non-deductive reasoning; in particular defeasible reasoning. In this paper, we discuss this possibility. Our thesis is that dialogical games are not intrinsically deductive and hermetic to defeasibility. Our argument is based on the distinction between the play level and the strategy level. In general, defeasibility can be grasped

---

[1]Following Peirce [14, CP 5.602], we may also take into account economic parameters of costs, risks and benefits.
[2]See e.g. Rahman and Keiff [17].

at the play level. The Inconsistency-Adaptive Dialogical Logic (IAD) of Beirlaen and Fontaine [3] also displays non-monotonicity at the strategy level, thanks to the implementation of specific rules.

Nonetheless, as we will see in the Section 2, it has been recently argued by Dutilh Novaes [7] that dialogical games are intrinsically indefeasible. But, according to Rahman et al. [18], this thesis relies on a confusion between the play level and the strategy level. This distinction fundamental in the construction of dialogues, and more generally the dialogical theory of meaning, is the subject of the Section 3. Finally, in the Section 4, we consider dialogical defeasibility and non-monotonicity in the context of IAD, in the light of these different levels of dialogical games. The dialogical rules and definitions are given in an Appendix (Section 6).

2 Dialogues and the BIO

Dialogues are games of argumentation between the Proponent (**P**) of a thesis and the Opponent (**O**). The game begins with **P** uttering the thesis that he has to defend against every possible criticism of **O**. Moves in the games, performed by means of questions or assertions, are either challenges against previously uttered statements or defences in response to challenges. The game is governed by two kinds of rules –the particle rules (see Appendix 6.2) and the structural rules (see Appendix 6.3)– that define the play level; i.e. they are definitory rules that indicate how to play, but not how to win. The particle rules provide the local meaning of logical constants in terms of interaction. They are abstract descriptions consisting of sequences of moves such that the first member is an assertion, the second is an attack and the third is a defence (when possible). Given that it is assumed that both players speak the same language (otherwise dialogues would not make sense), they are the same for both players. The structural rules determine the general organisation: how to begin, who plays, when, who wins, and so on. They provide the global meaning of the statements uttered in a dialogical game; that is, their meaning in a specific context of argumentation. For example, by means of the rules [**SR1c**] and [**SR1i**] we can distinguish between classical and intuitionistic games. By means of [**SR5.1**], we specify the use of the particle rule for negation in the context of IAD. One of the fundamental rules of dialogical logic is the Formal Rule [**SR2**], which says that only **O** can introduce new atomic formulas. An initial thesis φ uttered by **P** is claimed to be valid if and only if there is a **P**-winning strategy for φ; i.e. **P** can win the game no matter how **O** plays. For example, **P** wins the following game, but this does not mean that there is a winning strategy:

Example 1					
	O			**P**	
				$(p \lor q) \to (p \land q)$	0
1	$n := 1$			$m := 2$	2
3	$p \lor q$	0		$p \land q$	4
5	$?L^{\land}$	4		p	8
7	p	3		$?_{\lor}$	6

Explanation: In this play, **P**'s win depends on **O**'s (bad) choice of p (move 7). If **O** had chosen q, he would have won the play. Therefore, there is no winning strategy for **P**. The initial thesis is not valid.

In order to determine validity, it might seem that it must be assumed that **O** always performs the best choices. According to Dutilh Novaes [7, p. 602], "[w]hat starts as a strategic but not mandatory component of the dialogical game –putting forward indefeasible arguments– then becomes a constitutive structural element of the deductive method as such: only indefeasible arguments now count as correct moves in a deductive argument." She holds the thesis that the standard notion of logical truth has internalized in monological practices the role of the Opponent as an ideal interlocutor who seeks to defeat the argument by showing a case in which the premises are true but the conclusion is false. This is what she calls the built-in opponent conception (BIO) of logic and deduction.

Indeed, the standard view of logic is that logic is normative for correct thinking, by providing a criterion for deductive validity in terms of necessary truth preservation (NTP). But how should we understand NTP? Several attempts to link NTP with normative claims, by means of bridge principles, are unsatisfactory (see Dutilh Novaes [7, p. 591 ff]. She pretends that, rather than rules for correct thinking, logic has normative import for "specific situations of dialogical interaction" [7, p. 588]. These dialogical interactions can be represented by dialogical games beginning with the Proponent uttering the thesis and the Opponent trying to defeat the thesis by means of a countermove or by exhibiting a counter-model. If the Opponent cannot succeed in defeating the Proponent's argument, then the thesis is valid. According to the BIO conception of logic and deduction, it is this role of the Opponent that has been encapsulated, internalized, in the standard notion of deductive validity defined in terms of NTP.

According to Dutilh Novaes, the Proponent's thesis is valid if the Opponent cannot find a countermove to block the argument. So, although she identifies two components, one cooperative and one adversarial, dialogues are reduced to an inquiry into logical truths. By focusing on this inquiry into logical truths, the emphasis is put on the Opponent's role, trying to defeat the Proponent's argument. As she conceives dialogues, they assume that the Opponent always performs optimal moves. We are led to the conclusion that dialogues intrinsically involve a strategic component. Whereas the Proponent tries to establish the conclusion, the Opponent tries to block the establishment of the conclusion. This adversarial component accounts for the NTP; that is, the Opponent tries to defeat the argument, which is valid if and only if it cannot be defeated, if it is indefeasible. According to Dutilh Novaes [7, p. 597], NTP may have emerged as a strategic component in dialectical games. The BIO conception of deduction is the thesis according to which logic has internalized the Opponent in the sense that its role is now built in the framework itself. The traditional principles of deduction actually reflect rules for engaging in certain kinds of dialogical practices.

Games are thus reduced to an inquiry into logical truth and dialogues are intrinsically perceived from a strategic perspective: an argument is valid if it resists to the Opponent's attempts to defeat it. In the BIO conception of logic, it is the strategic role of the Opponent that has been internalized. Normativity is thus to be understood in terms of strategic recommendation. Indeed, the Opponent must play optimally and try to block the Proponent's inferential steps performed in order to derive the conclusion from the premises the Opponent has already granted. If they intrinsically involve a strategy component, then defeasibility cannot be properly approached in dialogical games. Nonetheless, as stressed by Rahman et al. [18, p. 284], this thesis relies on a confusion between the play level and the strategy level.

3 A Dialogical Theory of Meaning

Why should we pay a peculiar attention to the distinction, but also the links, between the play and the strategy levels? On the one hand, their distinction explains why dialogues do not intrinsically involve a strategic component. Meaning in dialogues arises from interactions at the play level, without being related to inferential requisites. At the play level, it makes no sense to say that **O**'s role is reduced to check the indefeasibility (or non-monotonicity) of a **P**'s move. On the other hand, the link between the play level and the strategy level shows how the latter stems from the former. We cannot think of the inferential level in dialogues without the more fundamental play level in which no strategic component is involved. In order to better understand how the dialogical theory of meaning is built, we refer to the Dialogues for Immanent Reasoning (DIR) of Rahman et al. [18], in which local and strategic reasons backing a statement are made explicit in the object language by incorporating features of the Constructive Type Theory (CTT).

The play level takes care of semantic issues. The strategy level is concerned with validity, through the notion of winning strategy. The play level is concerned with actual applications of the dialogical rules, namely the particle rules and the structural rules. These rules are normative for interaction and do not involve any strategic component. Particle and structural rules determine the meaning of statements in terms of rights and duties; i.e. the right to challenge a statement or to ask for reasons and the duty to answer such challenge or to give reasons. This provides a dialogical turn to Brandom's inferential approach to meaning [4] in terms of asking for and giving reasons. In addition, the particle rules also have a normative aspect with respect to the choices (see Appendix 6.2, the choice is for the challenger of a conjunction, but for the defender of a disjunction). The structural rules determine the general organization of dialogical games. Their correct application has no link with the notion of winning strategy. Indeed, knowing the meaning of an expression is knowing how to build a play for it, no matter who wins. It is only when linking plays to winning strategies that we may talk of validity and rules of inference. In a Wittgensteinian way, the play level reflects (i) the internal feature of meaning, and (ii) the meaning as mediated by language-games [18, p. 278]. The first point brings us to the necessity of a fully interpreted language, as the language of DIR which incorporates features of the CTT in order to make explicit the reasons backing the statements uttered by the players. The second point leads us to the notion of dialogue-definiteness and the notion of proposition as plays.

Having local reasons in the dialogical framework provides a structure in which the reasons asked for and given by the players actually appear in the object language. The Proponent can locally justify his statements by explicitly copying the Opponent's reasons for his own statements. Local reasons also allow to fully interpret contentual language within material dialogues.[3] This explains why there can be reasons put forward at the play level independently of what we might refer to as "inferential moves". They are reasons backing a statement at the play level, and the Proponent can copy the reasons brought forward by his Opponent in order to locally justify his statements even in the absence of a winning strategy, only in the context of an actual interaction, through what Rahman et al. call "equality in action".

A fully interpreted language is obtained by incorporating features of the CTT within DIR. In a nutshell, statements of the CTT are called judgments. If we consider the category *set* of sets, there are four forms of basic categoric judgments:

$$A : \textbf{\textit{set}} \ (A \text{ pertains to the category } \textbf{\textit{set}})$$

$$A = B : \textbf{\textit{set}} \ (A \text{ and } B \text{ are two equal elements of the category } \textbf{\textit{set}})$$

---

[3]See Chapter 10 in [18] for more details on material dialogues.

and for any set A:

$$a : A \ (a \text{ is an object of } A)$$

$$a = b : A \ (a \text{ and } b \text{ are two identical objects of } A)$$

The assertion that a proposition A is true is based on a judgment of the form "a is a proof of A". We can thus identify the category **prop** of propositions with the category **set** of sets, in accordance with the isomorphism of Curry-Howard [10], and consequently a proposition with the set of its proofs (hence the notion of proof-object). This led Ranta [19] to identify proof-objects of the CTT with winning strategies of Hintikka's Game-Theoretical Semantics.

Nonetheless, in relation to meaning as mediated by language-games, as mentioned in (ii) above, the notion of proposition in dialogical logic cannot be identified with strategies but with actual plays. If meaning is mediated by language-games, they must be actually playable by human beings. Following Lorenz [11, p. 258], a proposition becomes a dialogue-definite expression, an expression A such that there is an individual play about A, that can be said to be lost or won after a finite number of moves performed in accordance with the rules of interaction. To know the meaning of an expression A is to know how to build a play for A. And this is independent of the validity of A. Therefore, in dialogical games, propositions should not be identified with winning strategies, but with plays. That is why in DIR, statements are backed by two kinds of reasons: local reasons and strategic reasons. Local reasons are precisely those reasons by means of which a language can be fully interpreted at the play level, regardless the strategy level.

Local reasons are introduced in the object language of DIR as expression of the form $p : A$ where p is a local reason for the proposition A.The semantics of such expressions is provided by synthesis rules and analysis rules for local reasons; i.e. rules that explain how to compose the suitable reasons for a proposition A within a play, and how to separate a complex local reason into the elements required by the composition rule for A, respectively. Both are built on the formation rules for A. The formation rules being rules by means of which we verify whether the thesis is a well-formed expression. This device makes explicit the reasons backing a statement in a dialogical interaction. In formal dialogues, the simple fact that **O** gave a reason for an elementary proposition is sufficient for **P** to copy this reason for the same proposition.

In standard dialogues, reasons were left implicit. That is, a player uttering $A \wedge B$ was committed to utter A and to utter B if his argumentative partner made this request by challenging the conjunction. In DIR, there is first a formation play in which it must be shown that this conjunction is well-formed; i.e. that $A : \textbf{\textit{prop}}$

and that $B : \textbf{prop}$. Then, the reasons backing the player's utterance can be made explicit; i.e. the player must say he has a local reason for the left conjunct ($p_1 : A$) or for the second conjunct ($p_2 : B$), depending on the request of his adversary. In the end, the local reasons for elementary propositions are subject to the Formal Rule [**SR2**]. This is how DIR implements judgmental equality at the play level. Indeed, let us assume that $p : A$. Reflexivity statements of the form $p = p : A$ express the fact that if **O** states the elementary proposition A, then **P** can do the same on the basis of the same reasons. Intuitively, the idea can be expressed as follows: "My reasons for stating this proposition you are now challenging are exactly the same as the ones you brought forward when you yourself stated that very same proposition" [18, p. 8]. In short, this equality expresses at the object language level the fact that **P**'s defensive move rests on the authority **O** has previously asserted when producing his local reason.

Equality in action thus gives a dialogical turn to Sundholm's epistemic assumption. According to Sundholm [20, p. 556], validity involves the transmission of epistemic matters from premises to conclusion; i.e. upon the epistemic assumption that the premises are known, the conclusion is made evident. The notion of epistemic assumption appears when explaining what a valid inference is; i.e. an inference from the premises $J_1, ..., J_n$ to the conclusion J is valid if one can make the conclusion evident on the assumption that $J_1, ..., J_n$ are known. According to Martin-Löf[4], "known" cannot be taken in the sense of demonstrated; otherwise we would be explaining the notion of inference in terms of demonstration when demonstration is explained in terms of inference. MartinLöf's solution is to understand "known" in the sense of "asserted", "which is to say that others have taken responsibility for them, and then the question for me is whether I can take responsibility for the conclusion". Therefore, what must be known is not that the premises are known to be true, but that someone else take them as being such, and asserts them. In dialogical logic, this amounts to link judgmental equality to the Formal Rule. Then, there is a winning strategy for **X** only if **X** can base his moves leading to a win by endorsing himself the proposition whose justification is rooted in **Y**'s authority. But in dialogues, normativity is to be understood at the level of social interaction, not at the level of rules of inferences. Local rules are rules of interaction, and not rules of inference, telling rights and duties for the players. Inferences are built from these interactions when tight to strategies.

The CTT notion of proof-object actually finds its counterpart in the dialogical notion of winning strategy. It is only when linked to winning strategies that moves

---

[4]We refer to the transcription of Ansten Klev of Martin-Löf's lecture "Is Logic Part of a Normative Ethics?" held at the research unit FRE3593, Paris, May, 2015.

in a dialogue can be considered as inferential. Strategic reasons are a kind of re-capitulation of what can happen for a given thesis and show the entire history of the play by means of the instructions. They show an overview of the possibilities enclosed in the thesis. But the fundamental level of plays is also needed. This link clearly becomes apparent when a heuristic method to extract the strategy level from the multiplicity of possible plays enclosed by the initial thesis is spelled out. The strategy level is a generalization of the procedure which is implemented at the play level; it is a systematic exposition of all the variants. The strategy level allows to compare different plays on the same thesis. They need not be actually carried out by the players. They are only a perspective on the possibilities offered by the play level. Given that there is a **P**-winning strategy if and only if **P** has a way to win regardless of **O**'s choices, the **P**'s strategies are built on **O**'s choices. That is, each possible choice of **O** must be taken into account and dealt with in order to determine if **P** is able to win in all the different cases stemming from **O**'s possible choices. A heuristic method to build the **P**-winning strategies consists in taking into consideration **O**'s choices that entail a branching. For example, as in the Example 1, there is a branching when **O** defends a disjunction (move 3). There is another branching when **O** challenges **P**'s conjunction (move 5). All of these possibilities must be explored at the strategy level. Then, there is a winning strategy if **P** is able to win no matter the choices of **O**.[5] In order to build the strategies, we must apply every rule in order to extract every possibility of choice for **O**. Then, **P** will have a winning strategy if he is able to win each play opened by such choices. Therefore, strategies stem from the play level. We apply all the rules and then we consider all the different every possible play opened by the branchings.

The distinction between the play level and the strategy level is important because it dissociates semantic from inferential issues, while funding the latter on the former. The particle and the structural rules are not inferential rules, they are normative only for games of asking for and giving reasons; i.e. for the interaction between two argumentative partners. And **O**, as well as **P**, forms part of this interaction at the play level from which the meaning arises. His role cannot be reduced to verify the indefeasibility of the moves played by **P** or to check the validity of the initial thesis. His role is to take part in the execution of the rules by applying is right to ask for reasons and his duty to give reasons, on which **P** may rely to provide himself reasons backing his own statements. At the play level, we need not think of the players as optimal. There is room for error and defeasibility, although it is only from the perspective of the strategic level, we will be able to spot the best ways

---

[5]For more details on how to build a winning strategy, as well as the method of extensive forms of dialogues, see Chapter 5 in [18].

to play. Nothing prevents one player from playing badly, even when there exists a winning strategy. At the play level, we may also grasp limitations of computational skills, memory, information, and so on.

4 Non-Monotonicity in Dialogues

Given that the play level does not involve any strategic component, nothing precludes defeasibility in dialogical games. Moreover, there exists non-monotonic dialogical logics; e.g. the Inconsistency-Adaptive Dialogical Logic (IAD) of Beirlaen and Fontaine [3].[6]

Depending on the perspective on dialogues we adopt, defeasibillity and non-monotonicity can manifest themselves differently. On the one hand, we may speak of defeasibility of moves, at the play level. Given that the play level is normative, not for inferential moves, but for interactions, there is room for defeasibility of moves without changing the notion of validity. Moves might be retracted or defeated for various reasons; for example, if suboptimal plays or scant resources of the players (computational skill, memory, access to information, and so on) are taken into account. On the other hand, we may also speak of defeasible strategies, which yield non-monotonicity. Here, the notion of validity is fundamental; i.e. even by considering the strategy level, the link between the premises and the conclusions of an argument may be broken because of the introduction of new information. Nonetheless, strategies are not primitive, and non-monotonicity stems from the play level in which specific rules open new possibilities. In this last section, we discuss these issues in the context of IAD.

Let us begin by explaining how IAD works.[7] Standard dialogues are explosive: from an inconsistent set of premises, we can derive anything. But usually, the occurrence of inconsistencies in the course of an argumentative interaction is not a reason to infer random statements or to stop the process. That is why the argumentative partners may agree to play a different argumentative game with rules of paraconsistent logic. In such games, a specific structural rule [**SR5**] governing the use of the particle rule for negation (that remains unchanged) modifies the global meaning of negation. Indeed, **P** is not allowed to challenge a negation $\neg\varphi$ uttered by **O**, unless **O** has previously challenged an occurrence of the same negation before. We

---

[6]See also Fontaine and Barés [9] and Barés and Fontaine [1] for an application to abductive reasoning. Other dialogical approaches to defeasibility in dialogues have been proposed by Nzokou [13] and Dango [6] in order to study the use of proverbs in argumentative interaction of the African oral tradition. A dialogical approach of Belief Revison Theory based on Bonano's semantics can be found in Fiutek et al. [8].

[7]See the rules in Appendix 6.4, and Beirlaen and Fontaine [3] for more details.

may say that by challenging a negation $\neg\varphi$, **O** concedes that this negation behaves normally. This is enough to block explosion, but perfectly acceptable inferences like the disjunctive syllogism ($\neg\varphi, \varphi \vee \psi/\psi$) are now invalid.

That is why in IAD we reason as classically as possible, at least until an inconsistency is encountered. More precisely, **P** is now allowed to challenge a negation by means of a conditional move, even if **O** has not challenged an occurrence of the same negation before (See [**SR5.1**] in Appendix 6.4). A conditional move is a move by means of which **P** commits himself to defend an additional condition, of reliability in the case of IAD. A formula φ used to challenge a negation $\neg\varphi$ is reliable if $\varphi \wedge \neg\varphi$ does not pertain to a disjunction of abnormalities $Dab(\Theta)$ derivable from the set of premises.[8] Therefore, **O** is allowed to challenge a condition of reliability by introducing a $Dab(\Theta)$ containing $\varphi \wedge \neg\varphi$, and it is **P** who must show in a subdialogue that it is not derivable from the set of initial premises.

Let us illustrate the point by means of the following example:

Example 2							
		O			P		
d_1							
					$q[\neg p, p, p \vee q]$		0
1		$m := 2$			$n := 2$		2
3.1		$\neg p$					
3.2		p	0				
3.3		$p \vee q$					
5		p		3.3	$?_\vee$		4
		$- - -$		3.1	p	$\mathfrak{R}_p^\Sigma$	6
7		$?\mathfrak{R}_p^\Sigma(p \wedge \neg p)$	6		$\mathfrak{F}_\Sigma(p \wedge \neg p)$		8
$d_{1.1}$							
9		$p \wedge \neg p[\neg p, p, p \vee q]$	8		$- - -$		
					$\neg p$		10.1
11		$p \wedge \neg p$		9	p		10.2
					$p \vee q$		10.3
13		p		11	$?\wedge_L$		12
15		$\neg p$		11	$?\wedge_R$		14
		$- - -$		15	p		16
17		p	10.1		$- - -$		

---

[8]See Appendix (Section 6.4) for the definition of a a disjunction of abnormalities $Dab(\Theta)$.

Explanation: P challenges a negation by means of a conditional move (move 6). **O** challenges the condition by introducing $Dab(\{p \wedge \neg p\})$ (move 7). **P** answers by claiming that this inconsistency cannot be derived from the set of premises (move 8). **O** challenges **P**'s claim by taking the burden of proof of $p \wedge \neg p[\neg p, p, p \vee q]$ in a subdialogue.[9] **O** wins the subdialogue, thus showing that move 6 is not reliable. **P** loses the main dialogue.

By contrast, there would be a **P**-winning strategy for the initial thesis $q[\neg p, p \vee q]$, where no inconsistency is involved. The existence of a **P**-winning strategy for $q[\neg p, p \vee q]$ does not warrant the existence of a **P**-winning strategy for $q[\neg p, p, p \vee q]$. This is dialogical non-monotonicity, which is more generally defined in terms of dialogues that do not satisfy the following property:

Dialogical Monotony: If there is a **P**-winning strategy for $\varphi[\Sigma]$, then there is he also a **P**-winning strategy for $\varphi[\Sigma']$ (where $\Sigma \subseteq \Sigma'$).

Although non-monotonicity is a feature of the strategy level, it arises from specific rules implemented at the play level. What is fundamental is the use of negation, independently from the strategy level. The rules of IAD open new possibilities of plays which eventually yield non-monotonicity. We may understand the introduction of a Dab-formula in terms of Pollock's undercutting defeater [15, p. 485]; i.e. as a move breaking the relation between the premises and the conclusion. The link holds if the initial thesis is $q[\neg p, p \vee q]$, but it is broken if p is added to the initial set of premises. Nonetheless, from the strategy perspective, which is some kind of a recapitulation of the various possible plays, defeasibility does not appear. If **O** can defeat an argument by introducing a relevant Dab-formula, he will do it directly and win. If there is no such a Dab-formula, then the conditional move is reliable regardless his choices. Therefore, defeasibility of strategies, understood in terms of non-monotonicity, does not strictly speaking involve defeasible moves at the play level. non-monotonicity is manifested through different games with different initial theses; e.g., one with $q[\neg p, p \vee q]$, the other with $q[\neg p, p, p \vee q]$. But even in non-monotonic dialogues such as IAD, defeasibility needs not appear within the dialogue, whether it be at the play or at the strategy level. Defeasibility might appear at the play level (in suboptimal plays), but this would not be specific to IAD.

In IAD, the use of the negation rule grounds a kind of dynamic at the strategy level, since the existence of a **P**-winning strategy depends on the context of argumentation (the set of initial premises). But the rule itself is not dynamic, the rights and duties do not change in the course of the game. According to Batens [2, p. 461], non-monotonicity displays an external dynamics; i.e., a conclusion may be revised

---

[9]Notice that in the subdialogues **O** plays formally (he has the burden of proof) and that the rule are those of **LLD** with (SR5). See details in Appendix.

in the light of new information. This has nothing to do with suboptimal plays in dialogues. We may say that the players play optimally given a particular state of information, and then retract themselves if new information is added. In IAD, the existence of a **P**-winning strategy is determined since the beginning of a dialogical game, so that the strategy itself is not really defeated. But, external dynamics can be accounted for in terms of different dialogues with different initial theses (for example, one dialogical game for $\psi[\neg\varphi, \varphi \vee \psi]$ and another for $\psi[\neg\varphi, \varphi, \varphi \vee \psi]$. Therefore, from the strategy level perspective, such an external dynamic may be understood as a dynamic of passing from a dialogical game to another.

We may also think of external dynamics in terms of defeasibility in material dialogues of DIR, in which new assertions that are materially but not formally grounded are introduced. Indeed, local reasons can also be specified for material dialogues by means of "Socratic rules" which specify how to justify elementary assertions in relation to their proper content. Whereas in formal dialogues **P** is allowed to state an elementary proposition only on the ground that **O** has already stated it, the inquiry may go further in material dialogues. For example, in a first-order language, if **P** asserts that the ball is red, then **P** is committed to show that the ball is coloured. A language game establishes for each of the predicates of the game the suitable Socratic rule. Here, some kind of cooperation is involved, given that the players must finally agree on the use of the predicates involved in elementary propositions. In this case, defeasibility can be thought in relation to the propositional content, without arising from an adversarial component of dialogues. Various examples are put forward by Rahman et al. [16, p. 61 ff] in a dialogical framework applied to Islamic law. We may provide the following simpler example: **P** may justify the interdiction of drinking wine and driving because it is the product of fermented grapes. But **O** brings the case of vinegar as counterexample, and may introduce date wine as another example and justify its interdiction by its toxicity. This will yield a dynamic change and **P** will be invited to start again the game with the suggested justification.[10]

Batens [2, p. 462], also considers an internal dynamics, triggered by an analysis of already available information. In IAD, such an internal dynamics may be displayed at the play level. That is, in the course of suboptimal plays, one of the argumentative partners realizes after all that a conditional move is not reliable. The set of premises may be so huge that **O** is not immediately able to spot an inconsistency and to find the relevant Dab-formula to challenge a conditional move. But this has nothing to do with non-monotonicity, defined at the strategy level, and should not be understood in terms of inference. From a dialogical perspective, this is not a dynamics of proof, but a dynamics of interaction at the play level.

---

[10]We are thankful to an anonymous reviewer for this comment on material dialogues.

5 Conclusion

Both Dutilh Novaes, Martin-Löf, and Rahman et al., highlight the fact that logic is in first instance normative for dialogical interaction, rather than for correct thinking defined in terms of (deductive) inference. Nevertheless, in the absence of a clear conception of the distinction between the play level and the strategy level, Dutilh Novaes eventually holds the thesis that dialogues intrinsically involve a strategic component, and thus links again their foundations to the notion of inference. Following the insights of DIR of Rahman et al., we part with her on this issue: no strategic component is constitutive of dialogues at the play level. As a consequence, nothing prevents us from thinking of dialogues as an interactive framework for defeasible reasoning. It is even a requisite if we aim at considering dialogues as a general framework for human reasoning. Given that dialogical rules are normative for interaction rather than for correct (deductive) reasoning, or inference, they even constitute an interesting framework to deal with defeasible reasoning and non-monotonicity.

There are still different ways to understand defeasibility in dialogical games. Again, it seems that the different dialogical levels –play level and strategy level– provide tools to shed a new light on the issue. We can indeed speak of defeasible moves, at the play level. Defeasibility of moves has nothing to do with the notion of winning strategy, and subsequently with the notion of inference. But it may account for interesting features of argumentative interaction; e.g. resources limitations of the players when they try to spot an inconsistency in a set of initial premises. Defeasibility can also be related to the propositional content, in the context of material dialogues. Non-monotonicity is defined at the strategy level. In IAD, as in other monotonic dialogical approaches, the existence of a **P**-winning strategy for a given thesis is determined since the beginning by all the possible plays. Therefore, strategies are not strictly speaking defeasible. Non-monotonicity is accounted for in terms of different dialogues with different initial theses. But, as we explained it at the end of the previous section, such dialogues might be articulated in a dynamic framework displaying non-monotonicity explicitly.

6 Appendix - Definition and Rules

6.1 Basic Definitions

Let L be a propositional language, defined as follows: $\varphi := \varphi|\varphi \wedge \varphi|\varphi \vee \varphi|\varphi \to \varphi|\neg\varphi$

Lower case letters p, q, r, ... refer to atomic formulas in L. We use lower case Greek letters φ, ψ, χ, ... to refer to L-formulas, and upper case Greek letters Γ, Σ, Δ, ... to refer to finite sets of L-formulas.

X and **Y** (with **X** $\neq$ **Y**) are two player variables. **P** and **O** are two labels standing for the players of the games, the *Proponent* and the *Opponent* respectively.

The force symbol '!' is used for assertions and '?' for requests.

A move is an expression of the form **X**$-e$ where **X** is a player variable and e is either an assertion or a request.

We use $n := r_i$ and $m := r_j$ with $r_i, r_j \in \mathbb{N}*$ for the utterance of the rank the players choose according to the rule [**SR0**] below.

A *play* is a sequence of moves performed in accordance with the game rules. The initial thesis will be either a formula φ or an argument of the form $\psi[\varphi_1, ..., \varphi_n]$ which amounts to the claim that there is a winning strategy for the conclusion ψ given the concession of $\varphi_1, ..., \varphi_n$. The premises $\varphi_1, ..., \varphi_n$ are referred to as the initial concessions. In case the premise set is empty, the initial thesis is simply ψ. The *dialogical game* for a claim $\psi[\varphi_1, ..., \varphi_n]$ (respectively ψ) is the set $\mathcal{D}(\psi[\varphi_1, ..., \varphi_n])$ (respectively $\mathcal{D}(\psi)$) of all the plays with $\psi[\varphi_1, ..., \varphi_n]$ (respectively ψ) as the initial thesis.[11]

For every move M in a given sequence $\mathcal{S}$ of moves, $p_\mathcal{S}(M)$ denotes the position of M in $\mathcal{S}$. Positions are counted starting with 0. We will also use a function F such that the intended interpretation of $F_\mathcal{S}(M) = [m', Z]$ is that in the sequence $\mathcal{S}$, the move M is an attack (if $Z = A$) or a defence (if $Z = D$) against the move of previous position m'.

6.2   Particle Rules

Assertion	Attack	Defence
X-$!\varphi \wedge \psi$	**Y** - $?\wedge_L$	**X** - $!\varphi$
	or	or
	Y - $?\wedge_R$	**X** - $!\psi$ respectively
X - $!\varphi \vee \psi$	**Y** - $?\vee$	**X** - $!\varphi$ or **X** - $!\psi$
X - $!\neg\varphi$	**Y** - $!\varphi$	− − −
		No Defence
X - $!\varphi \rightarrow \psi$	**Y** - $!\varphi$	**X** - $!\psi$
X - $!\psi[\varphi_1, ..., \varphi_n]$	**Y** - $!\varphi_1$	**X** - $!\psi$
	...	
	Y - $!\varphi_n$	

---

[11]Where $\Sigma = \{\varphi_1, ..., \varphi_n\}$, we sometimes write $[\Sigma]$ instead of $[\varphi_1, ..., \varphi_n]$, for the sake of presentation.

6.3 (Standard) Structural Rules

[SR0][Starting Rule]

(i) If the initial thesis is of the form $\psi[\varphi_1, ..., \varphi_n]$, then for any play $\mathcal{P} \in \mathcal{D}(\psi[\varphi_1, ..., \varphi_n])$ we have:

   (ia) $p_{\mathcal{P}}(\mathbf{P}{-}!\psi[\varphi_1, ..., \varphi_n]) = 0,$

   (ib) $p_{\mathcal{P}}(\mathbf{O}{-}n := r_1) = 1$ and $p_{\mathcal{P}}(\mathbf{P}{-}n := r_2) = 2$.

(ii) If the initial thesis is of the form ψ, then for any play $\mathcal{P} \in \mathcal{D}(\psi)$ we have:

   (iia) $p_{\mathcal{P}}(\mathbf{P}{-}!\psi) = 0,$

   (iib) $p_{\mathcal{P}}(\mathbf{O}{-}n := r_1) = 1$ and $p_{\mathcal{P}}(\mathbf{P}{-}n := r_2) = 2$.

Clause (ia) (respectively (iia)) warrants that every play in $\mathcal{D}(\psi[\varphi_1, ..., \varphi_n])$ (respectively $\mathcal{D}(\psi)$) starts with $\mathbf{P}$ asserting the thesis $\psi[\varphi_1, ..., \varphi_n]$ (respectively ψ).

Clause (ib) (respectively (iib)) says that the players choose their respective repetition ranks among the positive integers.[12] Clerbout [5] (p. 791) showed that there is a $\mathbf{P}$-winning strategy when $\mathbf{O}$ chooses rank 1 if and only if there is a $\mathbf{P}$-winning strategy for any other choice of $\mathbf{O}$. In In IAD, $\mathbf{O}$ may need to choose rank 2 because of the dynamics of sections.

---

[12]A move M' performed by $\mathbf{X}$ in a dialogue is a repetition of a previous move M if (i) M' and M are two attacks performed by $\mathbf{X}$ against the same move N performed by $\mathbf{Y}$, or (ii) M' and M are two defences performed by $\mathbf{X}$ in response to the same attack N performed by $\mathbf{Y}$. The ranks guarantee the finiteness of plays by limiting the repetitions allowed in a dialogue.

[SR1c][Classical Development Rule] For any move M in $\mathcal{P}$ such that $p_{\mathcal{P}}(M) > 2$ we have $F_{\mathcal{P}}(M) = [m', Z]$ where $Z \in \{A, D\}$ and $m' < p_{\mathcal{P}}(M)$. Let r be the repetition rank of Player $\mathbf{X}$ and $\mathcal{P} \in \mathcal{D}\psi[\varphi_1, ..., \varphi_n]$ (respectively $\mathcal{D}(\psi)$) such that:

- the last member of $\mathcal{P}$ is a $\mathbf{Y}$-move,

- $M_0 \in \mathcal{P}$ is a $\mathbf{Y}$-move of position m_0,

- there are n moves M_1, ..., M_n of player X in $\mathcal{P}$ such that $F_{\mathcal{P}}(M_1) = F_{\mathcal{P}}(M_2) = ... = F_{\mathcal{P}}(M_n) = [m_0, Z]$ with $Z \in \{A, D\}$.

Let N be an $\mathbf{X}$-move such that $F_{\mathcal{P} \frown \mathcal{N}}(N) = [m_0, Z]$. We have $\mathcal{P} \frown N \in \mathcal{D}(\varphi)$ if and only if $n < r$.[a]

---

[a]"$\mathcal{P} \frown N$" denotes the extension of $\mathcal{P}$ with N.

Intuitionistic dialogical games are defined with a rule **[SR1i]**, by modifying **[SR1c]** so that the repetition ranks bound only the number of challenges, and players can defend only once against the last non-answered challenge.

[SR2][Formal rule] The sequence $\mathcal{S}$ is a play only if the following condition is fulfilled: if $N = \mathbf{P} - !\psi$ is a member of $\mathcal{S}$, for any atomic sentence ψ, then there is a move $M = \mathbf{O} - !\psi$ in $\mathcal{S}$ such that $p_{\mathcal{S}}(M) < p_{\mathcal{S}}(N)$.

This rule means that $\mathbf{P}$ can assert an atomic sentence ψ only if $\mathbf{O}$ previously asserted the same atomic sentence ψ.

[D1][X-terminal] Let $\mathcal{P}$ be a play in $\mathcal{D}(\psi[\varphi_1, ..., \varphi_n])$ (respectively $\mathcal{D}(\psi)$) the last member of which is an $\mathbf{X}$-move. If there is no $\mathbf{Y}$-move N such that $\mathcal{P} \frown N \in \mathcal{D}(\psi[\varphi_1, ..., \varphi_n])$ (respectively $\mathcal{D}(\psi)$) then $\mathcal{P}$ is said to be $\mathbf{X}$-terminal.

[SR3][Winning Rule for Plays] Player $\mathbf{X}$ wins a play $\mathcal{P} \in \mathcal{D}(\psi[\varphi_1, ..., \varphi_n])$ (respectively $\mathcal{D}(\psi)$) if and only if $\mathcal{P}$ is $\mathbf{X}$-terminal.

These rules determine the play level. Dialogical validity is defined by taking into account the the strategy level. The thesis of $\mathbf{P}$ is valid if and only if $\mathbf{P}$ has a winning strategy according to the following definition:

[D2][Winning]

1. A strategy of a player **X** in $\mathcal{D}(\psi[\varphi_1, ..., \varphi_n])$ (respectively $\mathcal{D}(\psi)$) is a function s_x which assigns a legal **X**-move to every non terminal play $\mathcal{P} \in \mathcal{D}(\psi[\varphi_1, ..., \varphi_n])$ (respectively $\mathcal{D}(\psi)$) the last member of which is a **Y**-move.

2. A **X**-strategy is winning if it leads to **X**'s win no matter how **Y** plays.

On the basis of the definition of winning strategy, we can define the notion of consequence for dialogical **CL** (classical logic); that is, a dialogical logic played with **[SR0]-[SR3]**, the so-called **CL**-rules:

[D3][CL-Consequence] $\Sigma \vdash_{CL} \psi$ (respectively $\vdash_{CL} \psi$) iff according to the **CL**-rules, there is a **P**-winning strategy for the thesis $\psi[\varphi_1, ..., \varphi_n]$ (respectively ψ).

A similar definition of consequence for dialogical logic **IL** (intuitionistic logic) is obtained by substituting the **IL**-rules to the **CL**-rules; i.e. by substituting **[SR1i]** to **[SR1c]**.

6.4 IAD

For the details of IAD, see Beirlaen and Fontaine (2016 REF). IAD is defined according to the following triple:

1. **LLD (Lower Limit Logic)** = Paraconsistent Dialogical Logic (**[SR0]-[SR3]** + **[SR5]**),

2. $\Omega =_{DF} \{\varphi \wedge \neg\varphi | \varphi$ is a formula$\}$,

3. Strategy = Reliability

Ω is a set of abnormalities. Where Θ is a finite subset of Ω, $Dab(\Theta)$ (for "disjunction of abnormalities) is the disjunction of the members of Θ. If Θ is a singleton, say φ, then $Dab(\Theta) = \varphi \wedge \neg\varphi$.

[SR5][Paraconsistent Negation Rule] The sequence S is a play only if the following condition is fulfilled: If there is a move $N = \mathbf{P}{-}!\psi - C - d$ in the sequence S such that:

(i) $n = p_S(N_1) = n_1$,

(ii) $F_S(N_1) = [m_1, A]$, and

(iii) $m_1 = p_S(M_1)$ such that $M_1 = O{-}!\neg\psi - C - d$.

Then, there is a move $M_2 = O{-}!\psi$ in S such that:

(i) $p_S(M_2) = m_2$ and $m_2 < n_1$,

(ii) $F_S(M_2) = [n_2, A]$, and

(iii) $n_2 = p_S(N_2)$ such that $N_2 = O{-}!\neg\psi - C - d$.

In IAD, even if $\mathbf{O}$ did not challenge another occurrence of the same negation before, $\mathbf{P}$ can challenge a formula $\neg\varphi$ previously uttered by φ by means of a conditional move; i.e. by assuming that φ is reliable:

[D7][Reliability] Let $\varphi[\Sigma]$ be the thesis of the Proponent. A formula ψ behaves reliably with respect to Σ iff there is no formula $Dab(\Theta)$ such that:

(i) $\psi \wedge \neg\psi \in \Theta$,

(ii) $\Sigma \vdash_{LLD} Dab\Theta$, and

(iii) $\Sigma \nvdash_{LLD} Dab(\Theta/\{\psi \wedge \neg\psi\})$.

[SR5.1][IAD Negation Rule] The sequence $\mathcal{S}$ is a play only if the following condition is fulfilled: If there is a move $N = \mathbf{P} - !\psi - C - d$ in the sequence $\mathcal{S}$ such that:

(i) $n = p_{\mathcal{S}}(N)$

(ii) $F_{\mathcal{S}}(N) = [m, A]$, and

(iii) $m = p_{\mathcal{S}}(M)$ such that $M = \mathbf{O} - !\neg\psi - \emptyset - d$

Then one of the following two conditions holds:

(i) N is performed by $\mathbf{P}$ in accordance with the **LLD** negation rule **[SR5]**, or

(ii) $N = \mathbf{P} - !\psi - \mathfrak{R}_{\psi}^{\Sigma} - d$ where $\mathfrak{R}_{\psi}^{\Sigma}$ abbreviates that ψ behaves reliably in view of the premise set Σ.

The reliability operator behaves as a logical constant whose meaning is given by the following particle rules:

Partice rule for the reliability operator $\mathfrak{R}$		
Assertion	Attack	Defence
$\mathbf{X} - !\varphi - \mathfrak{R}_{\varphi}^{\Sigma} - d_1$	$\mathbf{Y} - ?\mathfrak{R}Dab(\Theta)$ where $\varphi \wedge \neg\varphi \in \Theta$	$\mathbf{X} - !\mathfrak{F}_{\Sigma}(Dab(\Theta)) - \emptyset - d_1$
		Or $\mathbf{X}$ counter-attacks $\mathbf{X} - \mathfrak{I}_{\Sigma}(Dab(\Theta \backslash \{\varphi \wedge \neg\varphi\}) - \emptyset - d_1$ (where $(Dab(\Theta \backslash \{\varphi \wedge \neg\varphi\}) \neq \emptyset$

$\mathbf{Y}$ introduces a minimal disjunction of abnormality $Dab(\Theta)$ such that $\psi \in \Theta$. Then, either $\mathbf{X}$ claims that $Dab(\Theta)$ cannot be **LLD**-drawn from Σ; this he does by making use of the failure operator $\mathfrak{F}$ whose meaning is given by another particle rule. Or $\mathbf{X}$ claims that $Dab(\Theta)$ is not minimal (i.e. a smaller disjunction without $\varphi \wedge \neg\varphi$ is **LLD**-derivable); this he does by making use of the indispensability operator $\mathfrak{I}$ whose meaning is also given by another particle rule.

Particle rule for the failure operator $\mathfrak{F}$		
Assertion	Attack	Defence
$\mathbf{X} - !\mathfrak{F}_{\Sigma}\varphi - \emptyset - d_1$	$\mathbf{Y} - ?\varphi[\Sigma] - \emptyset - d_{1.i}$ $\mathbf{Y}$ opens a subdialogue $d_{1.i}$	$- - -$ No defence

The meaning of the $\mathfrak{I}$-operator is given by the following rule:

Particle rule for the indispensability operator $\mathfrak{I}$		
Assertion	Attack	Defence
$\mathbf{X}-!\mathfrak{I}_\Sigma\varphi - \emptyset - d_i$	$\mathbf{Y}-?\mathfrak{I}_\Sigma\varphi - \emptyset - d_i$	$\mathbf{X}-\varphi[\Sigma] - \emptyset - d_{i.j}$ $\mathbf{X}$ opens a subdialogue $d_{i.j}$

That is, $\mathbf{X}$ must show that a $Dab(\Theta)$ shorter than the one introduced by $\mathbf{Y}$ can be unconditionally derived from Σ.

When $\mathbf{Y}$ challenge the failure operator, he takes the burden of the proof of $\varphi[\Sigma]$ and must play under formal restriction, even if $\mathbf{Y} = \mathbf{O}$. That is why $\mathbf{Y}$ opens a subdialogue, in which he commits himself to defend $\varphi[\Sigma]$ by means of the **LLD**-rules.

We thus replace [**SR2**] by [**SR2.1**] and [**SR2.2**], and we add [**SR4.2**]:

[**SR2.1**][**Formal Restriction for Adaptive Dialogues**] If $\mathbf{X}$ *plays under formal restriction*, then the sequence Δ is a play only if the following condition is fulfilled: if $N = \mathbf{X}-!\psi - C_j - d$ is a member of Δ, for any atomic sentence ψ, then there is a move $M = \mathbf{Y}-!\psi - C_i - d$ in Δ such that $p_\Delta(M) < p_\Delta(N)$.

[**SR2.2**][**Application of Formal Restriction**] The application of the formal restriction is regulated by the following conditions:

(i) In the main dialogue d_1, if $\mathbf{X} = \mathbf{P}$, then $\mathbf{X}$ plays under the formal restriction.

(ii) If $\mathbf{X}$ opens a subdialogue $d_{1.i}$, then $\mathbf{X}$ plays under the formal restriction.

[**SR4.2**][**Adaptive LLD-Rule**] In a subdialogue $d_{1.i}$, only **LLD**-rules apply.

[**SR3.1**][**Winning rule for subdialogues**] A subdialogue $d_{1.i}$ is won by $\mathbf{X}$ if it is $\mathbf{Y}$'s turn and there are no more moves available to $\mathbf{Y}$. If $\mathbf{X}$ wins the subdialogue, we return to the main dialogue d_1 in which it is (still) $\mathbf{Y}$'s turn.

Finally, we define a notion of consequence for IAD:

[**D6**][IAD-**consequence**] $\Sigma \vdash_{IAD} \varphi$ (resp. $\vdash_{IAD} \varphi$) iff according to the IAD-rules there is a $\mathbf{P}$-winning strategy for the thesis $\varphi[\Sigma]$ (resp. φ).

References

[1] Cristina Barés Gómez and Matthieu Fontaine. Between sentential and model-based abductions: a dialogical approach. *Logic Journal of the IGPL*, 02 2020. jzz033.

[2] Diderik Batens. The need for adaptive logics in epistemology. In Shahid Rahman, John Symons, Dov Gabbay, and Jean-Paul Van Bendegem, editors, *Logic, epistemology, and the unity of science*, pages 459–485. Springer, Berlin, 2004.

[3] Mathieu Beirlaen and Matthieu Fontaine. Inconsistency-adaptive dialogical logic. *Logica Universalis*, 10(1):99–134, 2016.

[4] Robert Brandom. *Making it Explicit*. Harvard University Press, Cambridge, MA, 1994.

[5] Nicolas Clerbout. First-order dialogical games and tableaux. *Journal of Philosophical Logic*, 43(4):785–801, 2014.

[6] Adjoua Bernadette Dango. *Croyances et significations - Jeux de questions et réponses avec hypothèses*. College Publications, London, 2016.

[7] Catarina Dutilh-Novaes. Dialogical, multiagent account of the normativity of logic. *Dialectica*, 69(4):587–609, 2015.

[8] Virginie Fiutek, Helge Rückert, and Shahid Rahman. A dialogical semantics for bonanno's system of belief revision. In *Construction - Festschrift for Gerhard Heinzmann*, pages 315–334. College Publications, London, 2010.

[9] Matthieu Fontaine and Cristina Barés Gómez. Conjecturing hypotheses in a dialogical logic for abduction. In Dov Gabbay, Lorenzo Magnani, Woosuk Park, and Ahti-Veikko Pietarinen, editors, *Natural Arguments*, pages 379–414. College Publications, London, 2019.

[10] William Alvin Howard. The formulae-as-types notion of construction. In Jonathan Seldin and J. Roger Hindley, editors, *To H.B.Curry: Essays on Combinatory Logic, Lambda Calculus and Formalism*, pages 479–490. London Academic, London, 1980.

[11] Kuno Lorenz. Basic objectives of dialogue logic in historical perspective. *Synthese*, 127:255–263, 2001.

[12] Paul Lorenzen and Kuno Lorenz. *Dialogische Logik*. Wissenschqftliche Buchgesellschaft, Damstadt, 1978.

[13] Gildas Nzokou. *Logique de l'argumentation dans les traditions orales africaines*. College Publications, London, 2013.

[14] Charles Sanders Peirce. *Collected Papers of Charles Sanders Peirce*. Harvard University Press, Cambridge, 1931-1958.

[15] John L. Pollock. Defeasible reasoning. *Cognitive Science*, 11(4):481–518, 1987.

[16] Shahid Rahman, Muhammad Iqbal, and Youcef Soufi. *Inferences by Parallel Reasoning in Islamic Jurisprudence*. Logic, Argumentation & Reasoning. Springer, Cham, 2019.

[17] Shahid Rahman and Laurent Keiff. On how to be a dialogician. In D. Vanderveken, editor, *Logic, Thought and Action*, volume 2 of *Logic, Epistemology and the Unity of Science*, pages 359–408. Springer, 2005.

[18] Shahid Rahman, Zoe McConaughey, Ansten Klev, and Nicolas Clerbout. *Immanent*

Reasoning or Equality in Action - A Plaidoyer for the Play Level. Logic, Argumentation & Reasoning. Springer, Cham, 2018.

[19] Aarne Ranta. Propositions as games as types. *Synthese*, 76:377–395, 1988.

[20] Goran Sundholm. The neglect of epistemic considerations in logic: The case of epistemic assumptions. *Topoi*, 38:551–559, 2018.

[21] John Woods. *Errors of Reasoning. Naturalizing the Logic of Inference.* College Publications, London, 2013.

Received 16 March 2020

Beta Assertive Graphs:
Proofs of Assertions with Quantification

Francesco Bellucci
University of Bologna, Bologna, Italy
francesco.bellucci4@unibo.it

Daniele Chiffi*
Politecnico di Milano, Milan, Italy
chiffidaniele@gmail.com

Ahti-Veikko Pietarinen[†]
Tallinn University of Technology, Tallinn, Estonia
Research University Higher School of Economics, Moscow, Russia
ahti.pietarinen@gmail.com

Abstract

Assertive graphs (AGs) modify Peirce's Alpha part of Existential Graphs (EGs). They are used to reason about assertions without a need to resort to any *ad hoc* sign of assertion. The present paper presents an extension of propositional AGs to the Beta case by introducing two kinds of non-interdefinable lines. The absence of polarities in the theory of AGs necessitates Beta-AGs that resort to such two lines: standard lines that mean the presence of a certain method of asserting, and barbed lines that mean the presence of a general method of asserting. New rules of transformations for Beta-AGs are presented by which it is shown how to derive the theorems of quantificational intuitionistic logic. Generally, Beta-AGs offer a new non-classical system of quantification

Our thanks go to the reviewers of this journal as well as to the participants of the *Workshop on Assertion & Proof*, held in Lecce, Italy, in September 2019, for important comments, suggestions and questions raised. Jukka Nikulainen is to be acknowledged for his assistance in producing the graphs of the present paper, after having developed a fantastic LATEX package EGpeirce.sty for an easy typesetting Peirce's logical graphs, which now has the supplementary package Boxnotation.sty for the typesetting of *assertive graphs*, including its predicate extension.

*Paper prepared within the framework of the Dipartimento di Eccellenza project, Fragilità Territoriali (MIUR 2018–2022), DAStU, Politecnico di Milano.

†Paper prepared within the framework of the HSE University Basic Research Program and funded by the Russian Academic Excellence Project '5-100'.

by which one can logically analyse complex assertions by a notation which (i) is free from a separate sign of assertion, (ii) does not involve explicit polarities, and (iii) specifies a type-referential notation for quantification. These properties stand in important contrast both to standard diagrammatic notations and to standard, occurrence-referential quantificational notations.

Keywords: Existential/Assertive Graphs; Assertions; Quantifiers; Transformations; Intuitionistic Logic; Type vs. occurrence-referential notations.

1 Introduction

Intended interpretations of logical systems often motivate the meaning of logical constants and the justification of logic's fundamental principles. A good pre-theoretical justification of a logical system may be gained from models that are not mathematical in the strict sense, because one could then avoid a vicious circle of explaining knowledge of mathematical structures with other mathematical structures.

According to Gödel (1961), for example, we grasp abstract mathematical objects by means of a clarification of meaning that does not consist in formal definitions.[1] An instructive further example of this view is the assertion-based interpretation of intuitionistic logical constants, which derives from Heyting (1956). In it, conjunction, disjunction, implication, negation, etc. receive meaning via an informal notion of assertion. For example, "P and Q" can be asserted iff both "P" and "Q" can be asserted; "P or Q" can be asserted iff at least one of "P", "Q" can be asserted, and so on for other logical connectives.

Intended interpretations are thus reasonable guides by which one is to look for and assign an informal meaning to logical elements, thus helping to obtain intuitive models. To this effect, Dummett (1978, 214) had pointed out that "an intuitive model is a half-formed conception of how to determine truth-conditions for a given class of sentences. It is not an ultimate guarantee of consistency, nor the product of a special faculty of acquiring mathematical understanding. It is merely an idea in the embryonic stage, before we have succeeded in the laborious task of bringing it to birth in a fully explicit form".

These views provide the background motivation for the present paper. In many

---

[1]Gödel (1961, 383) was convinced that the phenomenological method may be used to conceptually clarify the meaning of abstract mathematical entities. For notice that he wrote that "now in fact, there exists today the beginning of a science which claims to possess a systematic method for such a clarification of meaning, and that is the phenomenology founded by Husserl. Here clarification of meaning consists in focusing more sharply on the concepts concerned by directing our attention in a certain way, namely, onto our own acts in the use of these concepts, onto our powers in carrying out our acts, etc.".

logical systems, an *ad hoc* sign of assertion, such as the one introduced by Frege in *Begriffsschrift* (Frege 1879), is commonly used (Frege 1879, Bellucci & Pietarinen 2017, Carrara, Chiffi & De Florio 2017). An assertion sign may then be allied with a sign that expresses its dual, *denial*, as in bilateralist systems (Rumfitt 2000). In actuality, however, there are systems that have no *ad hoc* sign for assertions at all. This may be the case even when assertions do play a major inferential role in such systems. Such is the case in the Existential Graphs (EGs), invented by Charles Peirce in the late 19th century (Peirce 2019). Instead of an explicit sign of assertion, these logical graphs have an *embedded* sign of assertion in their fundamental notation of *the sheet of assertion* (Peirce 2019, Roberts 1973, Bellucci & Pietarinen 2017, Pietarinen & Chiffi 2018).

Various forms of diagrams and graphical representations offer important advice in shaping the construction of intended models. Using the resources of Peirce's EGs, the present paper proposes an interpretation for acts of logical assertions involving quantification, represented in what is termed here the system of *Beta Assertive Graphs* (Beta-AGs). Beta-AGs is an extension of Alpha-AGs that was introduced in (Bellucci, Chiffi & Pietarinen 2018). It formalises a class of quantificational linguistic acts that are *assertive* in their very nature. It is an extension of Alpha-AGs in fashion analogous to Peirce's extension of propositional Alpha-EGs to quantificational Beta-EGs (Pietarinen 2015a,b).

In brief, the graphical formulas of Beta-AGs are constructed by means of (i) *standard lines* "——" meaning a particular (a certain) assertion, (ii) *barbed lines* "$\#$——" meaning a universal assertion (or a general method to assert), (iii) capital letters "F", "G",..., and boxed capital letters "$\boxed{F}$", "$\boxed{G}$",..., standing for assertions generically, (iv) a *connector* "+" between any assertion (such as "$F+G$") standing for disjunction, (v) a *blackened dot* "●" standing for the *absurdum*, (vi) a simple *juxtaposition* of assertions standing for conjunction (such as "$F\ G$"), and (vii) *cornerings* (doubly nested connected boxes) "$\boxed{\boxdot}$" standing for *implication*. In these cornerings, antecedents are denoted by whatever rests in the area within the outer rectangular shape, and consequents by whatever rests in the area within the inner area of the rectangular shape. The inner rectangle shares two of its sides with the outer one and is non-detachable from it.

The use of lines for quantification is motivated by the fundamental conceptual and semiotic distinction between *type-referential* and *occurrence-referential* logical languages, presented in the next section.

2   Why Lines? Occurrence-referential Languages

How many words are there in the sentence "Venice is always Venice"? In one sense, there are *three* words: "Venice", "is" and "always". This is the lexicographer's sense of the word "word". When a lexicographer says that the *Oxford English Dictionary* contains more than 300,000 words, what they mean is more than 300,000 word *types*. In another sense, "Venice is always Venice" contains *four* words, because in it the word type "Venice" occurs twice. This is an editor's sense of the word "word". When an editor says that a book should not exceed 120,000 words, what they mean is 120,000 word *occurrences*. The distinction between a linguistic type and an occurrence of a linguistic type is familiar in linguistics (Wetzel 2018).

Logical languages may differ as to whether the sameness and distinctness of the individuals of the universe of discourse are represented by means of variable types or variable occurrences. Peirce's Beta graphs that extend the propositional Alpha graphs with quantificational lines, and the standard notation for first-order logic with identity (hereafter, FOL=), illustrate this distinction well. Consider the following sentences in FOL=:

1. $\exists x(Fx \wedge Gx)$

2. $\exists x \exists y(Fx \wedge Gy)$

In (1), there are two occurrences of the variable type "x" within the scope of the quantifier, and these refer to the same individual. In (2) there is one occurrence of the variable type "x" and one occurrence of the variable type "y". These refer to distinct individuals (which may nonetheless coincide). In Beta graphs, (1) and (2) are represented as in Figs. 1 and 2, respectively.

Figure 1:

Figure 2:

In Beta graphs, the variable type is the "line of identity": a thick connector to the "hooks" imagined to occupy the peripheries of the predicate terms. In Beta graphs, the line of identity represents both individual existence (existential quantification) and identity. It represents the numerical identity of the individuals denoted by its terminal points. Since there is only one unique variable type (the line of identity), it is the distinctness of its occurrences that determine the distinctness of the objects denoted. Thus, in Fig. 1 there is one occurrence of the line type, and this means that reference is made to one single individual, of which it is asserted that it is both F

and G, just like in (1). By contrast, in Fig. 2 there are two occurrences of the line type (each disconnected portion of the line counting as a distinct occurrence), and this means that reference is made to two individuals (which may be the same), one of which is said to be F and the other of which is said to be G.

The difference may be stated in the following terms. A *type-referential notation* is a notation in which the sameness and the distinctness of individuals is represented by the *identity of the variable-type*: in such a language each occurrence of the variable-type within the scope of a quantifier refers to the same individual. An *occurrence-referential notation*, in contrast, is a notation in which the sameness and the distinctness of individuals is represented by the *identity of the variable-occurrence*: in such a language each occurrence of the variable-type within the scope of a quantifier refers to a possibly distinct individual (for details, see Bellucci & Pietarinen 2020). It is evident that FOL= is a type-referential notation, whereas Beta graphs exhibit an occurrence-referential notation.

In a type-referential notation like FOL= it is possible to represent syntactically distinct sentences such as (3), (4), and (5):

3. $\exists x \exists y (Fx \wedge Gy \wedge (x = y))$

4. $\exists x \exists y \exists z (Fx \wedge Gy \wedge (x = y) \wedge (y = z))$

5. $\exists x \exists y \exists z \exists t (Fx \wedge Gy \wedge (x = y) \wedge (y = z) \wedge (z = t))$

These three sentences are logically equivalent to (1). In Beta graphs, they all correspond to the graph in Fig. 1, because it is not possible in its language to write (1) without thereby at once also writing (3), (4), and (5).

Occurrence-referentiality thus means that one single Beta graph type corresponds to an equivalence class of formulas in FOL=. The distinction between type- and occurrence-referential notations impacts the possibilities of defining a translation from one language to another: in the case of standard Beta graphs and FOL=, each single Beta graph can always be translated to any one of the members of an equivalence class of sentences of FOL=.

3   Beta-Assertive Graphs (Beta-AGs)

In order to represent first-order quantification in the language of Assertive Graphs (AGs), this section defines Beta-AGs as a conservative extension of the theory of propositional, Alpha-AGs. Being a conservative extension, all the conventions and rules of inference that hold for the propositional level of AGs remain sound and unchanged in Beta-AGs.

First, the theory of Alpha-AGs is briefly reviewed; for more details, the reader is referred to Bellucci, Chiffi & Pietarinen (2018) and Chiffi & Pietarinen (2020).

3.1 Alpha-Assertive Graphs (Alpha-AGs)

The propositional theory of AGs is, in a nutshell, the following. The characteristic feature of AGs in general is that the graphs lack the standard notation of Existential Graphs (EGs), which are the continuous closed boundaries around any graph, typically termed the ovals or the cuts. In EGs, cuts signify both parentheses and contradictory negation. The absence of such cuts in AGs results from the specific design feature of its distinct notation, in which *polarities* of the areas of the graphs are not explicitly represented and in which the notation of boxes around graphs are not cuts on the sheet of assertion. For these reasons, there is no need to introduce a separate sign for negations.

The absence of cuts results in graphical formulas that are less cluttered, as multiple nestings of such boundaries are avoided. On the other hand, dispensing with polarities allows one to directly consider conditional forms of reasoning as expressed, as will be seen below, by means of the cornering of boxes. What is expressed by the cornering will accordingly be the main form of inference in AGs. This feature of having cornerings as a primitive notion means that AGs are somewhat closer to standard logical practice than EGs are.

3.1.1 Fundamental Conventions

Expressions of AGs are instances of graphs standing for assertions and their relations, recursively constructed from primitive assertions. All graph-instances are those that are *scribed* on a *sheet of assertion* (SA). SA is an unordered open-compact manifold. A blank SA is a sheet of assertion on which nothing (except boxes) appears. There are six fundamental conventions.

Convention 1. We always have a right to a blank SA.

Convention 2. We denote the assertion of a graph α by writing it enclosed within a *box*. So, if α is the proposition P of the language, this gives $\boxed{P}$, which also is a proposition of the same language.

Since graphs are scribed on the SA, anything on the SA is an assertion. A box around a graph is not necessary for that graph to be conceived as an assertion. The boxing is a *deictic device* that so to speak draws the interpreter's attention to the graph enclosed within it. It could mean stating "This is what I say: ___", "Have a look at ___", "Listen," etc. Hence the box is unlike Fregean sign of assertion, which in association with a content gives rise to an assertion (for details, see Carrara, Chiffi & Pietarinen 2020). Instead, the assertoric force is designated by the SA.

Convention 3. A *juxtaposition* is to assert more than one graph on the SA, at any non-overlapping position of the SA.

For example: $\boxed{P}$ $\boxed{Q}$. The meaning of juxtaposed graphs is that what their significations are is to be considered *independently* of the significations of the other, that is, without being influenced by what is scribed on any other position of the SA. Thus the juxtaposition of two graphs expresses the *conjunction* of two assertions of those graphs.

Since the boxes and be freely inserted and omitted around any graph, and since the conjunction of an assertion of two graphs is equivalent to the assertion of those two conjuncted graphs—a standard feature of the logic of assertions—the graph above is also equivalent to any of the following: $\boxed{P\,Q}$, $\boxed{P}$ Q, P $\boxed{Q}$, P Q, etc. (REMARK: Comma is not part of the language of graphs and does not appear on the SA.)

Since the SA is unordered (isotropic) in all directions, the following graphs are likewise examples of graphs equivalent to any of the graphs presented above in the context of Convention 3:

$$\frac{\boxed{P}}{Q}, \quad \frac{P}{\boxed{Q}}, \quad \quad d \quad \raisebox{-1ex}{Q}.$$

It is important to now notice that commutativity and associativity of assertions immediately result from this spatial representation of assertions on the SA and are not part of any explicit rule.

Convention 4. Two graphs juxtaposed on the SA not independently but *alternatively* asserted are connected by a *thin line* with a crossing bar: $\boxed{P}\!+\!\boxed{Q}$

This notation represents *disjunctive* assertions. (REMARK: The cross bar | added to the thin line of the notation for disjunction is both historical from the algebraic use of + and its variants for disjunction as well as to distinguish it from the quantificational line of the Beta extension.)

The next convention informs us of the fact that the box notation is nevertheless not a superfluous design feature of these graphs, and that its importance comes from the way in which we are to represent conditional assertions.

Convention 5. The sign of two nested boxed, one within the other and with the inner and outer boxes connected by the sharing of two adjacent sides, is called a *cornering*: $\boxed{P\,\boxed{Q}}$

Cornering represents an implicational relation between two assertions. The graph P that occurs on the *outer area* of the cornering is the antecedent and Q that occurs on the *inner area* of the cornering is the consequent of an implication from P to Q. This follows from these graphs being, just as existential graphs in general, interpreted in terms of what Peirce called the "endoporeutic principle" or "endoporeutic interpretation" (R 293, Peirce 1966), namely that one is to assign the semantic values to their constituents from the outside-in fashion: the graph in the outer area of the corner comes first, and thus makes the antecedent of the implication, while the graph in the inner area of the corner comes second, and is the consequent of the implication.

The notation of the cornering is not constructed from the simple nesting of two boxes. Cornering is, as noted, a primitive sign in which the two adjacent sides of the box that demarcate the inner area of the corner are welded with segments of the two adjacent sides of the box that demarcate the outer area of the cornering. Hence the graph above that expresses the implication of two assertions is not equivalent to $\boxed{P \boxed{Q}}$, which expresses the assertion of the conjunction of P with an assertion of Q.

Convention 6. An *absurdity* is indicated with a heavy dot '●', termed the *blot*.

REMARK. One could imagine the heavy dot to spread to that it would fill up the entire area in which it occurs. The idea of such blot resembles Peirce's own proposals (e.g., R S-30, 1906; see Pietarinen et al. 2020), according to which a blackening of an entire area leaves no room for any assertions and thus signals absurdity. For reasons of simplicity (that is, in order to avoid representing the entire sheet as blackened) and analyticity (that is, in order to represent juxtaposition of the blot with assertions), we take the blot to be a black dot bounded in the given nominal size.

This concludes our abridged presentation of the system of conventions for AGs.

3.1.2 The Language of Assertive Graphs

The set of well-formed (well-scribed) graphs of the language of AGs is defined as follows:

1. Atomic graphs h, j, s,... (of a denumerable set); the blank   , the blot ●, and their boxings $\boxed{h}$, $\boxed{j}$, $\boxed{s}$, $\boxed{}$, $\boxed{●}$ scribed on the SA, are well-formed graphs.

2. If H is a well-formed graph, then also its boxing scribed on the SA is a well-formed graph, thus: $\boxed{H}$.

3. If H is a well-formed graph and J is a well-formed graph, also the scribing of them at two different positions on the SA is a well-formed graph, thus: H J.

4. If H and J are well-formed graph, also their connection with a line $+$ is a well-formed graph, thus: $H+J$.

5. If H and J are well-formed graphs, then also the cornering, namely a graph in which J appears in the corner of H, scribed on the SA, is a well-formed graph, thus: $\boxed{H\boxed{J}}$.

By a graph we mean a well-formed (well-scribed) graph. Some examples of non-well-scribed graphs are

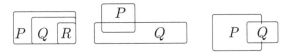

Clearly, the following is a well-formed graph whenever H is a well-formed graph: $\boxed{H\boxed{\bullet}}$. As the *blot* "$\bullet$" stands for the *absurdum*, a denial of the content of an assertion is expressed as a cornering in which the blot alone rests on the inside of the cornering while the assertion that is denied rests on the outer box of the cornering.

Since the SA is an unordered, open-compact space, the properties of commutativity, associativity and adjunction immediately follow from the properties of that space. Hence no separate rules are to be had in order to pronounce the equivalence of assertions such as $\boxed{H}$ $\boxed{J}$ and $\boxed{J}$ $\boxed{H}$ and $\genfrac{}{}{0pt}{}{\boxed{H}}{\boxed{J}}$ and $\genfrac{}{}{0pt}{}{\boxed{J}}{\boxed{H}}$, nor the equivalence of the assertions such as $\boxed{H}\mathbin{+}\boxed{J}\mathbin{+}\boxed{K}$ and $\boxed{J}\mathbin{+}\boxed{K}\mathbin{+}\boxed{H}$ and $\genfrac{}{}{0pt}{}{\boxed{K}\mathbin{+}\boxed{H}}{+\ \boxed{J}}$, and so on.

REMARK. When—as in the last example—graphs are connected with the line $+$, the ordering of these alternating connections is immaterial. The natural reason is that the graphical notation of logic is grounded on topological facts: for instance, the points at which disjunctive lines are connected to the graph are not fixed in any way. As long as the connections are preserved, any disjunct thus connected can freely move along the SA, including passing through other disjuncts. As in the above example, the disjunction between $\boxed{H}$, $\boxed{J}$ and $\boxed{K}$ means that these disjuncts are all connected to each other with the line order-independently.

3.1.3 Proofs in Assertive Graphs

The system of the logic of AGs is defined by graphical axioms and rules of transformation on the graphs of AGs. What follows is a concise presentation of these graphical axioms and rules.

Axioms of AGs. The blank space indicates a tautology and can appear anywhere on the SA.

Axiom I (The blank SA):   ,

Axiom II (Any graph implies a blank): $\boxed{H\ \Box}$

Axiom III (*Ex falso*): $\boxed{\bullet\ \boxed{H}}$

The sign '$\prec$' is used to denote the derivability relation for graphical expressions of the language of AGs. (REMARK: This sign was Peirce's favourite to denote logical consequence relation, and is adopted here accordingly.) Simply put, $G \prec H$ means that a graph G can be transformed into a graph H according to the rules of transformation.

Rules of Transformation. The following nine rules define the sound and complete set of transformation rules for the propositional part of AGs.[2]

1. Antecedent Separation /Antecedent Merging (As/Am):

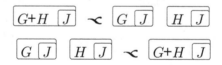

> That is, the disjunction of the antecedents of a cornering can be split into the juxtaposition of two cornerings with one (and not the same) of the two disjuncts as the antecedent and with the same consequent as the initial graph. Conversely, any two cornerings with the same consequent can be merged into a cornering with the disjunction of the antecedents of the initial graph and with the same consequent.

---

[2]Completeness can be shown by the Lindenbaum–Tarski construction, given that the underlying algebraic theory (Heyting algebra) is a variety and defines a congruence relation. A similar graphical intuitionistic system is defined in Ma & Pietarinen (2018), which includes the details of e.g. the admissible rules of the system. The present set of rules for AGs has to differ from that graphical intuitionistic system in order to compensate for the lack of polarities, and for that reason and also because there are more logical primitives in AGs, some additional rules have to be introduced.

2. Consequent Merging/Consequent Separation (Cm/Cs):

$$\boxed{G\ \boxed{H}}\quad \boxed{G\ \boxed{J}}\ \prec\ \boxed{G\ \boxed{H\ J}}$$

$$\boxed{G\ \boxed{H\ J}}\ \prec\ \boxed{G\ \boxed{H}}\quad \boxed{G\ \boxed{J}}$$

That is, the consequents of two cornerings with the same antecedent can be merged into the consequent of a cornering with the same antecedent as the initial cornerings. Conversely, the juxtaposed consequents in a cornering can be split into the juxtaposition of two cornerings with the same antecedent and with one (and not the same) of the two consequents.

3. Disjunct Contradiction (DC):

$$H \prec H{+}\bullet$$

$$H{+}\bullet \prec H$$

That is, any graph is equivalent to that graph disjuncted with the blot (*absurdum*) (the graphs here can just as well be boxed).

4. Cornering (Cr/UCr):

$$H \prec \boxed{\ \boxed{H}\ }$$

$$\boxed{\ \boxed{H}\ } \prec H$$

That is, any graph is equivalent to the cornering with that graph in its consequent and with a blank as its antecedent. This captures that if H is scribed on the SA then H follows from the assertion of a tautology. If H follows from the assertion of a tautology, then H holds. This latter clause is called the *uncornering rule* (UCr).

REMARK. An important feature of assertive graphs generically is that the boxes can be *cornered*. From a simple box containing any graph inside (including the blank graph), a cornering can be inferred, which contains the original graph in the inner area (consequent) of the cornering, and conversely. Cornered graphs represent conditional assertions, with antecedent (the outer) and consequent (the inner) areas.

5. Iteration/Deiteration (It/DeIt):

For the rule of iteration and its converse deiteration, first define the context of graphs. A *graphical context* is of the form K{ }, in which K is any graph-instance of the language of AGs, graph-instances enclosed within { } are said to be in the nest of K, and a single slot { } is the empty context. Let K{H} be the graph obtained from K{ } by substituting H for that slot. The two rules then are:

5.1 Iteration (It): If a graph G occurs on the SA or anywhere in the nest of graphs K, it may be scribed on any area (which itself is not part of G) which (i) is the same area on which G occurs or (ii) is in the nest of {G}:

$$\text{(i) } K\{G\} \prec K\{GG\}. \text{ (ii) } K\{GH\{J\}\} \prec K\{GH\{GJ\}\}.$$

The converse of (It) is deiteration (DeIt).

5.2 Deiteration (DeIt): Anything that is the result of an iteration may be deiterated, thus:

$$\text{(i) } K\{GG\} \prec K\{G\}. \text{ (ii) } K\{GH\{GJ\}\} \prec K\{GH\{J\}\}.$$

Conversely, anything resulting from deiteration can also be iterated.

The following examples apply the rule of iteration in AGs:

$$\boxed{H\ \square} \prec \boxed{H\ \boxed{H}}\ ; \ \boxed{H} \prec \boxed{H\ H}\ ; \ \boxed{H} \prec \boxed{H}\ \boxed{H}.$$

In the first example, H that lies on the antecedent area of the cornering is iterated into the consequent area. In the second example, the occurrence of H is iterated on the same area in which it occurs. In the third example, an occurrence of $\boxed{H}$ is likewise iterated on that same area. These three cases are all reversible by the application of the rule of deiteration.

6. Conjunction Elimination (CE):

$$G\ H \prec G$$
$$G\ H \prec H$$

That is, from the scribing of independently asserted G and H on the SA (excluding those cases in which G and H rest on the antecedent area of a cornering, unless at least one of them is a blot), it is possible to derive one of these assertions.

Naturally, the graphs could just as well be boxed:

$$\boxed{G}\ \boxed{H}\ \prec\ \boxed{G}$$
$$\boxed{G}\ \boxed{H}\ \prec\ \boxed{H}$$

Commutativity follows immediately from the spatial and non-linear nature of the language of AGs. Thus the second clauses in the above two pairs of rules are redundant.

7. Disjunction Introduction (DI):

$$G\ \prec\ G{+}H$$
$$G\ \prec\ H{+}G$$

That is, from the scribing of G on the SA it is possible to derive a disjunction of that graph with a graph H. The disjunction, as denoted by the connecting line, means "to be alternatively asserted". In a similar vein, from an assertion of G it is possible to derive the assertion of G or the assertion of H:

$$\boxed{G}\ \prec\ \boxed{G{\dashv}H}$$
$$\boxed{G}\ \prec\ \boxed{H{\dashv}G}$$

Again, disjuncts have no priority ordering on the topology of the SA.

8. Insertion in the Antecedent (InsA):

This rule also works with the contexts K{ } of graphs. The applicability of the rule of insertion in the antecedent (InsA) below is restricted to the antecedents of the cornerings whose immediate context in K (that is, the area on which the cornering is placed) is not an antecedent of another cornering. Then:

$$\mathrm{K}\{\boxed{G\,\boxed{H}}\}\ \prec\ \mathrm{K}\{\boxed{G\ J\,\boxed{H}}\}$$

That is, in any unoccupied position in the area of the antecedent of the cornering, which itself does not reside, as its immediate context, within an antecedent of a cornering, it is possible to insert any graph.

9. Deletion from the Consequent (DelC):

$$\boxed{G\,\boxed{H\ J}}\ \prec\ \boxed{G\,\boxed{H}}$$

That is, it is possible to delete any graph from the consequent of a cornering.

With these axioms and rules we can express all intuitionistically valid principles in AGs.

3.2   Quantification in Beta-Assertive Graphs

Since Beta-AGs maintain the constructive nature of AGs, two independent and primitive signs are needed to signal quantification. The *universal quantifier* is expressed as a line crossed by two *barbs* (*barbed line*) (Fig. 3a). The *particular (existential) quantifier* is expressed by a standard, *unbarbed line* (Fig. 3b).

Figure 3:   (a)   (b)

REMARK. Symbolising the two quantifiers separately stems from Peirce's early, 1882 notation for logical graphs. In a system of proto-graphs developed around 1882 (Peirce 1989, pp. 391–393, 394–399; see Roberts 1973, pp. 18–20; Bellucci & Pietarinen 2016) Peirce uses two kinds of "bonds", namely of lines, to express quantification: a plain line of connection represents an existentially quantified relative multiplication; a line crossed by a short mark represents universally quantified relative sum. This was long before Peirce finally managed, in mid-1890s, to introduce ovals and ditto the distinction between negative and positive areas (polarities), which rendered barbed lines superfluous. As the negation of a universal assertion is not a particular assertion, our system of intuitionistic Beta graphs recovers the necessity to notationally distinguish between two mutually indefinable quantifiers.

REMARK. The intended meaning of the barbed line is to denote having in possession "a general method to justifiably assert something" (Fig. 3a), while the intended meaning of the simple, unbarbed line is to denote having in possession "a specific, or a certain, method to justifiably assert something" (Fig. 3b). We may call both of these lines *lines of assertion*, in contradistinction to the lines of *identity* as coined by Peirce for classical, first-order Beta graphs.

This is not to say that the notion of identity would be detached from the meaning of the lines of assertion. Rather, it is merely to emphasise that it is the (methods of) assertions that are now explicitly being quantified by the outermost extremities of those lines. In fact no language of quantified graphical intuitionistic logic along the lines of EGs (pardon the pun) could as such be *without identity*—such as having its identity defined instead via a special two-place relation of equivalence—because by virtue of the continuity of the line, any such notation *also* incorporates the signification of numerical identity, in addition to the significations of universal and particular quantification, denotation of the binding scope of the values of quantification, and the predications that those values make.

Boxes have an important role in Beta-AGs just as they do in AGs. The box in

Fig. 4a means that "a content is asserted". The blank sheet (space of all assertions) means that a logical truth is asserted (Fig. 4b).

Figure 4: 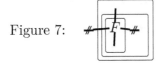 (b)

In Beta AGs, boxes are furthermore used as important graphical devices to disambiguate the logical orderings of quantifiers, showing what the priority scopes of the quantifiers are in addition to their binding. Boxes and lines may be combined as in Fig. 5a, which expresses "anything is asserted of F" (or, perhaps slightly more precisely, that "anything that is asserted means asserting F of it"). The graph of Fig. 5b, in turn, states that "something is asserted of F".

Figure 5:   ⊬─F⟩   ──F⟩
                (a)   (b)

The graph in Fig. 6 represents the FOL formula $\forall x \exists y F(x, y)$. This is consistent with Peirce's "endoporeutic" interpretation of graphs (that is, the interpretation that moves in the direction from the outermost part of the graph to the innermost parts, namely from the sheet of assertion inside the ever-more boxed or cornered areas, and not the opposite):

Figure 6:   ⊬─F─⟩

Nested quantifiers are expressed as in the graph of Fig. 7, corresponding to the formula of standard FOL $\forall x \exists y \forall u \exists z F(x, y, u, z)$:

Figure 7:   ⊬─F─#

The graph in Fig. 8 is an example of how to represent a quantification and disjunction, as in $\forall x (Gx \vee F)$:

Figure 8:   ⟨F⊢G⟩

The graphs in Figs. 5–8 are examples of how quantificational lines interact with boxes in AGs in the composition of complex formulas. The definition of well-formed (well-scribed) Beta-AGs is easily construed from these examples.

367

Notice, however, that the graph in Fig. 9, despite its apparent simplicity, is not a well-scribed graph of the language, as it does not disambiguate between the logical priority scopes of the two different lines of assertion:

Figure 9:   —F

Rather, this simple graph represents the Henkin–Hintikka formula $\begin{bmatrix} \exists x \\ \forall y \end{bmatrix} Fxy$. It denotes two independent, perhaps simultaneously made quantificational assertions (Pietarinen 2001, 2002). Such possibilities do not seem to have been explored in the literature.

4 Rules of Transformation in Beta-Assertive Graphs

The two major classes of irreversible rules of transformations, namely permissions to *insert* and permissions to *erase*, remain as they are in the theory of AGs (Chiffi & Pietarinen 2020). Unlike in EGs, these permissions are not defined in terms of the polarities of the areas (negative or positive). Instead, they take into account what is permissible within the antecedent and consequent areas of cornerings, respectively. Just like AGs, Beta-AGs are polarity-free; a design choice resulting from the absence of a primitive logical constant to designate negation, and the possibility of defining negation by the blot and the cornering in the intuitionistic fashion.

Moreover, the reversible *Beta-iteration* rule and its converse of *Beta-deiteration* behave just as they would when one is manipulating the standard lines of identity in classical Beta-EGs (Peirce 2019, Roberts 1973).

4.1 Rules of Transformation for Lines of Assertion

In addition to the rules of Alpha-AGs, however, in order to handle inferences with assertive quantificational lines several new rules are now needed.

In what follows, and just as is the case with ligatures (complexes of lines of identity) in the Beta part of EGs, any line depicted on antecedent areas as having a loose end may refer either to a continuous, branching or connected ligature.

Insertion Rules.

(**Ins 1.1**) Any two loose ends of an unbarbed line may be connected on the antecedent of cornering:

From [figure] we may infer [figure] .

(**Ins 1.2**) Any two loose ends of a barbed line can be connected on the antecedent of cornering:

From ⬚ we may infer ⬚ .

Two special cases of the insertion rule are:

(**Ins 1.2.1**) A barb '$/\!/$' can be inserted on a loose end of an unbarbed line on antecedent areas:[3]

From ⬚ we may infer ⬚ .

(**Ins 1.2.2**) A box can be added to a barbed line on a consequent area:

From ⬚ we may infer ⬚ .

Erasure Rules.

(**Er 2.1**) Any unbarbed line can be cut on a consequent area of a cornering (and on the sheet of assertion):

From ⬚ we may infer ⬚ .

(**Er 2.2**) Any barbed line can be cut on an antecedent area of a cornering:

From ⬚ we may infer ⬚ .

Two special cases of erasure are:

(**Er 2.2.1**) A barb can be erased from the barbed line whose outermost loose end rests on a consequent area of a cornering (or on the sheet of assertion):

From ⬚ we may infer ⬚ .

---

[3]REMARK. A line that has its extremal end on the boundary of the box *from the inside* is taken to have that end inside the box, that is, on the antecedent area. In contrast, a line whose innermost extremity lies on the boundary of the cut *from the outside*, is taken to penetrate into that inner area. These further conventions are analogous to the conventions concerning the behaviour of the line in standard Beta-EGs.

(**Er 2.2.2**) On an antecedent area of a cornering, a box can be removed from a barbed line that penetrates into it, thus:

From ⊞ we may infer ⊞ .

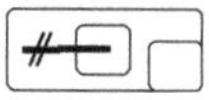

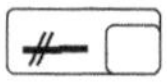

Axioms. In addition to the α-axioms of the previous section, the following β-axioms are at play.

β-(**Ax1**) An unattached, unbarbed line of assertion —— , a simple loop made out of such a line $\bigcirc$, and a simple dot **0** (an atrophied loop not to be confused with the blot of absurdity) can be inserted and erased anywhere on the sheet, including within boxes and cornerings.

β-(**Ax2**) An unattached barbed line of assertion ⊬—— , a simple barbed loop made out of such a line $\bigcirc$, and a barbed atrophied loop (a barbed dot)  can be inserted and erased anywhere on the sheet, including within boxes and cornerings.

4.2   Theorems and Examples

Next, some examples are given of how to derive the theorems of first-order intuitionistic logic in the system of the set of transformation rules of Beta-AGs as expounded above.

Theorems.

370

1.a) $\vdash_i \forall x \neg Fx \to \neg \exists x Fx$

(Ax)
(Cr)
(Ins)
(It)
β-(Ax1)
β-(It)
(Ins 1.1)
(Ins 1.2.1)

1.b) $\vdash_i \neg \exists x Fx \to \forall x \neg Fx$

(Ax)
(Cr)
(Ins)
(It)
(Ins 1.2.1)
β-(Ax1)
β-(It)
(Ins 1.2)

2. $\vdash_i \exists x \neg Fx \to \neg \forall x Fx$

From 1.a) (Ins 1.1), follow by

(Ins 1.2.1)

3. $\vdash_i \forall x Fx \to \exists x Fx$

(Ax)
(Ins)
(It)
(Ins 1.2.1)

4.a) $\vdash_i At \to \exists x Ax$

(Ax)
(Ins)
(It)

4.b) $\vdash_i \forall x Ax \to At$

(Ax)
(Ins)+(It)
(Er 2.2.1)

4.c) $\vdash_i \exists x(Ax \to B) \to (\exists x Ax \to B)$

4.d) $\vdash_i \forall x(B \to Ax) \to (B \to \forall x Ax)$

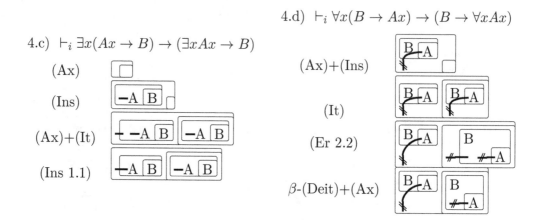

(Ax)

(Ins)

(Ax)+(It)

(Ins 1.1)

(Ax)+(Ins)

(It)

(Er 2.2)

β-(Deit)+(Ax)

Examples.

A couple of further examples illustrate the workings and the intuitionistic nature of the permissive transformations of Beta-AGs.

(A) $\vdash_i \exists x Fx \to \neg \forall x \neg Fx$

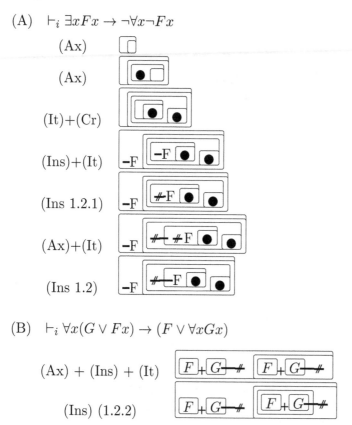

(Ax)

(Ax)

(It)+(Cr)

(Ins)+(It)

(Ins 1.2.1)

(Ax)+(It)

(Ins 1.2)

(B) $\vdash_i \forall x(G \vee Fx) \to (F \vee \forall x Gx)$

(Ax) + (Ins) + (It)

(Ins) (1.2.2)

(C) $\vdash_i \exists x (F \to Gx) \to (F \to \exists x Gx)$

(Ax) + (Ins) + (It)

(Er 2.1)

(Deit) + (Ax)

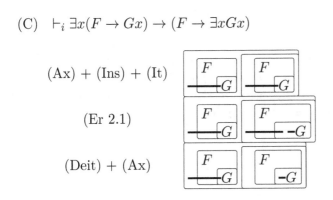

(D) $\nvdash_i \neg \forall x F x \to \exists x \neg F x$

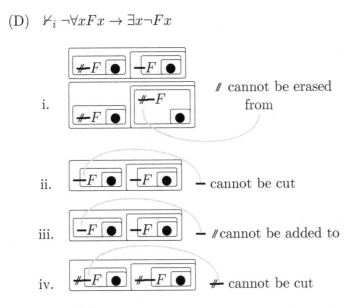

i. // cannot be erased from

ii. — cannot be cut

iii. — // cannot be added to

iv. — cannot be cut

The graph representing the formula (D) $\neg \forall x F x \to \exists x \neg F x$ is not a theorem of the system of Beta-AGs. Semantically, what is blocked is an attempt of moving negation over a general method of asserting in order to turn it into a particular assertion, because a negative of a universal does not result in a particular method of assertion ("absence of an assertion is not an assertion of absence").

As illustrated in the above diagrams (i)–(iv), one cannot erase a barb from the barbed line on an antecedent of a cornering, that one cannot cut a simple line on an antecedent, that a barb cannot be added to a simple line on an antecedent, and that a barbed line cannot be cut on an antecedent of a cornering. This exhausts the classes of possibilities.

In a similar vein, consider

(E) $\nvdash_i \neg\forall x \neg F x \rightarrow \exists x F x$

One may contrast this with what is the theorem of the system, namely (A) $\exists x F x \rightarrow \neg\forall x \neg F x$. Given the intuitionistic nature of the logic of Beta Assertive Graphs, the graph of the formula (E) is not a theorem of this system of Beta-AGs, since asserting that $\forall x \neg F x$ is not asserted will by itself not suffice to produce an instance of anything that would be asserted.

5 Conclusion

In this paper, the basic notational and inferential elements and rules of the system of Beta-AGs were sketched, thereby extending for the first time the assertive logic of AGs to the quantificational case.

The characteristic features of the system of Beta-AGs are the following.

1. Unlike standard FOL=, the language of Beta-AGs, just like the language of classical Beta-EGs, is occurrence-referential, meaning that it is the occurrence and not the type of a variable that determines sameness and distinctness of reference.

2. In Beta-AGs, two distinct notations for quantification—the lines of assertion—are needed, because just as in intuitionistic systems in general, these two quantifiers are not interdefinable. Beta-AGs maintain all the constructive features of AGs and result in a diagrammatic system for intuitionistic quantificational logic, represented by logical graphs with assertion-based interpretation of the main logical constants.

3. Since boxes are deictic notational devices used to express assertions and the grouping of formulas in lieu of parentheses, and since unlike in EGs there is no specific sign for negation (negation is defined as an implication of *absurdum*), there are no polarities, namely distinctions between negative and positive areas, in the language of AGs.

   Consequently, the insertion and erasure rules are more bountiful than in the classical case, which is necessitated by the need to circumvent the absence of polarities. While it might look like this results in a more cumbersome system of permissions than is the case in the classical EGs, it can be takes as the price to be paid when one desires to approximate and regiment well the pre-theoretic idea of intended models for a graphical logic that copes with intuitionistic assertions.

4. The graphical method of logic that explicitly represents and reasons about assertions manages to do that without any additional *ad hoc* sign of assertions. This is possible since the notion of assertion is embedded in the notion of the sheet of assertion (SA), which in the present system may be taken to be the canvass of all transformations, proofs or methods of asserting. The lines of assertions scribed on that canvass make explicit the quantificational logic of such assertions.

Completeness of the logic presented depends on the choices of interpretations of canonical models (such as presheaves) that are to correlate with that of the pre-theoretical notion of proofs and transformations. Other and related further lines of research may be devoted to investigate, from a graphical perspective of logic, the role of *admissible rules* in Beta Assertive Graphs (for example, (Cm/Cs) are the two admissible rules in the Alpha part of AGs), such as the presence of Markov's rule or the Independence-of-Premise rule, the corresponding principles of which nevertheless fail to hold intuitionistically.

The quantificational expressions of Beta-AGs are related to intuitionistic quantification, as noted in the present paper. They may be extended to cover classical propositional calculus by an addition of a rule of "an erasure of the coinciding corners", as shown to be the case for propositional AGs in Chiffi & Pietarinen (2020).

References

[1] Bellucci, F., Pietarinen, A.-V. 2017. Assertion and denial: A contribution from logical notations, *J. Appl. Logic* 25, 1–22.

[2] Bellucci, F., Pietarinen, A.-V. 2016. From Mitchell to Carus. Fourteen years of logical graphs in the making. *Tr. Peirce Society* 52, 539–575.

[3] Bellucci, F., Pietarinen, A.-V. 2020. Notational Differences. *Acta Analytica* 35, 289–314. https://doi.org/10.1007/s12136-020-00425-1

[4] Bellucci, F., Chiffi, D., Pietarinen, A.-V. 2018. Assertive graphs. *J. Applied Non-Classical Logic* 28(1), 72–91.

[5] Carrara, M., Chiffi, D., De Florio, C. 2017. Assertions and hypotheses: A logical framework for their opposition relations. *Logic J. IGPL*, 25(2), 131–144.

[6] Carrara, M., Chiffi, D., Pietarinen, A.-V. 2020. Some Logical Notations for Pragmatic Assertions. *Logique et Analyse* 251, 297–315.

[7] Chiffi, D., Pietarinen, A.-V. 2020. On the logical philosophy of assertive graphs. *Journal of Logic, Language and Information* 29, 375–397. https://doi.org/10.1007/s10849-020-09315-6

[8] Dummett, M. 1978. *Truth and Other Enigmas*. London: Duckworth.

[9] Frege, G. 1879. *Begriffsschrift.* Halle: L. Nebert. Translated as *Begriffsschrift, a Formula Language, Modeled upon that of Arithmetic, for Pure Thought.* In *From Frege to Gödel,* edited by Jean van Heijenoort. Cambridge, MA: Harvard University Press, 1967.

[10] Gödel, K. 1961. The modern development of the foundations of mathematics in the light of philosophy. Reprinted in Gödel, K. (1995). In S. Feferman, et al. (eds.), *Collected Works* (CWIII) (Vol.III, 374—387). Oxford: Oxford University Press.

[11] Heyting, A. 1956. *Intuitionism: An Introduction.* North Holland, Amsterdam.

[12] Ma, M., Pietarinen, A.-V. 2018. A Graphical Deep Inference System for Intuitionistic Logic, *Logique & Analyse* 245, 73–114. https://doi.org/10.2143/LEA.245.0.3285706

[13] Peirce, C.S. 1989. *Writings of Charles S. Peirce. Vol. 4.* Peirce Edition Project (ed.). Bloomington: Indiana University Press.

[14] Peirce, C.S. 2019. *Logic of the Future: Writings on Existential Graphs.* Volume 1: History and Applications. Pietarinen, A.-V. (ed.). Berlin: De Gruyter.

[15] Pietarinen, A.-V. 2001. Varieties of IFing, G. Sandu & M. Pauly (eds.), *Proceedings of the ESSLLI 2001 Workshop on Logic and Games,* Helsinki: University of Helsinki.

[16] Pietarinen, A.-V. 2002. Propositional Logic of Imperfect Information: Foundations and Applications, *Notre Dame Journal of Formal Logic* 42(4), 193–210.

[17] Pietarinen, A.-V. 2015a. Two Papers on Existential Graphs by Charles S. Peirce: 1. Recent Developments of Existential Graphs and their Consequences for Logic (R 498, R 499, R 490, S-36; 1906), 2. Assurance through Reasoning (R 669, R 670; 1911). *Synthese* 192, 881–922.

[18] Pietarinen, A.-V. 2015b. Exploring the Beta Quadrant. *Synthese* 192, 941–970.

[19] Pietarinen, A.-V., Chiffi, D. 2018. Assertive and existential graphs: A comparison. In P. Chapman, G. Stapleton, A. Moktefi, S. Perez-Kriz & F. Bellucci (eds.), *Diagrammatic Representation and Inference. Diagrams 2018* (pp. 565–581). Lecture Notes in Computer Science (Vol. 10871). Springer.

[20] Pietarinen, A.-V., Bellucci, F., Bobrova, A., Hayden, N., Shafiei, M. 2020. The Blot. In Pietarinen, A.-V., Chapman, P., Bosveld-de Smet, L., Giardino, V., Corter, J., Linker, S. (eds.), *Lecture Notes in Artificial Intelligence* 12169, Cham: Springer, 225–238.

[21] Roberts, D.D. 1973. *The Existential Graphs of C.S. Peirce.* The Hague: Mouton.

[22] Rumfitt, I. 2000. 'Yes and no'. *Mind* 109(436): 781–823.

[23] Wetzel, L. Types and Tokens. 2018. *The Stanford Encyclopedia of Philosophy* (Fall 2018 Edition), E.N. Zalta (ed.).

Received 17 July 2020

DLEAC and the Rejection Paradox

Massimiliano Carrara

Department of Philosophy, Sociology, Education and Applied Psychology (FISPPA) – Padua University

massimiliano.carrara@unipd.it

Andrea Strollo

Department of Philosophy, Nanjing University (China)

andreastrollo@nju.edu.cn

Abstract

In this paper we first develop a Dialetheic Logic with Exclusive Assumptions and Conclusions, DLEAC. We adopt the semantics of the *logic of paradox* (LP) extended with a notion of *model* suitable for DLEAC, and we modify its *proof theory* by refining the notions of *assumption* and *conclusion*, which are understood as speech acts. We introduce a new paradox – the *rejectability paradox* – first informally, then formally. We then provide its derivation in an extension of DLEAC contanining the *rejectability* predicate.

1 Introduction

Rejection is standardly considered a speech act that expresses an attitude of dissent. In the last years, some calculi – whose aim is to formalize such a notion, such as the *refutation* or *rejection calculi* – have been proposed.

For a general introduction to these calculi, specifically in propositional logic, see [19]. Let me review two examples: Skura's refutation calculi (developed in [16]) and Wansing's [21] natural deduction calculus. Skura's refutation calculi (see [16] but also [17], and [18]) is based on a Łukasiewicz-style refutation calculi for propositional logics (see on this [20]). Skura proposed a system for the modal logic of S4 in [17]. With the same purpose, H. Wansing, in his [21] (and in other papers), introduced a natural deduction calculus whose central idea was to begin with pairs comprising a set of assertions and a set of rejections, obtaining a similar pair by inference. Wansing's idea was to dualize the introduction and elimination rules for intuitionistic propositional logic with a primitive notion of dual proof to obtain a kind of *bi-intuitionistic* propositional logic that combines *verification* and its dual, i.e. *falsification*.

Here I propose a refutation calculus based on a dialetheic conception of *negation* and *refutation* or *denial*.

Dialetheic *negation* is not exclusive, whereas *denial* is: in a dialetheic framework A and $\neg A$ may both be true, but you cannot correctly assert and deny A. This is how Priest in [10] tries to recover the exclusivity of negation by introducing the notion of *rejection* or *denial*[1] as a *speech act*. He claims that while it is possible to *accept* both a sentence and its negation[2], one *cannot* accept and reject the same sentence. Assertion and rejection or denial are *incompatible* speech acts.

In this paper, I take the impossibility of accepting and rejecting the same sentence as primitive. In this way, I conceive the rejection of sentence A as a speech act that – in virtue of its very meaning – expresses the fact that A is *only* false. Similarly, the act of rejecting $\neg A$ expresses the fact that A is *only* true.

This dialetheic use of rejection suggests a theory of natural deduction, where the acts of *assuming* and *concluding* may be understood in an *ordinary* or *exclusive* mode. To assume a sentence in an ordinary mode amounts to supposing that it is *at least* true; to assume it in an exclusive mode amounts to supposing that it is *only* true.

To assume A in the ordinary mode then corresponds to the assertion of A, whereas to assume A in the exclusive mode corresponds to the rejection of $\neg A$.

Any sentence can be rightfully assumed in an *ordinary* or *exclusive* mode at will. Similarly, to prove a sentence in an ordinary mode proves that it is (at least) true (under certain assumptions); to prove it in an exclusive mode proves that it is only true.

Accordingly, concluding in the ordinary mode is to be understood as the assertion of the conclusion, and concluding in the exclusive mode as the rejection of the negation of the conclusion.

The acts of proving A and $\neg A$ in an exclusive mode are incompatible because they both indefeasibly lead to the rejection of some assumptions they depend on. Specifically, concluding A and $\neg A$ in an exclusive mode –independent of any hypothesis – cannot in principle be performed by any rational human being. In this way, I realize the dialetheic aim of taking exclusivity as extraneous to the meaning of logical negation and embedded in the speech acts of assuming and concluding. I am going to formalize such speech acts within a modified natural deduction, where they will be governed by *indefeasible rules*.

---

[1] For a general background on denial in non-classical theories, see [13, §3].

[2] On the thesis see, also, [8]. For a recent discussion of the topic see also [7].

The goal of this paper is to formulate the above-mentioned modified natural deduction, via a dialetheic logic with exclusive assumptions and conclusions DLEAC, where *exclusivity* is expressed *via* certain speech acts. Specifically, in DLEAC, exclusivity is expressed using the speech acts of assuming and concluding. In this paper, I adopt the semantics of the *logic of paradox* (LP)[3] extended with a notion of *model* suitable for DLEAC and I modify its *proof theory* by refining the notions of assumption and conclusion, which are understood as speech acts (I follow, in this part of the paper, [5]). In the second part of the paper, I introduce a new paradox – the *rejectability paradox* – first informally, then formally; I give its derivation in an extension of DLEAC.

2   The Basics of DLEAC

Let me first introduce the basic elements of DLEAC, specifically its syntax and semantics.

Let L be a language of first-order logic with identity (FOL =) with individual constants and predicates of any *ariety*. For the sake of simplicity, I omit function symbols in L. I adopt the semantics for LP extended with a new, generalized notion of the model.

Let me briefly review the semantics for LP.[4]

A dialetheic interpretation of the propositional logic consists of an evaluation v that assigns to each atomic formula a member of the set $\{\{1\}, \{0\}, \{0, 1\}\}$. The v is extended to the complex formulas using the following clauses:

($\vee$) $v(A \vee B) = \{1\}$ if either $0 \notin v(A)$ or $0 \notin v(B)$;
$v(A \vee B) = \{0\}$ if $1 \notin v(A)$ and $1 \notin v(B)$;
$v(A \vee B) = \{0, 1\}$ otherwise.

($\wedge$) $v(A \wedge B) = \{1\}$ if $0 \notin v(A)$ and $0 \notin v(B)$;
$v(A \wedge B) = \{0\}$ if either $1 \notin v(A)$ or $1 \notin v(B)$;
$v(A \wedge B) = \{0, 1\}$ otherwise.

($\neg$) $v(\neg A) = \{1\}$ if $v(A) = \{0\}$;
$v(\neg A) = \{0\}$ if $v(A) = \{1\}$;
$v(\neg A) = \{0, 1\}$ otherwise.

A sentence A *is true* if $1 \in v(A)$, *is false* if $0 \in v(A)$;
A *is exclusively true* if $0 \notin v(A)$, *is exclusively false* if $1 \notin v(A)$.

---

[3]For a general background on LP, see [1], [2], [9], [15], [3].
[4]For details see [11, sez. 5.2, 5.3].

This semantics is extended in a similar way to first order logic with identity. I simplify, making the assumption that there is a name in the language L for every object of the domain D of quantification.

An evaluation v assigns to every individual constant a member of the domain D, and assigns to every unary predicate P two subsets of D: the extension P^+ and the counter-extension P^-, possibly overlapping, with the only constraint that $P^+ \cup P^- = $ D. Then:

$$v(Pa) = \{\,1\,\} \;\; \text{if } a \in P^+ - P^-$$
$$v(Pa) = \{\,0\,\} \;\; \text{if } a \in P^- - P^+$$
$$v(Pa) = \{\,0,1\,\} \;\text{if } a \in P^+ \cap P^-$$

Similarly for predicates of degree > 1.
The constraints for the identity sign $(=)$ are the following:

$(=)^+ = \{(a,a) : a \in \mathsf{D}\}$, while $(=)^-$ is arbitrary with the only constraint that $(=)^+ \cup (=)^- = \mathsf{D}$.

The clauses for the universal and existential quantifiers are analogous to those of conjunction and disjunction, respectively.

I extend the semantics of LP by introducing a notion of model suitable for DLEAC.

Let S be any set of sentences of a first order language L, some of which may be starred (i.e. marked by a star *). Observe that stars * do not belong to the object language L.

A model M of S is an LP-interpretation in which all sentences of S are true and the starred ones are exclusively true.

A sentence A (a starred sentence A^*) is a *semantic consequence* of a set S of possibly starred sentences, in symbols $S \models A(*)$, if it is true (exclusively true) in every model of S.

3   DLEAC: Deductive rules

Let A, B, C... be formulas of a first order language L, and let Γ be a finite set of possibly starred formulas.

A sequent is an expression of the form:

$$\Gamma: C \; (*),$$

to be read: "From the assumptions in Γ, one can infer the conclusion C (in an ordinary or exclusive mode)."

The non-starred formulas in Γ are assumed to be in an ordinary mode, and the starred ones in an exclusive mode. Similarly, the conclusion C can be understood in an ordinary or in an exclusive mode.

3.1 Basic deductive rules for DLEAC

In this section I list the primitive inference rules (I follow [5]). When stars occur in parentheses ( ) the deductive rule holds in the double form:

- with all stars in parentheses at work

- with all stars in parentheses deleted.

Reflexivity:

$$A(*)\colon A(*)$$

$$A*\colon A$$

The informal reading of the first rule is the following: From the assumption that A is only true (at least true), it follows that A is only true (at least true). The informal reading of the second rule is: From the assumption that A is only true it follows that A is (at least) true.

Weakening:

$$\frac{\Gamma : A(*)}{\Gamma\,\Delta : A(*)}$$

Cut:

$$\frac{\Gamma : A(*),\ \Delta\,A(*)\colon B}{\Gamma\,\Delta : B}$$

381

$$\frac{\Gamma : A(*), \, \Delta \, A(*) : B^*}{\Gamma \, \Delta : B^*}$$

Conjunction:

$$I\wedge \frac{\Gamma : A(*), \, \Delta : B(*)}{\Gamma \, \Delta : A \wedge B(*)}$$

$$E\wedge \frac{\Gamma : A \wedge B(*)}{\Gamma : A(*)}$$

$$E\wedge \frac{\Gamma : A \wedge B(*)}{\Gamma : B(*)}$$

Disjunction:

$$I\vee \frac{\Gamma : A \, (*)}{\Gamma : A \vee B \, (*)}$$

$$E\vee \frac{\Gamma A : C(*), \, \Delta \, B : C(*), \, \Lambda : A \vee B}{\Gamma \, \Delta \, \Lambda : C(*)}$$

$$E\vee \frac{\Gamma A^* : C(*), \, \Delta \, B^* : C(*), \, \Lambda : A \vee B^*}{\Gamma \, \Delta \, \Lambda : C(*)}$$

Double negation:

$$A \, (*) : \neg\neg A(*)$$

$$\neg\neg A \, (*) : A(*)$$

Introduction of absurd (IA):

$$\frac{\Gamma : A^*, \Delta : \neg A}{\Gamma\Delta : A \wedge \neg A^*}$$

The informal justification of IA is the following: From A and $\neg A$ follows $A \wedge \neg A$. Furthermore, since A is only true, it cannot be a *dialetheia*; therefore $\neg A$ also cannot be a *dialetheia*. As a result, neither of the conjuncts of $A \wedge \neg A$ can also be false, and therefore, $A \wedge \neg A$ is *only* true.

Since $\neg(A \wedge \neg A)$ is a *dialetheic logical law*, the conclusion $A \wedge \neg A^*$ is an *authentic absurd*, i.e. a conclusion unacceptable even by a dialetheist. Since, dialethically, $A \wedge \neg A$ might be true, it does not count as an absurd. For this reason, by an *absurd*, I mean a formula $A \wedge \neg A$ that is *only true*.

Reductio ad absurdum (RAA):

$$\frac{\Gamma A^*:\ B \wedge \neg B^*}{\Gamma : \neg A}$$

$$\frac{\Gamma A:\ B \wedge \neg B^*}{\Gamma : \neg A^*}$$

Informally, RAA works in this way: If the assumption that A is true (only true) leads to the *authentic absurd*, it cannot be true (only true), hence it is *only false* (at least false).

The rules for the quantifiers are analogous to those of conjunction ($\wedge$) and disjunction ($\vee$). The rules for identity are as follows:

Introduction of identity (I $=$):

$$: x = x$$

Elimination of identity (E $=$):

$$x = y, Px : Py$$

$$E = \frac{\Gamma A^*\colon \neg(t = t)^*}{\Gamma \colon \neg A}$$

$$E = \frac{\Gamma A\colon \neg(t = t)^*}{\Gamma \colon \neg A^*}$$

Observe that, according to the semantics of identity $(=)$, a sentence having the form $(t = t)$[5] cannot be *exclusively false*.

3.2 Derived deductive rules for DLEAC

In this section, I introduce some derived rules of DLEAC:

Material conditional:

$$\frac{\Gamma A \ (*)\colon B(*)}{\Gamma \colon \neg A \vee B}$$

$$\frac{\Gamma A^*\colon B}{\Gamma \colon \neg A \vee B}$$

$$\frac{\Gamma A\colon B(*)}{\Gamma \colon \neg A \vee B(*)}$$

Elimination of absurd (Ex absurdo quodlibet) (EA):

$$EA \ \frac{\Gamma \colon A \wedge \neg A^*}{\Gamma \colon B^*}$$

Notice that EA is a derived rule.

Modus ponens (MPP):

$$MPP \ \frac{\Gamma \colon A^*, \ \Delta \colon \neg A \vee B}{\Gamma \Delta \colon B}$$

---

[5]where 't' is an individual constant or a variable.

$$\text{MPP1} \ \frac{\Gamma : A, \Delta : \neg A \vee B^*}{\Gamma\Delta : B^*}$$

For an example of how **DLEAC** works, here is the derivation of MPP1:

1	1.	A	Assumption
2	2.	$\neg A \vee B^*$	Assumption
3	3.	$\neg A^*$	Assumption
1, 3	4.	$A \wedge \neg A^*$	IA
1, 3	5.	B^*	EA
6	6.	B^*	Assumption
6	7.	B^*	*Reflexivity*
1, 2	8.	B^*	E∨

Following **LP**, the material conditional is not a genuine conditional because, in general, it does not permit the validity of MPP.[6] In this approach, the validity of MPP is appropriate under a starred assumption. This way, we obtain the following reading of the quasi-validity of MPP for a dialetheist: MPP is appropriate when at least one of the two premises is starred.

The following are other derived rules of **DLEAC**.

De Morgan rules:

$$\frac{\Gamma : \neg(A \wedge B)(*)}{\Gamma : \neg A \vee \neg B \ (*)}$$

$$\frac{\Gamma : \neg A \vee \neg B \ (*)}{\Gamma : \neg(A \wedge B)(*)}$$

$$\frac{\Gamma : \neg(A \vee B)(*)}{\Gamma : \neg A \wedge \neg B(*)}$$

$$\frac{\Gamma : \neg A \wedge \neg B(*)}{\Gamma : \neg(A \vee B)(*)}$$

The Law of non-contradiction:

---

[6]For an extended discussion of this topic see [4].

$$\Gamma : \neg(A \wedge \neg A)$$

The Law of the excluded middle:

$$\Gamma : (A \vee \neg A)$$

4   The Completeness of DLEAC

Let S be any set of possibly starred sentences.

I suggest that S is *dialetheically consistent* ($\mathsf{d} - consistent$) if no conclusion of form $(A \wedge \neg A)^*$ is derivable from S.

Theorem 1. If S is $\mathsf{d} - consistent$, then it has a *model* $\mathcal{M}$.

Proof. Let S be $\mathsf{d} - consistent$. Extend the language L to a language L' with an infinite sequence of new individual constants $c_1, c_2, ..., c_n, ...$ Let

$$A_1, A_2, ..., A_n, ...$$

be a sequence of all L'-sentences. I inductively define the sequence:

$$S_0, S_1, ..., S_n, ...$$

of sets of (possibly starred) L'-sentences as follows:

1. $S_0 = S$;

2. $S_{n+1} = S_n$ if A_{n+1} is derivable from S_n and is not an existential sentence;

3. $S_{n+1} = S_n \cup \{B(c)(*)\}$ if $A_{n+1} = \exists x B(x)$ and $S_n \vdash \exists x B(x)(*)$, where c is the first constant not occurring in S_n nor in A_{n+1};

4. $S_{n+1} = S_n \cup \{\neg A_{n+1}(*)\}$ if A_{n+1} is not derivable from S_n.

Let us consider the following definition:

$$S_\omega = \cup_{n \in \mathsf{N}} \, S_n$$

One can prove by induction that each S_n is $\mathsf{d} - consistent$, so that S_ω is $\mathsf{d} - consistent$.

Consider, for example, 3. Suppose, by reduction, that S_{n+1} is inconsistent. If $S_{n+1} = S_n \cup \{B(c)\}$, then $S_n \vdash \neg B(c)^*$ and hence $S_n \vdash \forall x \neg B(x)^*$, against the $\mathsf{d} - consistency$ of S_n. If $S_{n+1} = S_n \cup \{B(c)^*\}$, then $A_{n+1} = \exists x B(x)^*$. Then $S_n \vdash \neg B(c)$ and hence $S_n \vdash \forall x \neg B(x)$, against the $\mathsf{d} - consistency$ of S_n.

$\square$

S_ω is *deductively complete*: for any L'-sentence, if not $S_\omega \vdash A$, then $S_\omega \vdash \neg A^*$.

An interpretation I of L' can be defined as follows. Take the set D of all individual constants as domain. Evaluation v can be defined as follows:

$1 \in v(A)$ *iff* $S_\omega \vdash A$, $0 \in v(A)$ *iff* $S_\omega \vdash \neg A$, for every atomic L'-sentence.

One can prove, by induction on the complexity of a sentence A, that $v(A) = \{1\}$ *iff* $S_\omega \vdash A^*$, $v(A) = \{0\}$ *iff* $S_\omega \vdash \neg A$ *, $v(A) = \{0, 1\}$ *iff* $S_\omega \vdash A$ and $S_\omega \vdash \neg A$.
It follows that I is a model of S_ω and hence of S.

Completeness. If $S \models A(*)$ then $S \vdash A(*)$

Proof. Let $S \models A$. Suppose, by reduction, that it is not the case that $S \vdash A$. Then $S \cup \{\neg A^*\}$ is $\mathsf{d} - consistent$ and hence has a model where $\neg A$ is only true, against the hypothesis. Similarly if $S \models A^*$.

$\square$

5 Extending a theory with the truth predicate

In this section, I show (I refer to what is done in [5]) that any dialetheic interpretation of a first order language L can be extended to an interpretation of a language L' capable of expressing its own truth predicate.

Let L be a first order language with predicates and individual constants (for simplicity, I ignore functions). Let I be any interpretation of L and D its domain of quantification. Extend L with a new predicate symbol T and infinitely many individual constants. Extend D to D' by adding all L'-sentences to D. Let I' map the new constants 1-1 onto D' so that any member of D' has an L'-name. If A is an L'-sentence, we indicate by $\lceil A \rceil$ its name.

I' puts all sentences in the counter-extension of the L-predicates and the members of D in the counter-extension of T. As shown (in [5]), it is possible to fix the interpretation of T in such a way that it turns out to be the truth predicate of I', so that, for all L'-sentences A, A and $T(\lceil A \rceil)$ have the same truth values.

> *Theorem 2.* There is an extension of I to an interpretation I' of L' such that, for every $L' - sentence$ A, A and $T\lceil A \rceil$ have the same truth values, while the values of the $L - sentences$, relativized to D are unchanged.

An evaluation v' is a *sub-evaluation* of v, in symbols $v' \subset v$, if v' is obtained from v by suppressing a truth value of some atomic dialetheias.

> *Lemma.* If a sentence has a unique v-value, this is also the unique v'-value, for any $v' \subset v$.

Proof. The proof is obtained by an induction on the complexity of the sentence. □

Proof of the theorem 2. The following sequence can be defined by transfinite induction:

$$v_0 \supset v_1 \supset ... \supset v_\alpha, ... \text{ (for all ordinals).}$$

They are evaluations of sentences of form $T(\lceil A \rceil)$ for all L'-sentences A:

$$v_0(T(\lceil A \rceil)) = \{0, 1\} \text{ for all } A;$$

$$v_{\alpha+1}(T\lceil A \rceil) \text{ is defined by cases:}$$

(i) $v_{\alpha+1} (T\lceil A \rceil) = v_\alpha(A)$ if $v_\alpha(A)$ is a singleton, while $v_\alpha (T\lceil A \rceil)$ is not;

(ii) $v_{\alpha+1} (T\lceil A \rceil) = v_\alpha (T\lceil A \rceil)$ otherwise.

$$v_\beta (T(\lceil A \rceil)) = \cap_{\alpha<\beta} v_\alpha (T(\lceil A \rceil)) \text{ for } \beta \text{ limit.}$$

One can prove, by transfinite induction, that:

> for all α, if $v_\alpha (T\lceil A \rceil)$ is a singleton, then $v_\alpha (T\lceil A \rceil) = v_\alpha(A)$.

1. $\alpha = 0$. Trivial

2. $\alpha = \beta + 1$. Let v_α $(\mathsf{T}\lceil A\rceil)$ be a singleton. We distinguish the two cases above (i) and (ii):

(i) v_α $(\mathsf{T}\lceil A\rceil) = v_\beta$ (A) and, since v_β (A) is a singleton and v_α is a sub-evaluation of v_β, by the *lemma* v_α $(A) = v_\beta(A)$.

(ii) v_α $(\mathsf{T}\lceil A\rceil) = v_\beta$ $(\mathsf{T}\lceil A\rceil)$. Therefore, v_β $(\mathsf{T}\lceil A\rceil)$ is a singleton. By the induction hypothesis, v_β $(\mathsf{T}\lceil A\rceil) = v_\beta(A)$ and, by the lemma, $v_\alpha(A) = v_\beta(A)$.

3. α limit. If v_α $(\mathsf{T}\lceil A\rceil)$ is a singleton, then v_α $(\mathsf{T}\lceil A\rceil) = v_\beta$ $(\mathsf{T}\lceil A\rceil) = v_\beta(A)$, for some $\beta < \alpha$. By the lemma, $v_\alpha(A) = v_\beta(A)$.

$\square$

Observe that if $v_\beta \neq v_\alpha$ with $\beta < \alpha$, there is some sentence A such that $v_\beta(A) = \{0, 1\}$, while $v_\gamma(A)$ is a singleton for all $\gamma > \beta$. And since only countably many sentences can satisfy–for some ordinal–this condition, it follows that at least from the first uncountable ordinal on, the sequence becomes stationary. If δ is such an ordinal, v_δ is clearly the required evaluation.

Consider then the following familiar rules for a truth predicate.

Primitive Tarki's rules:

$$\frac{\Gamma : A \ (*)}{\Gamma : \mathsf{T}(\lceil A\rceil)(*)}$$

$$\frac{\Gamma : \mathsf{T}(\lceil A\rceil)(*)}{\Gamma : A \ (*)}$$

From the above-mentioned rules, it follows that, semantically, $\mathsf{T}(\lceil A\rceil)$ and A possess the same truth-values.

Derived Tarki's rules:

$$\frac{\Gamma : \neg\mathsf{T}(\lceil A\rceil)(*)}{\Gamma : \mathsf{T}(\lceil \neg A\rceil)(*)}$$

$$\frac{\Gamma : T(\lceil \neg A \rceil)(*)}{\Gamma : \neg T(\lceil A \rceil)(*)}$$

The guiding idea of the evaluation constructed above suggests that, when Tarski's rules fail to determine a unique value for a sentence of form $T(\lceil A \rceil)$, there is no reason to arbitrarily choose one of the two truth values for $T(\lceil A \rceil)$, which will, therefore, be evaluated as a dialetheia.

Conservativity. The extension of any theory by means of the predicate T with Tarski's rules is *conservative.*

Proof. It is derivable from *Completeness* and *Theorem 1.*   □

6   The refutation paradox in DLEAC

Let us go back to the speech acts of *assertion* and *denial*. Classically, to deny A is equivalent to asserting $\neg A$: A is correctly denied iff $\neg A$ is correctly asserted,[7] but the dialetheic denial of A is *stronger* than the assertion of $\neg A$. As Littman and Simmons observed in [6], because the dialetheist appeals to "non-standard relations between assertability and deniability", a full account of these notions is required. Specifically, any such account would need to deal with apparent paradoxes that turn on the notion of *assertability* and/or *deniability.*

In their paper Littman and Simmons proposed a paradox concerning assertion, *the assertability paradox* ([6], 320). Take a sentence α having the form:

(α) α is not assertable.

They argue that (α) is a dialetheia. Here is the proof they give:

Proof.

Suppose (α) is *true.* Then what it says is the case. So (α) (i.e. (α) α is not assertable) is not assertable. But we have just asserted (α). So (α) is assertable–and we have a contradiction.

Suppose, on the other hand, that (α) is *false.* Then what (α) says is not the case, and (α) is assertable. So we may assert: (α) is not assertable. Again, we have a contradiction.

□

---

[7][14] calls this the *denial equivalence.*

If α is a dialetheia, then it is both assertable and not assertable. But how is it possible to both assert and not assert a sentence? This seems to be impossible also for a dialetheist. While acknowledging that certain sentences can be both true and false, a dialetheist cannot admit that a sentence is assertable and not assertable: "there seems to be no room for manoeuvre. So, the dialetheist will need to say more" ([6], 320).

Note that there is a problem in the proof of the assertability paradox: The mere supposition that α is *true does not imply its assertability*. Indeed, assertability implies the recognition, not just the mere supposition, of the truth of (α).

In the following text, I propose a revised version of the assertability paradox called *the amended assertability paradox*. To amend the argument, a reasoning by which the truth of (α) can be recognized is provided. The reasoning is given by the following proof.

Let us prove dialetheically that α is true by distinguishing the following two cases:

- (1) Assume that (α) is false. Then, its negation is true, so (α) is assertable and then it is true.

- (2) Assume that (α) is true. Then it is true.

According to the Law of the excluded middle – as formulated in classical first order logic – (α) is true. In this way, we have a proof – not just a supposition – of the truth of (α), and we can assert it. So (α) is assertable, in opposition to what(α) claims and is false. Therefore it is a dialetheia. That is the *revised assertability paradox*.

Now, again, let us repeat our question: If α is a dialetheia, it is both assertable and not assertable, but how is it possible to both assert and not assert a sentence?

Is this a *real problem* for a dialetheist? The quick answer is, 'No'. Why should it be a problem for a dialetheist to admit that a sentence is *both* assertable and *not* assertable? Once the exclusivity of logical negation has been rejected, the non assertability of a sentence does not exclude its assertability, even if it is far from clear what it means that a certain sentence *is and is not assertable*.

Here I am not interested in giving a philosophical answer to the above questions. They concern the philosophical *status* of these speech acts, and this is not the place to discuss them.

Priest, in ([12]) proposed something similar to the *assertability paradox*: the *irrationalist paradox*. Let I be a sentence having the form:

(I): it is not rational to accept I.

You can both accept and and reject (I).

Priest's derivation is as follows ([12] 121).

Proof.

Let Rat be an operator expressing *rational acceptance* and $R = \neg\mathsf{Rat}(R)$, Priest derives R from the schema

$$(P) \quad \neg\mathsf{Rat}(A \wedge \neg\mathsf{Rat}(A)),$$

as follows:

$$\frac{\dfrac{\dfrac{\neg\mathsf{Rat}(R \wedge \neg\mathsf{Rat}(R))}{\neg\mathsf{Rat}(R \wedge R)}}{\neg\mathsf{Rat}(R)}}{R}.$$

That is, R, and hence $\neg\mathsf{Rat}(R)$, is deducible from P: $P \vdash R$.

Assuming that rational acceptance (Rat) is closed under single-premise deducibility, and that P is rationally acceptable (*and it seems to be: if someone believes A, and, at the same time, believes that it is not rationally permissible to believe A, that would seem to be pretty irrational – not something that is itself rationally permissible*), we have $\mathsf{Rat}(P) \vdash \mathsf{Rat}(R)$, and we have a contradiction.

$\square$

Priest calls this kind of paradox a *rational dilemma*. He observes that a dialetheist cannot rule out a priori the occurrence of rational dilemmas:

> Arguably, the existence of dilemmas is simply a fact of life ([10], 111).

Moreover, he maintains that the irrationalist's paradox is much more problematic for a classicist than for a dialetheist. For the latter it is not irrational to believe both a sentence α and that it is irrational to believe α, if such a belief is also rational, an option clearly closed to a fan of classical logic. This argument is in line with the observation to the *assertability paradox* done before: if negation is non-exclusive, a dialetheist can rightly assert a non-assertable sentence, if she has recognized that it is also true.

Again, it is not the aim of this paper to debate Priest's argument on *the irrationalist paradox*.

I would like just to expand the *revised assertability paradox* in a *rejection direction*. In what follow I first informally introduce *the rejectability (or deniability) paradox*, then I give its derivation in an extension of **DLEAC** with the *rejectability* predicate.

Let R be a sentence having the form:

(R) the sentence R is rejectable.

You can both accept and reject (R).

Proof.

Assuming that (R) *is true*, then it is rejectable. So, there is a state of knowledge in which one can *reject* it. In such a state, one recognizes that what (R) says is true, so that one is in a position to *assert* (R). So you *both reject and accept R*. Thus, the assumption of (R) leads to a state (of knowledge) in which *one can both assert and reject (R), and that is dialetheically inacceptable*.

It then follows that (R) *cannot be true*. But, then, we can *reject* it, recognize its truth and *assert* it, which, again, is in opposition of Priest's thesis of *the impossibility of accepting and rejecting the same sentence*. Again: *dialetheically inacceptable*.

□

Notice – *en passant* – that, in contrast with the *assertability paradox*, the *deniability paradox* goes against the dialetheist thesis that *assertion* and rejection are *incompatible* speech acts.

Since, as in the amended assertability paradox, evidence of truth or falsity must be available, the agent is assumed to be able to recognize them by the reasoning displayed in the informal proof. The agent is then assumed to have some minimal logical, semantical, and pragmatical skills.[8] Such requirements are fairly minimal and are ordinarily satisfied by normal agents in normal circumstances. If a dialetheist tried to avoid the paradox by rejecting such an assumption, she would make dialetheism a viable view only for limited cognitive agents. Such an extreme move would be like invoking the ghost of Tarski at the cognitive level, by limiting the cognitive resources of an agent instead of the expressive capacity of the language.

In the next part of this section I logically argue that, in a dialetheic logic, such as **DLEAC** (which is compatible with dialetheism), expanded with a rejectability predicate (which is used in natural languages with some intuitive derivation rules for the introduction and elimination of rejection) a strong absurd can be derived.

---

[8]In particular, she must be able to understand (R) and truth and falsehood predicates, be capable of rejecting and asserting, having elementary logical skills, and some ability to reflect on her own reasoning processes.

Let 'R' be the rejectability predicate $R(\ulcorner A \urcorner)$ is to be read: *A is rejectable*; or more explicitly: *an ideal rational human can reach indefeasible reasons for excluding the truth of A.*

While assuming an ideal agent with such epistemic capacities might be problematic in general, in the context of (R) it is justified. As the informal argument of the paradox shows, evidence for truth and falsity of (R) is indeed attainable. Generalizing the inference rules of (R) to other contexts would be possible at the price of introducing complications unnecessary for present purposes.[9]

The inference rules for R are:

$$\text{ER} \ \frac{\Gamma : R(\ulcorner A \urcorner)}{\Gamma : \neg A^*}$$

$$\text{IR} \ \frac{: \neg A^*}{: R(\ulcorner A \urcorner)}$$

Let k be a sentence of form $R(\ulcorner k \urcorner)$:

1	1.	$R(\ulcorner k \urcorner)$	Assumption
1	2.	$\neg R(\ulcorner k \urcorner)^*$	ER
1	3.	$R(\ulcorner k \urcorner) \wedge \neg R(\ulcorner k \urcorner)^*$	IA
	4.	$\neg R(\ulcorner k \urcorner)^*$	RAA
	5.	$R(\ulcorner k \urcorner)$	IR
	6.	$R(\ulcorner k \urcorner) \wedge \neg R(\ulcorner k \urcorner)^*$	IA

Notice that (6) is a *strong absurd* in **DLEAC**.

On the contrary, observe that – according to *conservativity* as it was formulated in Sec.5 – it is impossible to use the liar paradox to obtain an *absurd*. Indeed, while you obtain a formula saying that the liar is false, there is no formula saying that it is *only false*, as is shown in the following proof.

Let A be a sentence of form $\neg T(\ulcorner A \urcorner)$. You have that $T(\ulcorner A \urcorner)$ is a dialetheia.

---

[9]For example, (IR) should also require that evidence is available for $\neg A^*$, which is the case, however, in the case of (R). To be fully complete, the proof should also incorporate the reasoning through which the agent recovers such evidence.

1	1.	$T(\ulcorner A \urcorner)$	Assumption
1	2.	$\neg T(\ulcorner A \urcorner)$	Tarski
	3.	$\neg T(\ulcorner A \urcorner) \vee \neg T(\ulcorner A \urcorner)$	*Material conditional*
	4.	$\neg T(\ulcorner A \urcorner)$	Assumption
	5.	$\neg T(\ulcorner A \urcorner)$	*Reflexivity*
	6.	$\neg T(\ulcorner A \urcorner)$	E∨
	7.	$T(\ulcorner A \urcorner)$	Tarski
	8.	$\neg T(\ulcorner A \urcorner) \wedge T(\ulcorner A \urcorner)$	I∧

Observe that–even using the starred assumptions–you can not get an *absurd*, as in the following proof:

1	1.	$T(\ulcorner A \urcorner)^*$	Assumption
1	2.	$\neg T(\ulcorner A \urcorner)^*$	Tarski
1	3.	$\neg T(\ulcorner A \urcorner) \wedge T(\ulcorner A \urcorner)^*$	IA
	4.	$\neg T(\ulcorner A \urcorner)$	RAA
	5.	$T(\ulcorner A \urcorner)$	Tarski
	6.	$\neg T(\ulcorner A \urcorner) \wedge T(\ulcorner A \urcorner)$	I∧

7   Conclusion

In the first part of the paper I exposed **DLEAC**, a dialetheic logic in which *exclusivity* is expressed *via* the speech acts of assuming and concluding. An expansion of **DLEAC** with a predicate for rejection and some intuitive derivation rules for its *introduction* and *elimination* led to a strong absurd, a problem for a dialetheist.

8   Acknowledgements

I wish to thank the referees, Filippo Mancini and Enrico Martino for their comments to a preliminary version of the paper. This research has been supported by the BIRD Project: "Razionalità, Normatività e Decisioni" (FISPPA-UNIPD).

References

[1] F. G. Asenjo. A calculus of antinomies. *Notre Dame Journal of Formal Logic*, 16:103–105, 1966.

[2] F. G. Asenjo and J. Tamburino. Logic of antinomies. *Notre Dame Journal of Formal Logic*, 16:17–44, 1975.

[3] Jc Beall. *Spandrels of Truth*. Oxford University Press, Oxford, 2009.

[4] M. Carrara and E. Martino. Logical consequence and conditionals from a dialetheic perspective. *Logique et Analyse*, 57(227):359–378, 2014.

[5] M. Carrara and E. Martino. DLEAC: A dialetheic logic with exclusive assumptions and conclusions. *Topoi*, 38(2):379–388, 2019.

[6] G. Littman and K. Simmons. A critique of dialetheism. In G. Priest, J. C. Beall, and B. Armour-Garb, editors, *The Law of Non-Contradiction: new philosophical essays*, pages 314–335. OUP, New York, Oxford, 2004.

[7] J. Murzi and M. Carrara. Denial and disagreement. *Topoi*, 34(1):109–119, 2015.

[8] T. Parsons. Assertion, denial and the liar paradox. *Journal of Philosophical Logic*, 13:136–52, 1984.

[9] G. Priest. The logic of paradox. *Journal of Philosophical Logic*, 8:219–241, 1979.

[10] G. Priest. *Doubt Truth to be a Liar*. Oxford University Press, Oxford, 2006.

[11] G. Priest. *In Contradiction*. Oxford University Press, Oxford, 2006. Expanded edition (first published 1987 Kluwer-Dordrecht).

[12] G. Priest. Hopes fade for saving truth. *Philosophy*, 85:109–40, 2010.

[13] D. Ripley. Negation, denial and rejection. *Philosophy Compass*, 6(9):622.629, 2011.

[14] D. Ripley. Embedding denial. Forthcoming in *Foundations of logical consequence*, O. T. Hjortland and C. Caret (eds), Oxford University Press, 2014.

[15] R. Routley. Dialectical logic, semantics and metamathematics. *Erkenntnis*, 14:301–31, 1979.

[16] T. Skura. Refutation calculi for certain intermediate propositional logics. *Notre Dame Journal of Formal Logic*, 33(4):552–560, 1992.

[17] T. Skura. Refutations and proofs in S4. In Heinrich Wansing, editor, *Proof Theory of Modal Logic*, pages 45–51. Springer, 1996.

[18] T. Skura. A refutation theory. *Logica Universalis*, 3(2):293–302, 2009.

[19] T. Skura. Refutation systems in propositional logic. In Franz Guenthner Dov M. Gabbay, editor, *Handbook of Philosophical Logic*, pages 115–157. Springer, 2011.

[20] J. Łukasiewicz. *Aristotle's Syllogistic From the Standpoint of Modern Formal Logic*. Garland Pub., 1957.

[21] H. Wansing. Falsification, natural deduction and bi-intuitionistic logic. *Journal of Logic and Computation*, 26:425–450, 2016.

Received 20 March 2020

Asserting Boo! and Horray! Pragmatic Logic for Assertion and Moral Attitudes

Daniele Chiffi*

DAStU - Politecnico di Milano
chiffidaniele@gmail.com

Abstract

The present paper puts forward a formal and pragmatic treatment of the Frege-Geach problem and other related problems usually associated with logical connectives for sentences expressing moral attitudes within an extension of logic for pragmatics (LP), a logic for acts of assertion. First, we present the Frege-Geach problem, showing its relevance for distinguishing asserted from non-asserted contexts in logical inferences. Second, we introduce the basic elements of LP, underlying its capacity to clearly disambiguate asserted from unasserted contexts. Third, we extend LP into ELP, a pragmatic logic for expressive sentences, and provide a suitable way to deal with the Frege-Geach problem. Fourth, our framework is compared with classical expressivist systems and we clarify how in ELP it is possible to provide a plausible answer to the remaining logical problems regarding expressive sentences. Finally, we show the relation between assertion and attitudes in our perspective and suggest an outline to implement ELP as a logical basis for different kinds of expressivist systems.

1 Introduction

Expressivists believe that moral sentences are conventional devices for expressing positive or negative attitudes towards their objects. Expressivist views on moral

*This research has been supported by Fundação para a Ciência e a Tecnologia, project *Values in Argumentative Discourse* (PTDC/MHC-FIL/0521/2014). A special thanks goes to the following colleagues who gave me many useful comments on early versions of this paper: Sergio Galvan, Erich Rast, Javier González de Prado Salas, Francesco Orsi, Fabrizio Macagno, Dina Mendonça, Dima Mohammed, Marcin Lewinski. Moreover, I would like to thank Massimiliano Carrara, Ciro De Florio, Ahti-Veikko Pietarinen and Francesco Bellucci for the uncountable and stimulating discussions on the logical nature of assertions. The present paper is written in memory of Carlo Dalla Pozza (1942-2014) and based on his research in pragmatic logic.

sentences have captured the interest of many philosophers working in metaethics, logic, and philosophy of language, specifically in relation to the justification of inferences with sentences expressing (non-cognitive) attitudes. In virtue of this, the justification of a basic form of inference as *modus ponens* in presence of (non-cognitive) attitudes – i.e., the Frege-Geach problem – has received much attention since the seminal work of Geach [23]. We will mainly focus on the logical aspects of sentences expressing attitudes and their structure. Then, we will show how to formally treat expressive sentences in a formal pragmatics framework. We will base our investigations on the formal framework called "Logic for Pragmatics"(LP) [12] [13] – a logical language originally devoted to providing a fine-grained treatment of illocutionary acts. Traditionally, the Frege-Geach problem has been interpreted as a challenge for distinguishing asserted from non-asserted contexts. This is why it seems worth investigating the significance of a possible application of a logic for assertions in order to deal with sentences expressing attitudes. The proper aim of this paper is to introduce a pragmatic logic for expressive attitudes (in an extension of LP) so as to overcome some difficulties – specifically, the ones associated to the connectives of statements involving attitudes – envisaged in the metaethical literature. We will show that the presence of a variety of connectives for different types of formulas will be useful for dealing with complex expressive sentences. The structure of the paper is the following. First of all, we introduce in Section 2 the connection, envisaged by Frege, between the illocutionary force of assertion and *modus ponens*, which will turn out to be fundamental in the formal treatment of expressive inferences. Then, in Section 3 the pragmatic logic for assertions LP is introduced. In this logical system it is possible to give a proper treatment of Frege's views on the inferential role of assertion, specifically in the case of *modus ponens* [12] [13]. In Section 4, a pragmatic logic for expressive attitudes (ELP) is presented in order to properly deal with logical inferences involving expressive sentences. In Section 5, classical attempts to solve the Frege-Geach problem are briefly analysed and compared with our perspective. Some remarks of philosophical interest are discussed in the Conclusion.

2 Assertion and *Modus Ponens*

Frege [20] noticed that the proper treatment of *modus ponens* has to take into account the concept of assertion, i.e., an inference occurs between assertions, not between propositions. In this context, a judgement is the acknowledgement of the truth of a proposition, and assertion is its external counterpart [19]. Consider the *modus ponens* rule from a Fregean perspective:

(1) the assertion of α,

(2) the assertion of $(\alpha \rightarrow \beta)$,

(3) therefore, the assertion of β.

Notice that in (2) the occurrence of β is not asserted, because only the whole conditional is asserted. Hence, we can say that β in (2) is unasserted (as well as α), while in (3) β is finally asserted. It has been noticed by Russell [36] that if one places *modus ponens* in an unasserted framework, then two possible problems can come up, especially if one interprets *modus pones* in the following way: α, $\alpha \rightarrow \beta$; therefore β. But if one interprets β in the second premise in the same way of the conclusion, then what we have to prove is already known, hence *modus ponens* is useless, while if β in the second premise has a different meaning from β in the conclusion, then the argument suffers from a fallacy of equivocation. An elaborated assertive framework for *modus ponens* may solve these problems and explains the inferential role of Frege's assertion sign '⊢', which indicates the external counterpart of an internal judgement. This Fregean turnstile is a composed sign constituted by a horizontal stroke, which refers to the propositional content of the judgement, expressing what is judgeable in principe, and a vertical stroke expressing that the assertion of the content has been executed [20]. We will now focus on the case of *modus ponens* with expressive attitudes.

Geach [23] noticed that our moral judgements can resemble the formal behaviour of assertions in a moral argument. Consider the following argument:

- (4) Tormenting the cat is wrong.

- (5) If tormenting the cat is wrong, then getting your little brother to torment the cat is also wrong.

- (6) Therefore, getting your little brother to torment the cat is wrong.

One can see that in (4) and (6) the idea of "being wrong" expresses a genuine attitude towards the world, while (5) seems to indicate an unasserted context where there is no genuine moral position towards tormenting the cat. If one holds an expressivist position, it is easy to handle (4) and (6) but not (5) since neither the antecedent nor the consequent are asserted, but instead the whole conditional is asserted. This problem is called the "Frege-Geach problem" in metaethics or, more generally, the "embedding problem". Since the Frege-Geach problem is about the distinction between asserted and non-asserted contexts and ther role for logical inferences, we will deal with this problem within the framework of the pragmatic logic for assertions, which is introduced in the next section. Those who are already acquainted with LP may skip it.

3 Logic for Pragmatics: Assertion

LP is a logical language mainly inspired by Frege. However, in Frege's system there are no connective working on assertions in order to create complex assertions. Therefore, in a certain sense LP is an extension of Frege's logical system. Frege distinguishes the concept of thought from the concept of judgement. The thought expressed in a (propositional) content has a truth value assigned, while the judgement is the acknowledgement of truth by a thought. The concept of thought (proposition) is a semantic concept, while the concept of judgement (or assertion) is a pragmatic one. The propositions can be either true or false, while the judgements can be justified (J) or unjustified (U). A well-formed elementary formula in LP is obtained by prefixing an operator of pragmatic mood (indicated by "⊢" for assertions) to a proposition (radical formula). Sentential formulas (which we indicate with δ) are built from the elementary formulas using the following pragmatic connectives ∩ (conjunction), ∪ (disjunction), ⊃ (implication), ∼ (negation), ≡ (equivalence). The specification of the illocutionary-force operator shows the existence of a relation between the user of a symbol and the symbol itself. Notably, an assertion in LP is justified iff there exists any (informal) proof of the content of the radical formula [13]. The pragmatic language LP is the following:

The pragmatic language LP, which is the union of the set of radical formulas RAD and the set of sentential formulas SENT, can be defined recursively:

RAD $\qquad \gamma ::= p|\neg\gamma|\gamma_1 \wedge \gamma_2|\gamma_1 \vee \gamma_2|\gamma_1 \rightarrow \gamma_2|\gamma_1 \leftrightarrow \gamma_2|$

SENT $\qquad$ i) Elementary sentential formulas $\theta ::= \vdash \gamma$

$\qquad\qquad$ ii) Sentential $\delta ::= \theta| \sim \delta|\delta_1 \cap \delta_2|\delta_1 \cup \delta_2|\delta_1 \supset \delta_2|\delta_1 \equiv \delta_2|.$

Radical formulas of LP have a truth value (true or false) and sentential formulas have a justification value ("J" justified or "U" unjustified) defined in terms of the intuitive notion of *proof* or *conclusive evidence*. The semantics of LP is classical and provides only the interpretation of the radical formulas, by assigning them a truth value and interpreting propositional connectives as truth functions in the standard way. The semantic rules for radical formulas are thus the usual classical Tarskian ones and specify the truth conditions (only for radical formulas) through an assignment function σ, thus regulating the semantic interpretation of LP. Let γ_1, γ_2 be radical formulas and $1 =$ true and $0 =$ false; then:

1. $\sigma(\neg\gamma_1) = 1$ iff $\sigma(\gamma_1) = 0$

2. $\sigma(\gamma_1 \wedge \gamma_2) = 1$ iff $\sigma(\gamma_1) = 1$ and $\sigma(\gamma_2) = 1$

3. $\sigma(\gamma_1 \vee \gamma_2) = 1$ iff $\sigma(\gamma_1) = 1$ or $\sigma(\gamma_2) = 1$

4. $\sigma(\gamma_1 \rightarrow \gamma_2) = 1$ iff $\sigma(\gamma_1) = 0$ or $\sigma(\gamma_2) = 1$

5. $\sigma(\gamma_1 \leftrightarrow \gamma_2) = 1$ iff $\sigma(\gamma_1) = \sigma(\gamma_2)$.

A distinguished feature of LP are pragmatic connectives. These connectives have a meaning due to the BHK (Brouwer, Heyting, Kolmogorov) interpretation of intuitionistic logical constants. The illocutionary force of assertion plays a key role in determining the *pragmatic* component of the meaning of an elementary expression, together with the *semantic* component expressed in radical formulas.

Justification rules regulate the pragmatic evaluation π, specifying the justification conditions for the assertive formulas in function of the σ-assignments of truth values for their radical sub-formulas:

JR1 – Let γ be a radical formula. $\pi(\vdash \gamma) = J$ iff a proof exists that γ is true, i.e. that σ assigns to γ the value 1. $\pi(\vdash \gamma) = U$ iff no proof exists that γ is true.

JR2 – Let δ be an assertive formula. Then, $\pi(\sim \delta) = J$ iff a proof exists that δ is unjustified. i.e., that $\pi(\delta) = U$.

JR3 - Let δ_1 and δ_2 be assertive formulas. Then:

1. $\pi(\delta_1 \cap \delta_2) = J$ iff $\pi(\delta_1) = J$ and $\pi(\delta_2) = J$;
2. $\pi(\delta_1 \cup \delta_2) = J$ iff $\pi(\delta_1) = J$ or $\pi(\delta_2) = J$;
3. $\pi(\delta_1 \supset \delta_2) = J$ iff a proof exists that $\pi(\delta_2) = J$ whenever $\pi(\delta_1) = J$;
4. $\pi(\delta_1 \equiv \delta_2) = J$ iff $\pi(\delta_1 \supset \delta_2) = J$ and $\pi(\delta_2 \supset \delta_1) = J$.

The Soundness criterion (SC) is the following:

Let be $\gamma \in RAD$, then $\pi(\vdash \gamma) = J$ implies that $\sigma(\gamma) = 1$.

SC states that if an assertion is justified, then the content of assertion is true.

It is evident from the justification rules that sentential formulas have an intuitionistic-like formal behaviour and can be translated into the modal system S4, where $\Box\gamma$ means that there is an (intuitive) proof (or conclusive evidence) for γ. It is worth noting that the justification rules are partial, since they do not allow determining in all cases the justification value of a complex sentential formula when all the justification values of its components are known. For instance,

NR1 $\pi(\delta) = J$ implies $\pi(\sim \delta) = U$,

NR2 $\pi(\delta) = U$ does not imply $\pi(\sim \delta) = J$,

NR3 $\pi(\sim \delta) = J$ implies $\pi(\delta) = U$,

NR4 $\pi(\sim \delta) = U$ does not imply $\pi(\delta) = J$.

PRAGMATIC VALIDITY FOR ASSERTIVE FORMULAS: A formula δ is pragmatically valid iff for every Tarskian semantic interpretation σ and for every pragmatic function of justification π, then $\pi(\delta)=J$.

LP has both an intuitionistic fragment (ILP) and a classical fragment (CLP).

The axioms of ILP are the following:

A1. $\qquad \delta_1 \supset (\delta_2 \supset \delta_1)$

A2. $\qquad (\delta_1 \supset \delta_2) \supset ((\delta_1 \supset (\delta_2 \supset \delta_3)) \supset (\delta_1 \supset \delta_3))$

A3. $\qquad \delta_1 \supset (\delta_2 \supset (\delta_1 \cap \delta_2))$

A4. $\qquad (\delta_1 \cap \delta_2) \supset \delta_1; (\delta_1 \cap \delta_2) \supset \delta_2$

A5. $\qquad \delta_1 \supset (\delta_1 \cup \delta_2); \delta_2 \supset (\delta_1 \cup \delta_2)$

A6. $\qquad (\delta_1 \supset \delta_3) \supset ((\delta_2 \supset \delta_3) \supset ((\delta_1 \cup \delta_2) \supset \delta_3))$

A7. $\qquad (\delta_1 \supset \delta_2) \supset ((\delta_1 \supset (\sim \delta_2)) \supset (\sim \delta_1))$

A8. $\qquad \delta_1 \supset ((\sim \delta_1) \supset \delta_2)$

The intuitionistic fragment is obtained by limiting the language of LP to complex formulas that are valid with atomic radicals. This is an intuitionistic fragment ILP [13]. *Modus ponens* rule for ILP is the following:

[MPP'] $\qquad$ if $\delta_1, \delta_1 \supset \delta_2$, then δ_2

where δ_1 and δ_2 contain atomic radicals.

The classical fragment CLP of LP can also be characterized. CLP is composed by the set of *sfs* without pragmatic connectives.

Axioms for CLP are the following:

Ai $\vdash (\gamma_1 \to (\gamma_2 \to \gamma_1))$
Aii $\vdash ((\gamma_1 \to (\gamma_2 \to \gamma_3)) \to ((\gamma_1 \to \gamma_2) \to (\gamma_1 \to \gamma_3))$
Aiii $\vdash (\neg\gamma_2 \to \neg\gamma_1) \to ((\neg\gamma_2 \to \gamma_1) \to \gamma_2))$

Modus ponens rule for CLP is:

[MPP]
$(1)^* \vdash \gamma_1$
$(2)^* \vdash (\gamma_1 \to \gamma_2)$
$(3)^*$ therefore, $\vdash \gamma_2$

A modal translation of LP can be given by a function $()^*$ from assertive formulas to the corresponding modal ones in the system S4, where $\Box\gamma$ means that there is an *intuitive proof* or *conclusive evidence* for γ:

$(\vdash \gamma)^* \qquad \Box\gamma$

$(\sim \delta)^* \qquad \Box\neg(\delta)^*$

$(\delta_1 \cap \delta_2)^* \quad (\delta_1)^* \wedge (\delta_2)^*$

$(\delta_1 \cup \delta_2)^* \quad (\delta_1)^* \vee (\delta_2)^*$

$(\delta_1 \supset \delta_2)^* \quad \Box((\delta_1)^* \to (\delta_2)^*)$

$(\delta_1 \equiv \delta_2)^* \quad \Box((\delta_1)^* \leftrightarrow (\delta_2)^*)$

Connectives for radical and sentential formulas are related by these bridge principles:

(a) $(\vdash \neg\gamma) \supset (\sim\vdash \gamma)$

(b) $((\vdash \gamma_1) \cap (\vdash \gamma_2)) \equiv (\vdash (\gamma_1 \wedge \gamma_2))$

(c) $((\vdash \gamma_1) \cup (\vdash \gamma_2)) \supset (\vdash (\gamma_1 \vee \gamma_2))$

(d) $(\vdash (\gamma_1 \to \gamma_2)) \supset (\vdash \gamma_1 \supset \vdash \gamma_2)$

(e) $(\vdash (\gamma_1 \leftrightarrow \gamma_2)) \supset (\vdash \gamma_1 \equiv \vdash \gamma_2)$

Bridge principles (a)–(e) show the formal relations between pragmatic connectives and connectives in the radicals. (a) states that from the assertion of not-γ the non-assertability of γ can be inferred. (b) expresses that the conjunction of two assertions is equivalent to the assertion of a conjunction; (c) states that from the disjunction of two assertions the assertion of a disjunction is inferable. Formula (d) expresses that from the assertion of a classical material implication follows the pragmatic implication between two assertions. Formula (e) shows that from the assertion of a biconditional follows the equivalence of assertions.

Let us consider the two variants of *modus ponens* again:

$(1)^* \vdash \gamma_1$

$(2)^* \vdash (\gamma_1 \to \gamma_2)$

$(3)^*$ therefore, $\vdash \gamma_2$

In virtue of bridge principle (d), from $(2)^*$ follows that $(\vdash \gamma_1 \supset \vdash \gamma_2)$. This means that the following variant of the *modus pones* rule can be justified:

$(1)^\circ \vdash \gamma_1$

$(2)^\circ (\vdash \gamma_1) \supset (\vdash \gamma_2)$

$(3)^\circ$ therefore, $\vdash \gamma_2$

Thus we can say that $(2)^*$ is a stronger claim compared to $(2)^\circ$; the former is the material implication, the latter is the inference in an argument when we use "therefore", as it was observed by Russell [36], who named the problem of justification of *modus pones* in presence of unasserted formulas as the "embedding problem". As we will see, this will turn out to be particularly relevant for our view on the Frege-Geach problem.

4 Pragmatic Logic for Expressive Sentences

In this section we will explore the possibility to apply our pragmatic framework to expressive sentences in order to better understand their logical form and inferential relations. Two expressive and pragmatic operators are introduced in order to extend LP, namely: $H^P!$ that expresses a positive attitude towards the content of the radical formula and the operator $B^P!$ that expresses a negative attitude toward a content. $H^P!$ and $B^P!$ have an intuitionistic-like formal behaviour as the assertion sign and other pragmatic connectives in LP. As pointed out by Reichenbach [33], assertions and other linguistic features of language can be represented as terms in "pragmatic capacity" that do not obey to truth-functional connectives and cannot be iterated.

Given the intuitionsitic-like formal behaviour of pragmatic operators for expressive statements, they are not fully inter-definable, notably $H^P!\gamma \supset B^P!\neg\gamma$ and $B^P!\gamma \supset H^P!\neg\gamma$, but their readings from right to left do not hold. Moreover, we indicate by B and H the *descriptions* of expressive attitudes and, for this reason, they can be true or false B and H are inter-definable because they behave like alethic modalities (as *forbidden* and *obligatory* respectively) and are connected by classical connectives in radical formulas. In fact, the the description of an expressive judgment can be true or false, while a genuine moral sentence is in this framework justified or not on the basis of a set of assumed principles.[1]

---

[1]In philosophy of law there is, for instance, the distinction between the *logic of normative propositions* (deontic logic), in which sentences can be true or false, and the proper *logic for the*

The language of this extension of LP is termed ELP (Pragmatic Logic for Expressive Sentences) and it is the following:

Descriptive signs: propositional letters: p, q, r, ..

Logical signs for radical formulas: $\neg, \wedge, \vee, \rightarrow, \leftrightarrow, H, B$

Logical signs for sentential formulas: the sign of pragmatic illocutionary force $\vdash$, $H^P!$ and $B^P!$

Pragmatic connectives: $\sim$ pragmatic negation, $\cap$ pragmatic conjunction, $\cup$ pragmatic disjunction, $\supset$ pragmatic implication, $\equiv$ pragmatic equivalence.

Formation Rules (FRs):

Radical formulas (*rfs*) are recursively defined by the following FRs:

FR1 (atomic formulas): every propositional letter is a *rf*.

FR2 (molecular formulas): (i) Let γ be a *rf*, then $\neg\gamma$ is a *rf* .

(ii) Let γ_1 and γ_2 be *rfs*, then $\gamma_1 \wedge \gamma_2$, $\gamma_1 \vee \gamma_2$, $\gamma_1 \rightarrow \gamma_2$, $\gamma_1 \leftrightarrow \gamma_2$, $H\gamma$, $B\gamma$ are *rfs*.

Sentential formulas (sfs) are recursively defined by the following FRs:

FR3 (elementary formulas): Let γ be a *rf*, then $\vdash \gamma$, $H^P!$ and $B^P!$ are *sfs*.

FR4 (complex formulas):

(i) Let λ be a *sf*, then $\sim \lambda$ is a *sf* .

(ii) Let λ_1 and λ_2 be *sfs*, then $\lambda_1 \cap \lambda_2$, $\lambda_1 \cup \lambda_2$, $\lambda_1 \supset \lambda_2$, $\lambda_1 \equiv \lambda_2$, are *sfs*.

As we have already pointed out, descriptions of attitudes in our framework are represented by the operators H and B, which behave formally like O (obligatory) and F (forbidden) respectively in the modal logic KD and belong to radical formulas as they can be true or false. They are different from the genuine expressive operators $H^P!$ and $B^P!$ that if associated to radical formulas may form sentential formulas which can be justified or unjustified. Hence, we have to apply the classical connectives (used in radical formulas) to B and H. This distinction is important in order to switch from the assertion of a descriptive ethical statement to a genuine

---

expressivist conception of norms in which inferences between normative jugdements can be justified or unjustified [1], [30]. For the expressivist conception, "norms are the result of the *prescriptive* use of language".

expressive statement and *vice versa*, even if, as we will see later on, this is not always possible for more complex formulas[2]. Let us consider the following bridge principles connecting assertions and attitudes:

(12) $\vdash B(\gamma) \equiv B^P!(\gamma)$

(13) $\vdash H(\gamma) \equiv H^P!(\gamma)$[3]

Principles (12) and (13) require some clarification. Let us focus on (12), but a similar discourse can be replicated for (13). In our pragmatic language the notion of justification indicates the existence of a proof (or conclusive evidence). Indeed, if $B^P!(\gamma)$ is justified, this is equivalent to say that there is a proof (or conclusive evidence) of proposition $B(\gamma)$. However, this is the same condition justifying $\vdash B(\gamma)$. This reasoning is behind the justification of (12) and (13). For a detailed discussion of pragmatic bridge principles, see [2], [32]. Moreover, (12) and (13) do *not* mean that whatever assertive formula with the description of an attitude in the radical is formally equivalent with a sentential formula prefixed with an expressive operator like $H^P!$ or $B^P!$. Principles (12) and (13) mean that the justification conditions of the two elementary formulas are pragmatically equivalent[4] . The relation between assertive and expressive formulas is, in fact, *shaped* by the logical structure. For instance, since disjunction in the radical does not entail the pragmatic disjunction but only *vice versa* (see the bridge principle (d)), therefore from $\vdash (H\gamma_1 \vee H\gamma_2)$ cannot be derived the formula $(\vdash H\gamma_1) \cup (\vdash H\gamma_2)$ (or, if (13) is accepted, the formula $(H^P!\gamma_1) \cup (H^P!\gamma_2))$[5].

In the literature there exist other *hybrid* or *ecumenical* forms of expressivism in which moral sentences express states of mind consisting of *both* descriptive beliefs and genuine attitudes[6]. The expressive elementary formula stating the justification of

---

[2]Since LP has a modal translation in the modal system S4 and expressive operators have a translation in KD, then the modal translation of ELP can be expressed in a bimodal system in which there is a *fusion* of KD and S4. Consistency and completeness results deriving from the fusion of this kind of simple propositional modal systems can be shown to persist [7].

[3]Principles (12) and (13) might be weakened. For instance, by imposing that $\vdash B(\gamma) \supset B^P!(\gamma)$. The other sense of implication could be less acceptable according to different philosophical views.

[4]One can imagine different forms of bridge principles connecting the assertion (or a different illocutionary act) of attitudes and genuine expressivist attitudes similarly to what happens in the *logic of normative propositions* with bimodal systems equipped with both normative and alethic connectives. On normative bridge principles, see [21], [22].

[5]We want to have the operator of assertion explicit in our system in order to clarify the inferential and pragmatic structure of sentences expressing attitudes and to be able to speak with a uniform structure of expressive and non-expressive sentences. Nonetheless, nothing prevents us from presenting the language of ELP with only the expressive operators of approval and disapproval and no explicitation of the assertion sign (and without thus proposing bridge principles like (12) and (13)).

[6]See, for instance, [34] and [6]. For a similar view on aesthetic sentences, see also [17].

$H^P!(\gamma)$ can be *translated* in the descriptive multimodal formula $\Box H\gamma$ that indicates the existence of conclusive evidence for $H\gamma$, and given the factivity of the box of S4, then it is possible to derive $H\gamma$ from $\Box H\gamma$. The following multimodal translation justifies the descriptive readings of $H\gamma$ and $B\gamma$ and the choice of (12) and (13) as bridge principles. Given the following translation function, in which $(\ )**$ is a function from attitudes to multimodal formulas (where the box is as S4 modality and H a deontic modality) we have that[7]:

λ	$\lambda**$
$(\vdash H^P!\gamma)$	$\Box H\gamma$
$(\sim H^P!\gamma)$	$\Box\neg\Box(H\gamma)$
$(H^P\gamma_1 \cap H^P\gamma_2)$	$(\Box H_1) \wedge (\Box H_2)$
$(H^P\gamma_1 \cup H^P\gamma_2)$	$(\Box H_1) \vee (\Box H_2)$
$(H^P\gamma_1 \supset H^P\gamma_2)$	$\Box((\Box H_1) \rightarrow (\Box H_2))$
$(H^P\gamma_1 \equiv H^P\gamma_2)$	$\Box((\Box H_1) \leftrightarrow (\Box H_2))$

The multimodal translation above is obtained as a special case (only with elementary sentences) of the following general translation:

λ	$\lambda**$
$(\vdash H\gamma)^{**}$	$\Box H\gamma$
$(\sim \lambda)^{**}$	$\Box\neg(\lambda)^{**}$
$(\lambda_1 \cap \lambda_2)^{**}$	$(\lambda_1)^{**} \wedge (\lambda_2)^{**}$
$(\lambda_1 \cup \lambda_2)^{**}$	$(\lambda_1)^{**} \vee (\lambda_2)^{**}$
$(\lambda_1 \supset \lambda_2)^{**}$	$\Box((\lambda_1)^{**} \rightarrow (\lambda_2)^{**})$
$(\lambda_1 \equiv \lambda_2)^{**}$	$\Box((\lambda_1)^{**} \leftrightarrow (\lambda_2)^{**})$

It is now evident that our translations provide a ground for supporting (12) and (13)[8].

Finally, we have that an analogous of the necessitation rule holds. Namely, if γ is a truth-functional consequence of $\gamma_1, \gamma_2.., \gamma_n$, then $H^P!(\gamma)$ follows from $H^P!(\gamma_1)$, $H^P!(\gamma_2), .. H^P!(\gamma_n)$.

The validity of the moral inference is related to the Frege-Geach problem. In fact, we can interpret the *original example* of the Frege-Geach problem in our pragmatic and expressive language, i.e., we can indicate (4), (5) and (6) in ELP with the expressive operator B^P!

---

[7] For a negative attitude, the translation can be obtained in a similar way.

[8] However different bridge principles expressing alternative forms of relations between the assertion (or different illocutionary acts) involving attitudes and genuine moral attitudes may be conceived, based on different philosophical insights.

(14) $B^P!(\gamma_1)$,
(15) $\vdash ((B\gamma_1) \to (B\gamma_2))$,
(16) therefore, $(B^P!\gamma_2)$.

Note that in ELP it is possible to handle the three statements of Geach's original example for the Frege-Geach problem in a uniform way. ELP is a suitable logic for treating expressive moral judgements. If we compare the assertive language LP with its extension ELP, we can observe many similarities between logical assertions and the expressive operators. Nevertheless, notice that the justified assertion of a proposition implies the acknowledgment of the truth of the proposition, while this is not the case for the expressive operators. That is why we need to extend LP with ELP in order to take into account the specific features of attitudes and relating them to the inferential structure of assertions. We will show that the Frege-Geach problem may receive a suitable formal treatment in ELP. In fact, if (14) (15) (16) hold in ELP, then it is possible to derive in ELP (a better form of expressive) *modus ponens*[9]

(17) $B^P!(\gamma_1)$,
(18) $B^P!(\gamma_1) \supset B^P!(\gamma_2)$,
(19) $B^P!(\gamma_2)$.

Note that if we want to apply (12) to (17) - (18) - (19) we can rewrite the inference above in the following way (with the explicitation of the assertion sign):

(17)° $\vdash B(\gamma_1)$,
(18)° $\vdash B(\gamma_1) \supset \vdash B(\gamma_2)$,
(19)° $\vdash B(\gamma_2)$.

So, in our pragmatic language it is possible to express material implication in the radical formula and pragmatic conditional among sentential sentences (and their logical relation).

The analysis of negation in ELP is even more articulated. There is the negation in the radical formula "¬", the pragmatic negation $\sim$, and the justification values of unjustified "U". The richness in the expressive logical power of ELP makes it particularly suitable to formulate different forms of negations and connectives for asserted and unasserted contexts and to deal with the major logical problems of expressivism as we will see in the next Section. Indeed, the main idea behind ELP is to provide a pragmatic tratment of connectives in order to clarify the logical structure of the expressivist conception of moral sentences.

---

[9]See, also, [14]. The revision of that paper has been published very recenty [15].

5 A Comparison with Other Systems

In this section we will focus on the seminal contributions to the expressivist structure of sentences involving attitudes and compare them with our perspective[10]. We will consider (i) Blackburn's proposal based on higher-order attitudes, (ii) Gibbard's norm expressivism, and (iii) Schroeder's general attitude of *being for*.

5.1 Blackburn's Higher-Order Approach

Blackburn developed a logic for expressive attitudes in what he calls the "quasi-realist" framework [3],[4]. He wants to prove the validity of judgements and inferences, without any appeal to the concept of truth, referring to a possible substitute for this notion in moral theory. He argues that the validity (or invalidity) of our moral judgements has to depend only on the connections between attitudes (or their failures). Blackburn called his views "quasi-realistic" and clarified their relationships with the classical projectivism in the following manner:

"Projectivism is the philosophy of evaluation which says that evaluative properties are projections of our own sentiments (emotions, reactions, attitudes, commendations). Quasi-realism is the enterprise of explaining why our discourse has the shape it does, in particular by way of treating evaluative predicates like others, if projectivism is true. It thus seeks to explain, and justify, the realistic-seeming nature of our talk of evaluations - the way we think we can be wrong about them" [3, p. 180].

Blackburn aims to discover the 'deep structure' of our moral arguments, i.e., they are not a description of the world, but they obey to some patterns of validity as for descriptive statements, although they are not "truth apt" (i.e., they do not aim to achieve the truth). Indeed, he presents a logical system, which seeks to explain the particular structure of moral arguments. He introduces in this logic two operators $H!$ (Horray!-operator, which stands for a high-order positive attitude) and $B!$ (Boo!-operator, which stands for a high-order negative attitude). It is possible to define the language of Blackburn's system, which is usually termed as "Higher-Order Approach" (HOA) in the following way:

HOA ::= $\gamma| \neg\gamma |\gamma_1 \wedge \gamma_2| \gamma_1 \vee \gamma_2 |\gamma_1 \to \gamma_2| B!\gamma | H!\gamma |$.

The main difference with our pragmatic system (presented in the previous section) is that Blackburns's operators can be iterated and can work as second-level

---

[10]Further contemporary views on expressivism are not investigated in this paper and we hope to be capable to do it in a future work. For instance, [29].

attitudes referring to attitudes (or beliefs) of the first level indicated within squared brackets. Thus, the sentence (5) can be translated as:

$H!([B!$ tormenting the cat$] \Rightarrow [B!$ getting your little brother to torment the cat$])$

And $\Rightarrow$ indicates "the involvement of one mental state with another" necessary for a conditional [4]. Thus, the *modus ponens* for expressive sentences in Blackburns' framework will be:

- (20) $B!$ tormenting the cat,

- (21) $H!([B!$ tormenting the cat$] \Rightarrow [B!$ getting your little brother to torment the cat$])$,

- (22) therefore, $B!$ getting your little brother to torment the cat.

But the sign $\Rightarrow$ in this system is not a sign with a clear logical meaning; hence, the problem of the logical validity of the inference is still open in this system. As a matter of fact, if one endorses the premises, and not the conclusion, then no logical contradiction arises, only a moral failure, which does not affect the rationality of the moral agents [50].

The $H!$ operator is definable contextually in this way: "the subject A accepts $H!\gamma$" is equivalent with "A hoorays that γ". However, Unwin [46] has noted some syntactical problems in Blackburn's system. Such problems can be easily handled in ELP.

Problem 1. (*Negations*)

The expression "A accepts $H!\gamma$" has three different negated forms, whereas "A hoorays that γ" admits only two forms:

N1. A does not accept $H!\gamma \equiv$ A does not hooray that γ
N2. A accepts *not* $H!\gamma \equiv$???
N3. A accepts $H!(not\,\gamma) \equiv$ A hoorays that *not* γ

N1 and N3 are unproblematic, while N2 admits no clear formulation in HOA. From a pragmatic perspective:

N1°. A does not accept $H!\gamma \equiv ((H^P!\gamma) = U)$
N2°. A accepts *not* $H!\gamma \equiv ((\vdash \neg H\gamma) = J)$

Notice that from $(\vdash \neg H\gamma) = J$, it is possible to derive, in virtue of the bridge principle (a), that $((\sim\vdash H\gamma) = J)$. From the latter formula, by principle (13) we have $((\sim H^P!\gamma) = J)$, a formula without any connective in the radical and thus a better way to formalize expressive attitudes.

N3°. A accepts $H!(not\,\gamma) \equiv ((H^P!\neg\gamma) = J)$

It is worth remarking N2 cannot be translated as $B^P!\neg\gamma$ because of the failure of the complete inter-definability of our expressive operators [11].

Problem 2. (*Disjunctions*)
Unwin [46] has showed a second problem in Blackburn's framework for expressive disjunctions, namely:

D1. either A accepts $H!\gamma_1$ or A accepts $H!\gamma_2$
D2. A accepts $(H!\gamma_1 \vee H!\gamma_2)$
D3. A accepts $H!(\gamma_1 \vee \gamma_2)$

The expression D2 cannot be properly handled in Blackburn's system, while in ELP it is possible to express D1, D2 and D3. Namely,

D1°. $(H^P!\gamma_1) = J$ or $(H^P!\gamma_2) = J$
D2°. $(H^P!\gamma_1 \cup H^P!\gamma_2) = J$
D3°. $H^P!(\gamma_1 \vee \gamma_2) = J$

The disjunction in D1° can be interpreted as a metalinguistic disjunction in ELP of two justified formulas because it connects formulas already equipped with a justification value. Since the metalanguage of LP is classical, also the metalanguage of ELP is classical. Therefore, disjunction in D1° is classical and principles like *excluded middle* can be shown, for instance, to hold. In D2°, disjunction is pragmatic and hence it shows an intuitionistic-like formal behaviour in ELP. This form of disjunction is difficult to express in HOA, as observed by Unwin [46]. Finally, disjunction in D3° is classical since it is a connective in the radical formula, so it is not metalinguistic as in D1°[12].

---

[11]If we want to deal with a normative language without having the negation in the radical and without using the pragmatic negation $\sim$, it is possible to extend ELP with the illocutionary act of denial, indicated by $\dashv$. In such an extended pragmatic framework, the justification of $\vdash \neg\gamma$ would be sufficient to justify $\dashv \gamma$. Indeed, if we have conclusive evidence for $\neg\gamma$, then we can justifiably deny γ. However, from the justification of $\dashv \gamma$ does not follow the justification of $\vdash \neg\gamma$. The resources required for the justification of an act of denial are not always sufficient, in fact, to justify the assertion of the opposite content. In a sense, it is not assumed Frege's *classical denial equivalence thesis* between the assertion of negation and denial [35]. So, if we the act of denial is included as a logical sign, from $((\vdash \neg H\gamma) = J)$ follows that $((\dashv H\gamma) = J)$, which is a weaker assumption compared to $((\sim\vdash H\gamma) = J)$ [16]. A further extension of LP has been proposed to deal with the illocutionary act of hypothesis ([8]; [11]; [9]).

[12]Suppose that a judge must decide whether to accept or not the positive or negative evidence towards a crime hypothesis (Hyp). If γ_1 represents positive evidence and γ_2 negative evidence towards Hyp, then we have that:

- D1° expresses a situation in which the judge accepts that from the lack of positive (negative)

411

Problem 3. (*Mixed formulas*)

A formula like $(H!\gamma_1) \vee (\gamma_2)$ raises problems to an expressivist account because the disjunction connects a purely descriptive proposition γ_2 with a formula indicating an expressive judgement [27]. Even if we cannot straightly write mixed formulas like the previous one in ELP for they are not well-formed, there are ways to deal with this issue in our pragmatic perspective. On the one hand, it is correct to state that truth-functional connectives may not be applied to sentential formulas and pragmatic connectives may not be applied to radical formulas. Nonetheless, it is possible to connect a sentential formula expressing a moral attitude with assertive formulas not expressing any moral attitude in virtue of the principles (12) and (13). For instance, in ELP a formula like $(H^P!\gamma_1) \cup (\vdash \gamma_2)$ is formally equivalent with $(\vdash H\gamma_1) \cup (\vdash \gamma_2)$. From the latter, by bridge principle (c), it is finally possible to infer $\vdash (H\gamma_1 \vee \gamma_2)$, but not *vice versa*. This shows that there are specific constraints for complex formulas in ELP on the possibility to switch from sentential formulas expressing attitudes to the assertion of (the description of) those attitudes. This is not an ad-hoc strategy, since bridge principles (a)-(e) hold not only in ELP but also in LP. Moreover, this means that the switch from descriptive operators B and H and their pragmatic versions $H^P!$ and $B^P!$ is a delicate issue in case of complex formulas.

Problem 4. (*Iteration of operators*)

A formula like $H!(H!\gamma)$ cannot be expressed in ELP since it would not be a well-formed formula[13]. No iteration of pragmatic operator is possible and this is a general feature of any pragmatic operator. Pragmatic operators (also when asserting attitudes) can be applied only to radical formulas[14].

---

evidence *per se*, he or she may accept that the existence of negative (positive) evidence is inferable.

- D2° indicates a context in which the judge accepts that from the lack of positive (negative) evidence *per se*, he or she may not accept that the existence of negative (positive) evidence is inferable.

- D3° represents a minimal situation in which the judge has to accept the positive evidence or the negative evidence.

[13]One may argue that pragmatic operators like the assertion sign cannot be iterated, while attitudes may be iterated, for instance, in a multi-agent context. For instance, we want to say things like: "Tom approves that Tim blames his father's behaviour". Somehow, we agree on this point. Therefore, it could be possible, in line of principle, to formalize iterated (or nested) non-cognitive attitudes in a multiagent framework extending ELP, in which formulas are associated to different agents. However, this task is out of the scope of this paper.

[14]Blackburn [4] developed a deontic view in which there are two operators for a positive attitude ($H!$) and a tolerance-related attitude respectively ($T!$), which behave like the operators of obligatory

Thus, the main problems associated to Blackburn's proposal cannot be replicated in ELP. Even if there are *prima facie* some similarities between Blackburn's and our pragmatic operators, their expressive and inferential power is different. Conditional statements in ELP are not required to have different levels of attitudes to solve the question of unasserted propositions in the Frege-Geach problem. Our perspective is more similar to the original view of Peter Geach and prevents the unwanted iteration of pragmatic operators. Moreover, in ELP it is possible to express a greater variety of complex sentences compared to HOA and, as we have just seen, many of them are required to make some remarkable logical distinctions for a deep analysis of expressive views on moral sentences.

5.2   Gibbard's Norm Expressivism

Other attempts to solve the Frege-Geach problem have been provided by Allan Gibbard. We will not go investigate all the main features of this interesting theory, but only those that are relevant for the expressivist problems associated with connectives. In [24], a logical framework was proposed in order to deal with normative sentences with a possible-world semantics in which worlds are conceived as "factual-normative worlds" indicated by (w, n), which is a combination of a possible world with a set of general norms. The main idea of Gibbard is that our normative judgements depend on a set of norms, and they are considered rational when based on the acceptance of a system of norms. Acceptance is to be intended as a non-cognitive attitude. He states that "a person's normative judgements all told on a given matter will typically depend on his acceptance of more than one norm, and the norms he accepts may weigh in opposing directions... Our normative judgements thus depend not on a single norm, but on a plurality of norms that we accept as having some force, and on the ways we take some of these norms to outweigh or override others" ([24], pp. 86-7). And in any system of norms we have that: "we can characterize any system N of norms by a family of basic predicates 'N- forbidden', 'N-optional', and 'N-required'. Here 'N-forbidden' simply means 'forbidden by system of norms N', and likewise for its siblings. Other predicates can be constructed from these basic ones; in particular 'N-permitted' will mean 'either N-optional or N-required' " ([24], p. 87). These N-corresponding versions of normative predicates are *descriptive* and therefore may occur in embedded contexts.

---

and permissible (and are therefore intertefinable) equipped with a semantics based on Hintikka's notion of "deontic alternatives" [28]. However, it has been pointed out that this second attempt of Blackburn can be hardly considered expressivist [46], since normative propositions are merely true or false. This view is a theory for normative propositions (as deontic logic) rather than a logic for the expressive conception of norms.

In Gibbard's system we have that a normative judgement S holds at a factual-normative world $< w, n >$ if and only if $S*$, the sentence which results from S by replacing all of its normative predicates with the descriptive predicates which N-correspond to them, is true in the possible world w [24]. However, the substitution of genuine normative statements with descriptive ones can be problematic for expressivists [31] and, as we have seen, this is especially evident in presence of complex normative sentences.

In this system, the meaning of a sentence depends on what is ruled out in every world. So, (4) rules out all the normative-factual possible worlds in which tormenting the cat is not wrong, (5) rules out the intersection of the set of normative factual worlds in which "tormenting the cat is not wrong" and the set of normative factual worlds in which "it is not wrong to get your little brother to torment the cat". Therefore, (4) and (5) rule out all the set of normative worlds in which "it is not wrong to get your little brother to lie" and what (6) rules out is included in the world ruled out by the premises. But, as observed by Sinnott-Armstrong [43], this procedure can be applied to non-normative statements as well, even if it seems that this property was a *desideratum* by Gibbard's himself in order to have a uniform treatment of expressivist and descriptive sentences. Consider the following argument:

(23) I lie,

(24) If I lie, then my little brother lies,

(25) therefore, my little brother lies.

If we apply Gibbard's technique, we see that the worlds ruled out by (25) are included in the worlds ruled out by (23) and (24). This fact explains the validity of the argument. In any case, in Gibbard's system *the description of a norm* and the *expressive dimension of a norm* are interchangeable and – as we have seen – this fact may be complicated in the case of complex normative formulas. The relation between a description of an expressive attitude and a genuine expressive attitude is a delicate issue and requires to be somehow carefully investigated. Moreover, the classical concept of validity turns out to be not adequate for the genuine expressive attitudes. Unwin [47] observed that in Gibbard's logic a sentence simply holds in a factual normative world or else not, without any further qualification. This fact explains the problem of negations in Gibbard's logic, that we have already seen in the case of Blackburn's system. In fact, consider the following negated expressions:

N(1) A does *not* accept that it is obligatory that γ.

N(2) A accepts that it is *not* obligatory that γ.

N(3) A accepts that it is obligatory that *not* γ.

One can see that N(2) can be handled neither in Gibbard's system nor in Blackburn's framework, while ELP allows for such kind of distinction as already shown. Unfortunately, the problem regarding expressivist negations remains in the new formulation of Gibbard's logic in later works [25], [26], as it has been pointed out by Dreier [18]. In the new formulation of his theory, Gibbard argues that normative judgements must be based on some plans. The idea that in any specific situation a particular action is the thing to do if it expresses a plan to do it in that situation. However, one might claim for the opposite view, i.e. it may be the case that we judge something good or wrong and then we act according to a plan, aiming to achieve a good or wrong target and not the other way around. Gibbard states that a plan is what rules out all the alternatives that a moral agent has, permitting at least one. If a plan holds for every contingency, then we have what Gibbard calls "hyperplan". He assumes that the external negation in N(2) corresponds to a disagreement with someone's attitude. Let us consider the following sentence within Gibbard's framework.

(26) Miss Manners thinks one must write thank you notes by hand.

As we have seen, this type of sentence allows three different negated formulas. Consider the following negated sentences:

- n1. Miss Manners does *not* believe that one must write thank you notes by hand.

- n2. Miss Manners believes that it is *not* so that one must write thank you notes by hand.

- n3. Miss Manners believes that one must *not* write thank you notes by hand.

Thus, in n1 a precise plan is lacking, while in n3 there is the idea of planning something about the negation of something. n2 "leaves open the possibility that she is indifferent among the ways of writing thank you notes, so that, according to her, not writing thank you notes by hand is permissible but so is writing them by hand" ([18], p. 720). In Gibbard's system, n2 is not formalized properly, while this can be done in ELP. We have already seen that n2 can be formalized in ELP in such a way such that we do not logically infer anything specific about Miss Manners' attitudes towards $\neg\gamma$. Miss Manners is free to attribute any genuine expressive value towards $\neg\gamma$.

The pragmatic and illocutionary component of expressive sentences is made clear in ELP and given the intuitionistic-like structure of assertion it incorporates this means that the justification functions are only partial. Hence, there is room in ELP for situations like n2 in which an agent is free to attribute any expressive attitude towards a propositional content.

5.3 Schroeder's *Being For*

Mark Schroeder introduced a full expressivist semantics for moral sentences based on a general non-cognitive attitude that he termed *being for* [37]. Such attitude "is bearing a very general positive attitude" and it is expressed in normative languages. However, he noticed that "everything I do in this paper can be done with a basic attitude that is negative" ([37], p. 589) and this is relevant for our reading of his views. One of the main reasons for introducing the attitude of *being for* relies in the fact that Schroeder wants to solve the problems of expressivism with negation and presents a semantics in which: (i) inconsistency (or better to say in this case "discordance") is explained in terms of conflict among different attitudes towards the same content (A-type inconsistency) rather than as a conflict emerging from the application of the same attitude to opposite contents (B-type inconsistency)[15]; and (ii) it is possible to solve the problems of expressivism with negation. B-type inconsistency seems more suitable for cognitive attitudes like beliefs, whereas A-type inconsistency seems to handle in a better way non-cognitive attitudes.

Being for is a placeholder for some existing attitudes to be possibly filled in later [49]. In this perspective disapproval is interpreted, for instance, as "*being for* blaming for". Schroeder seems to implicitly assumes that his theory is mainly based (at least *prima facie*) on a classical framework. In fact, classical principles like *excluded middle* are explicitly said to hold in his semantics [37] and used, for instance, in the justification of Geach's original *modus ponens* argument[16]. Specifically, if N is an normative sentence meaning "being for n", then the following semantics is assigned to the following formulas:

$N = $ *being for* n
$\backsim N = $ *being for* not n
$N_1 \& N_1 = $ *being for* n_1 and n_2
$N_1 \lor N_1 = $ *being for* n_1 or n_2

The remaining connectives of material implication and biconditional can be easily reconstructed in a similar way. The general attitude of *being for* is used to create a new place to handle the three cases of negations for expressive sentences. Let us consider:

(27) Jon is for blaming for murdering

---

[15]See ([48]).

[16]However, in [39] it is said that moral sentences do not necessarily need to obey *excluded middle*.

n[1] Jon is *not* for blaming for murdering

n[2] Jon is for *not* blaming for murdering

n[3] Jon is for blaming for *not* murdering

Schroeder justifies *modus ponens* (4-6) in his semantics in the following manner:

(28). *being for* (blaming tormenting the cat)

(29). *being for* (not blaming tormenting the cat *or* blaming getting your little brother to torment the cat)

(30). *being for* (blaming getting your little brother to torment the cat)

Notice that (29) presupposes a classical view on the conditional as a material implication and cannot be a felicitous expressivist reading of (5) according to Skorupski [44]. Schroeder's reply is that his semantics is about purely normative language [40]; but if this so, then it still remains the problem of how to deal with sentences that are not fully normative. And many of the problems for expressivism come, indeed, from sentences that are not fully normative[17]. If we want to deal with sentences that are not fully normative, then Schroeder's semantics is not completely capable of differentiating things like *being for* a disjunction from a disjunction of *being for* attitudes. Skorupski noted in general in Schroeder's semantics that there are some problems with the connectives involved in the construction of expressive sentences [44], [45], (of course, of a language not fully normative). The *crux* of the matter is, in fact, related to disjunction as we have seen in (29). Since Schroeder uses disjunction in his solution to the Frege-Geach problem, then Schroeder's semantics does not seem to have all the resources for providing an expressivist treatment for those sentences which are not *fully* normative. Schroeder observes that also in a cognitive framework, if we substitute *being for* with a belief operator in (28) (29) (30) we should face a similar problem. What is not fully underlined in this discussion is the inferential role of assertion, which is fundamental in the Frege-Geach problem and many other logical questions related to expressivism. Moreover, Skorupski ([45], p. 484) noticed regarding negation that "the question is not about the particular case of negation and 'wrong', where Schroeder has offered a solution specific to a particular case, but about the way negation can exchange scope with normative predicates in general".

Since Schroeder agreed that "there will be several ways of implementing" his basic ideas, we will try to provide one possible implementation in our formal pragmatic structure. As noted by Schroeder himself, "if the problem arises from a lack of

---

[17]ELP can differentiate fully normative sentences, in which there is no connective in the radical formulas, from not fully normative sentences in which there are both pragmatic connectives for sentences expressing attitudes and connectives for radical formulas.

structure, there can be only one solution: to add structure" ([38], p. 61) and this is what we are going to do. However, we will make use of the idea of a general attitude of *being for* and adopt our pragmatic framework rather his semantics that seems to suffer from the criticisms of Skorupski's observations.

Let us consider the symbol $\trianglerighteq$, staying for the general attitude of *being for* and added in the language of ELP. Strictly speaking, $\trianglerighteq$ may be substituted with a specific non-cognitive attitude, e.g. it can be an *approval, disproval,* etc. If Bl stands for blaming and M for murdering, a pragmatic reading of (27) would be:

(31) $(\trianglerighteq Bl(M)) = J$

Let us consider now how negation interacts with normative predicates in our pragmatic framework. Now it is evident that we can locate negation in three different places in order to express n_[1], n_[2], and n_[3], namely,

n_[1] $(\trianglerighteq Bl(M)) = U$

n_[2] $(\trianglerighteq \neg Bl(M)) = J$

n_[3] $(\trianglerighteq Bl\neg(M)) = J$

Notice that in the formula n_[1] there is no nested negation (so it is a fully normative statement), while this is not the case for the last two formulas. So, our analysis allows to clearly differentiate internal and external forms of negations.

Finally, it is possible to use the general attitude of *being for* in order to provide a justification of *modus ponens* in ELP extended with this general attitude.

(32) $\trianglerighteq \gamma_1,$
(33) $(\trianglerighteq \gamma_1) \supset (\trianglerighteq \gamma_2),$
(34) $\trianglerighteq \gamma_2.$

Notice that (35) is a genuine pragmatic conditional with an intuitionistic-like formal behaviour which is indeed not equivalent to the antecedent of the conditional disjuncted with the consequent, since the logical structure of our pragmatic language is not classical but intuitionistic-like. It is now evident that *modus ponens* holds when $\trianglerighteq$ is substituted in a uniform way with a specific attitude, otherwise it may lead to implausible results[18]. However, if we do not introduce explicitly the assertion sign and its relation with attitudes, than we cannot say that (32)-(34) is a suitable treatment of the Frege-Geach problem, since there is no unasserted formula. In virtue of this, we indicate by $\triangleright$ the descriptive counterpart of $\trianglerighteq$ in the radical formula

---

[18] A similar point is made in [39]. It is also clear that when $\trianglerighteq$ is not substituted in a uniform way, then the structural rules of *weakening* and *contraction* may fail. But if this is true, then the logic governing this general attitude is not classical.

and also assume that a bridge principle connecting assertion and *being for* exists as in the case of other attitudes, namely:

(35) $\vdash \triangleright p \equiv \underline{\triangleright}\, p$[19]

The original example for the Frege-Geach problem (4-6) can be now expressed as follows:

(36) $\underline{\triangleright}\, \gamma_1$,
(37) $(\vdash \triangleright \gamma_1) \supset (\vdash \triangleright \gamma_2)$,
(38) $\underline{\triangleright}\, \gamma_2$.

From (38) (39) (40), in which $\triangleright$ is unasserted in the major premise, it is possible to infer the proper *modus ponens* for attitudes (34) (35) (36) according to expressivist views.

Finally, it is possible to disambiguate in our framework the attitude towards a disjunction from the disjunction of two attitudes (and of course the same can be done for all other connectives):

(39) $\underline{\triangleright}\, (\gamma_1 \vee \gamma_2)$
(40) $(\underline{\triangleright}\, \gamma_1) \cup (\underline{\triangleright}\, \gamma_2)$

Our interpretation of the general attitude of *being for* seems to clarify the role of this attitude in unasserted contexts.[20] Notice that our pragmatic approach is not *per se* an *alternative* to solve all the problems that face expressivism. The original motivations of logic for pragmatics were not meta-ethical. Nonetheless, our logical structure may contribute to clarify the inferential properties and the well-formed expression of sentences involving attitudes.

6 Conclusion

After having discussed the inferential role of assertion for the justification of *modus ponens* and its relevancy for the Frege-Geach problem, we have introduced the basic

---

[19]Like (12) and (13), also in this case one may imagine different forms for this bridge principle.

[20]Our perspective may thus clarify the role of asserted and unasserted contexts which has been at the origins of the Frege-Geach problem. In addition, the pragmatic logic for assertions has been used for many other applications. This makes our pragmatic treatment of sentences expressing attitudes not an ad-hoc strategy. Finally, as observed by Charlow [10], we also believe that the Frege-Geach problem is, in fact, not a problem for expressivism *per se* but it is common to many different languages. Similarly, Schwartz and Hom [42] have also recently pointed out that the negation problem can be replicated in non-espressive contexts. The adoption of our formal pragmatic structure seems to clarify that the logical basis of the problems that in the literature have been associated with expressivism can be replicated in other languages. This is made particularly evident when the assertion sign is made explicit.

elements of the pragmatic logic LP. Then, we have extended the pragmatic language of LP into a pragmatic language for expressive attitudes ELP. We analysed how to deal with the issue of asserted and unasserted contexts for conditionals, showing the merit of adopting our pragmatic view to address the Frege-Geach problem. The remaining problems related to other connectives for expressive sentences may receive a logical treatment in ELP. So, ELP seems to be a formally adequate structure for the analysis of expressive sentences and their combinations, in which the pragmatic acts of assertion (of descriptive contents) and genuine attitudes are logically related. This seems particularly coherent with hybrid forms of expressivism.[21]

Finally, we showed how our logical and pragmatic perspective can be modified and used to investigate different versions of expressivism.

References

[1] Alchourrón, C.E., Bulygin, E. (1981). The expressive conception of norms. In R. Hilpinen (ed.), *New Studies in Deontic Logic*. Dordrecht: Reidel.

[2] Bellin, G., Dalla Pozza, C. (2002). A Pragmatic Interpretation of Substructural Logics, (pp. 139-163). In W. Sieg, R. Sommer, C. Talcott (eds.), *Reflections on the Foundations of Mathematics. Essays in Honor of Solomon Feferman* - ASL Lecture Notes in Logic, 15.

[3] Blackburn, S. (1984). *Spreading the Word*. Oxford: Oxford University Press.

[4] Blackburn, S. (1988). Attitudes and contents. *Ethics*, 98(3): 501-517. Reprinted in Blackburn (1993), pp. 182-197.

[5] Blackburn, S. (1993). *Essays in Quasi-Realism*. Oxford: Oxford University Press.

[6] Boisvert, D.R. (2008). Expressive Assertivism. *Pacific Philosophical Quarterly*, 89: 169-203

[7] Carnielli, W., Pizzi, C. (2008). *Modalities and Multimodalities*. Berlin: Springer.

[8] Carrara, M., Chiffi, D., De Florio, C. (2017). Assertions and hypotheses: A logical framework for their opposition relations. *Logic Journal of the IGPL*, 25(2), 131-144.

[9] Carrara, M., Chiffi, D., De Florio, C. (2019). Pragmatic logics for hypotheses and evidence. *Logic Journal of the IGPL*, https://doi.org/10.1093/jigpal/jzz042.

---

[21] Our pragmatic proposal seems to also be on the right track regarding more general expressivist requirements. According to Woods [49], in fact, for the development of an adequate non-cognitivist treatment of moral thought it is *necessary* to: (i) assign a content to both embedded and unembedded occurrences of expressions (and their appropriate relations); (ii) legitimate the logical relations between the expressions so analyzed; (iii) accommodate intuitive data about legitimate and illegitimate embeddings; (iv) accommodate mixed cases (not fully normative sentences). Moreover, he pointed out that it would also be *desirable* to have an account which is: (v) compositional, (vi) finitely specifiable, in the sense that we can lay down a set of rules for generating complex expressions and (vii) general, i.e. we know how to connect any form of complex expression without accommodating further manners of composition.

[10] Charlow, N. (2014). The problem with the Frege-Geach problem. *Philosophical Studies*, 167(3): 635-665.

[11] Chiffi, D., Pietarinen, A.-V. (2018). Abductive inference within a pragmatic framework. *Synthese*, DOI: 10.1007/s11229-018-1824-6.

[12] Dalla Pozza, C. (1991). Un'interpretazione pragmatica della logica proposizionale intuizionistica. In G. Usberti (ed.). *Problemi fondazionali nella teoria del significato*. Firenze: Leo S. Olschki.

[13] Dalla Pozza, C., Garola, C. (1995). A pragmatic interpretation of intuitionistic propositional logic. *Erkenntnis*, 43(1): 81-109.

[14] Dalla Pozza, C. (2008). A Pragmatic logic for the expressive conception of norms and values and the Frege – Geach problem. Lecce: Siba.

[15] Dalla Pozza, C, Garola, C., Negro, A., Sergio, D. (2020). A Pragmatic Logic for Expressivism, *Theoria*, https://doi.org/10.1111/theo.12240

[16] De Florio, C. (2020). Reflections on Logics for Assertion and Denial. *Journal of Applied Logics*, forthcoming.

[17] de Prado Salas, J.G., Milić, I. (2018). Recommending beauty: semantics and pragmatics of aesthetic predicates. *Inquiry*, 61(2): 198-221.

[18] Dreier, J. (2006). Disagreeing (about) What to Do: Negation and Completeness in Gibbard's Norm-Expressivism. *Philosophy and Phenomenological Research*, 72(3): 714-721.

[19] Dummett, M. (1973). *Frege: Philosophy of Language*. London: Duckworth.

[20] Frege, G. (1879). *Begriffsschrift, eine der arithmetischen nachgebildete Formelsprache des reinen Denkens*, Halle a. S.: Louis Nebert. Translated as Concept Script, a formal language of pure thought modelled upon that of arithmetic, by S. Bauer-Mengelberg in J. van Heijenoort (ed.)(1967). *From Frege to Gödel: A Source Book in Mathematical Logic*, 1879-1931. Cambridge, MA: Harvard University Press.

[21] Galvan, S. (1988). Underivability Results in Mixed Systems of Monadic Deontic Logic. *Logique et Analyse*, 31(121): 45 – 68.

[22] Galvan, S. (2001). The Principle of Deontic Reflexivity and Kantian Axiom. *Logique et Analyse*, 44(176): 329-347.

[23] Geach, P. (1964). Assertion. *Philosophical Review*, 74(4): 449-465.

[24] Gibbard, A. (1990). *Wise Choices, Apt Feelings: A Theory of Normative Judgment*. Oxford: Clarendon Press.

[25] Gibbard, A. (2003). *Thinking How to Live*. Cambridge, MA: Harvard University Press.

[26] Gibbard, A. (2006). Précis of *Thinking How to Live*. *Philosophy and Phenomenological Research*, 72(3): 687 – 698.

[27] Hale, B. (1993). Can there be a logic of attitudes?, (pp. 337- 363). In J. Haldane & C. Eright (eds.), *Reality, Representation, and Projection*. Oxford: Oxford University Press.

[28] Hintikka J. (1969). Deontic logic and its philosophical morals (pp. 184-214). In *Models for Modalities*. Synthese Library, vol 23. Springer, Dordrecht.

[29] Horgan, T., Timmons, M. (2006). Cognitive Expressivism. In T. Horgan & M. Timmons (eds.) (pp. 255-298). *Metaethics After Moore*. Oxford: Oxford University Press.

[30] Kalinowski, J. (1985). Sur l'importance de la logique déontique pour la philosophie du droit. *Rivista internazionale di filosofia del diritto*, 62: 212-226.

[31] Miller, A. (2013). *Contemporary Metaethics: An Introduction*. Malden, MA: Polity Press, second edition.

[32] Negro, A., Sergio, D. (2013). Pensare attraverso la prova. I fondamenti del linguaggio pragmatico di Carlo Dalla Pozza (pp. 163-204). In A. Di Giorgio & D. Chiffi (Eds.), *Prova e giustificazione*. Turin: Giappichelli.

[33] Reichenbach, H. (1947). *Elements of Symbolic Logic*. New York: McMillan.

[34] Ridge, M. (2006). Ecumenical Expressivism: Finessing Frege. *Ethics*, 116: 302-336.

[35] Ripley, D. (2011). Negation, denial, and rejection. *Philosophy Compass*, 6(9): 622–629.

[36] Russell, B. (1903). *Principles of Mathematics*. New York: Norton.

[37] Schroeder, M. (2008a). *Being for: Evaluating the semantic program of expressivism*. Oxford: Oxford University Press.

[38] Schroeder, M. (2008b). How expressivist can and should solve their problem with negation. *Noûs*, 42(4): 573-599.

[39] Schroeder, M. (2009). Hybrid Expressivism: Virtues and Vices. *Ethics*, 119(2): 257-309.

[40] Schroeder, M. (2012). Skorupski on Being For. *Analysis*, 72(4): 735-739.

[41] Schroeder, M. (2013). Tempered Expressivism, (pp. 283- 313). In R. Shafer-Landau, *Oxford Studies in Metaethics*, Volume 8. Oxford: Oxford University Press.

[42] Schwartz, J., C. Hom (2015). Why the negation problem is not a problem for expressivism. *Noûs*, 49(4): 824-845.

[43] Sinnott-Armstrong, W. (2000). Expressivism and Embedding. *Philosophy and Phenomenological Research*, 61(3): 677-693.

[44] Skorupski, J. (2012). The Frege-Geach objection to expressivism: still unanswered. *Analysis*, 72(1): 9-18.

[45] Skorupski, J. (2013). Reply to Schroeder on being for. *Analysis*, 73(3): 483-487.

[46] Unwin, N. (1999). Quasi-Realism, Negation and the Frege-Geach Problem. *The Philosophical Quarterly*, 49(196): 337-352.

[47] Unwin, N. (2001). Norms and Negation: A problem for Gibbard's logic. *The Philosophical Quarterly*, 51(202): 60-75.

[48] van Roojen, M. (1996). Expressivism and irrationality. *The Philosophical Review*, 105(3): 311-335.

[49] Woods, J. (2018). The Frege-Geach Problem, (pp. 226- 242). In T. McPherson and D. Plunkett, *Routledge Handbook of Metaethics*. New York: Routledge.

[50] Wright, C. (1988). Realism, antirealism, irrealism, quasi-realism. *Midwest Studies in Philosophy*, 12: 25-49.

Received 30 October 2019

Towards Depth-Bounded Natural Deduction for Classical First-Order Logic

Marcello D'Agostino[*]
Università degli Studi di Milano
marcello.dagostino@unimi.it

Costanza Larese
Università degli Studi di Milano
costanza.larese@unimi.it

Sanjay Modgil
King's College London
sanjay.modgil@kcl.ac.uk

Abstract

In this paper we lay the foundations of a new proof-theory for classical first-order logic that allows for a natural characterization of a notion of *inferential depth*. The approach we propose here aims towards extending the proof-theoretical framework presented in [6] by combining it with some ideas inspired by Hintikka's work [18]. Unlike standard natural deduction, in this framework the inference rules that fix the meaning of the logical operators are symmetrical with respect to assent and dissent and do not involve the discharge of formulas. The only discharge rule is a classical dilemma rule whose nested applications provide a sensible measure of inferential depth. The result is a hierarchy of decidable depth-bounded approximations of classical first-order logic that expands the hierarchy of tractable approximations of Boolean logic investigated in [11, 10, 7].

We wish to thank the participants in the workshop "Assertion and Proof" held in Lecce on 12-14 September 2019, and the participants in the 2nd Workshop on Logic and Information held in Milan on 28 November 2019. We also wish to thank an anonymous reviewer for helpful comments.

[*]Funded by the Italian Ministry of University (MUR) as part of the the PRIN 2017 project n. 20173YP4N3.

1   Introduction

In the Manifesto of the Vienna Circle, written in 1929, Rudolph Carnap, Hans Hahn and Otto Neurath claimed that logic is analytic and tautological:

> Logical investigation leads to the result that all thought and inference consists of nothing but a transition from statements to other statements that contain nothing that was not already in the former (tautological transformation). [...] The scientific world-conception knows only empirical statements about things of all kinds, and analytic statements of logic and mathematics. [5, pp. 308, 311]

In a series of papers collected in the volume *Logic, Language-Games and Information* published in 1973, Hintikka attacked the logical empiricists' thesis. Starting from the Church-Turing result that classical first-order logic is undecidable (1936), Hintikka argues that there is a class of polyadic first-order logical truths that are synthetic and informative. He formulates a theory of distributive normal forms for classical first-order logic, on the basis of which he defines two objective and non-psychological notions of information content. The former, which he calls "depth information", is equivalent to Bar-Hillel and Carnap's semantic information and is not increased by deductive reasoning, thus justifying the traditional claim that logic is tautological. The latter, which he calls "surface information", might be increased by deductive reasoning and is computable, thus vindicating the idea that logic is informative.

The non-trivial deductive reasoning that does increase surface information is, according to Hintikka, to be regarded as synthetic. But, of course, the terms "analytic" and "synthetic" are given a meaning that is different from the one used in the Vienna Circle and that is based on an original interpretation of the distinction put forward by Kant in his first *Critique*. According to Hintikka, a derivation is synthetic if at least one of its steps introduces *new individuals*; a derivation is analytic if all its steps merely discuss the individuals which we have already introduced. Inferences can be synthetic at any degree k and whether or not a sentence follows from a given set of premises by means of a k-degree synthetic inference is *decidable* for every fixed k.[1]

Despite Hintikka's rejection of the idea that all logical inferences are analytic, his approach still classifies as analytic a wide class of inferences that includes not only many valid inferences of polyadic predicate logic, but also the entire set of valid inferences of propositional logic and monadic predicate logic. As a result, his work provides only a *partial* vindication of the idea that logical deduction is informative.

---

[1] For a thorough discussion of Hintikka's view and its comparison with Kant's analytic/synthetic distinction, see [22].

These doubts find an important confirmation in the theory of computational complexity: if the decision problem for Boolean logic is (most probably) intractable[2], that is to say, undecidable in practice, how is it possible to maintain that propositional logic is analytic and uninformative?

This observation was at the origin of the approach of Depth-Bounded Boolean Logics (DBBLs) [11, 10, 7]. In this approach the standard semantics of the Boolean operators is replaced by a weaker "informational semantics" whereby the meaning of a logical operator $\star$ is fixed by specifying the sufficient and the necessary conditions for an agent a to *actually possess* the information that a $\star$-sentence φ is true (respectively false), and be therefore in the disposition to *assent to* (respectively *dissent from*) it, solely in terms of the information that a actually possesses about the immediate components of φ. An inference is analytic if its conclusion can be established in terms of the actual information that is implicitly contained in its premises according to this weaker explanation of the logical operators. Synthetic inferences are those that essentially require the introduction of "virtual information", i.e., information that we do not actually possess, but must be temporarily assumed in order to reach the conclusion (as, for example, in the discharge rules of natural deduction). In this approach a propositional inference can be synthetic at any given degree k (the "depth" of the inference), depending on the nested use of virtual information; moreover, whether or not a sentence follows from a given set of premises by means of a k-degree synthetic inference is *tractable* for every fixed k.

Our main purpose in this paper is to lay the foundations for a unified treatment of classical first-order logic that brings together the main insights of the two approaches outlined above. More specifically, our main aim is to extend to the standard quantifiers the informational semantics for the Boolean operators in order to obtain a general view of the analytic/synthetic distinction and of the classification of inferences in terms of their depth.

The main contributions of this paper are: (i) a natural characterization of the intuitive "surface" meaning of quantifiers along the same lines as the characterization given for the Boolean operators in DBBLs; (ii) the definition of a suitable first-order extension of the propositional natural deduction system of [7][3] and of the associated notion of inferential depth, in such a way that k-depth inference is decidable. Typical technical results such as soundness, completeness and subformula property are stated but their proofs are omitted.

---

[2]See, for example, [14].

[3]See also [6] for a thorough proof-theoretical investigation.

2 Is logical inference "tautological"?

What do "analytic" and "tautological" mean? The affinity between the notion of analyticity and the discipline of logic is not something natural and atemporal, but rather the result of a precise historical development. This development, which concerns both the analytic/synthetic distinction and the conception of logic, might be summarized in three turning points. First, according to Kant, a judgment is analytic if and only if the concept of the predicate is (covertly) contained in that of the subject [20, A6-7/B10-11]. However, although Kant uses traditional logic as an instrument to define the notions, such as that of containment, that lie at the basis of his definition of analyticity, in his first *Critique* (1781, 1787²) he is not interested in determining whether logic itself is analytic. It is with Frege's *Foundations of Arithmetic* (1884), that represents our second step, that the relationship between logic and analyticity becomes stronger. Frege holds that a truth is analytic if and only if it can be proved with help of logical laws from definitions only [13, §3, p. 4]; as a result, logical truths, being provable through logical laws, are analytic. Interestingly enough, however, Frege explicitly rejects the idea that truths or inferences that are analytic in this sense are uninformative. With the logical empiricist movement, we reach the third step: logical truths are assumed to be analytic and they are used in order to catch the rest of analytic truths and inferences. Even W.V.O. Quine, despite his thorough criticism of the sharp analytic/synthetic distinction made by the Vienna Circle, in his *Two dogmas of empiricism* (1951), maintained that logical truths can after all be safely classified as "analytic" [24, p. 23].[4]

Now, if we assume, as the logical empiricists do, that logical deduction is analytic — and thus its conclusions result from some kind of analysis that unfolds the meaning of the logical operators —, then we seem to be obliged to conclude that it must be trivial, that is, uninformative and tautological, at least on the basis of the standard theory of semantic information [9]. This appears to be the side effect of the paradox of analysis [21, p. 323], which states that analysis cannot be sound and informative at the same time: for if it is sound, the analyzed and the analyzandum are equivalent and analysis cannot be augmentative; and if it is informative, then the analyzed and the analyzandum are not equivalent and the analysis is incorrect. The logical empiricists were bold enough to accept the "triviality" of deductive reasoning as a consequence of their commitment to the principle of analyticity of logic.

---

[4]On this point see also [8], Section 3.

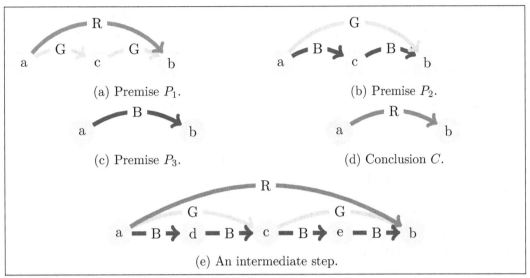

(a) Premise P_1.   (b) Premise P_2.

(c) Premise P_3.   (d) Conclusion C.

(e) An intermediate step.

Figure 1: Configurations of individuals involved in the argument from premises P_1, P_2 and P_3 to conclusion C.

3   Hintikka on "synthetic" logical inference

In order to convey the main idea underlying Hintikka's approach, while avoiding technicalities, consider the following example, which is a simplified version of the case first presented in [4] and then discussed in [17, p. 86 ff.], that illustrates a kind of reasoning that is synthetic according to Hintikka's sense of the term. Consider the argument from the premises P_1, P_2 and P_3 to the conclusion C:

$$P_1 : \forall x \forall y (Rxy \rightarrow \exists z (Gxz \wedge Gzy))$$
$$P_2 : \forall x \forall y (Gxy \rightarrow \exists z (Bxz \wedge Bzy))$$
$$P_3 : \forall x \forall y ((Bxy \wedge Cx) \rightarrow Cy)$$
$$C : \forall x \forall y ((Rxy \wedge Cx) \rightarrow Cy).$$

As Figure 1 suggests, P_1 says that whenever two points are connected through a red arrow, then there exists a third point, which is interpolated through green arrows. Similarly, P_2 says that whenever two points are connected through a green arrow, then there exists a third point, which is interpolated through blue arrows. P_3 says that whenever two points are connected through a blue arrow and the former is colored, then also the latter is colored. Similarly, C says that whenever two points are connected through a red arrow and the former is colored, then also the latter is colored.

What is the reasoning that leads us from the premises to the conclusion? We could start from premise P_1 and say that whenever two points, a and b, are connected through a red arrow, then there exists a third point, call it c, which is interpolated through green arrows. Then, we could use premise P_2 and reason as follows. Since a and c are connected through a green arrow, then there is another individual d, which is linked to a and c through blue arrows; similarly, since also c and b are connected through a green arrow, then there is a fifth point e, which is linked to c and b by blue arrows. Then, given the premise P_3, if a is colored, then also d is colored (because they are connected through a blue arrow); for the same reason and given that d is colored, then also c is colored; again, since c is colored then also e is colored; last, we get that b is also colored. In this way and since we didn't assume anything about the instantiating individuals, we reach the general conclusion that the colored marker ink spreads along red arrows too.

According to Hintikka's theory, this argument is synthetic, because some of its intermediate steps introduce new individuals into the argument. In particular, the intermediate step depicted in Figure 1e makes use of individuals d and e, which do not enter the configurations of the premises and conclusion of the argument, but, at the same time, are in certain relations with the other individuals.

As mentioned in the Introduction, this approach still classifies as analytic and informationally trivial all the inferences of propositional logic and of the monadic predicate calculus. This has been widely regarded as unsatisfactory especially in light of the development of the theory of NP-completeness according to which the decision problem for propositional logic is most likely to be intractable. The tension between the (probable) intractability of Boolean logic and its alleged informational triviality seems very similar to the tension that motivated Hintikka in arguing that the undecidability of first-oder logic is at odds with the philosophical claim that its inferences are analytic and tautological.

4 Depth-Bounded Boolean Logics

Standard formalizations of classical logic cannot capture the essential difference between these two inferences:

$$\frac{\begin{array}{c} P \vee Q \\ Q \to R \\ \neg P \end{array}}{R} \qquad \frac{\begin{array}{c} P \vee Q \\ P \to Q \end{array}}{Q}$$

The argument to establish the soundness of the first inference is the following:

$$
\begin{array}{ll}
1 & P \vee Q \\
2 & Q \to R \\
3 & \neg P \\
\hline
4 & Q \ (\text{from 1 and 3}) \\
5 & R \ (\text{from 2 and 4})
\end{array}
$$

Notice that, here, at each step we are using information that we actually possess. On the other hand, a typical argument for the second example would run as follows:

$$
\begin{array}{cc}
1 & P \vee Q \\
2 & P \to Q \\
\end{array}
$$

3.1	*Suppose that P*	3.2	*Suppose that ¬P*
	Q (from 2 and 3.1)		Q (from 1 and 3.2)

$$Q$$

The sense in which the conclusion of the first argument is "implicitly contained" in the premises is different from the sense in which the conclusion of the second argument is. In the latter we make essential use of information that we do not *actually possess* and is not even implicitly contained in the information that we actually possess. This is what we call "virtual information". We simulate information states that are richer than the actual one and consider the two possible outcomes of the process of acquiring such information.

In Gentzen's Natural Deduction the use of virtual information is associated with a technical device, known as "discharging of assumptions":

$$
\begin{array}{ccc}
\Gamma & \Delta, [P] & \Lambda, [Q] \\
\vdots & \vdots & \vdots \\
P \vee Q & R & R \\
\hline
 & R &
\end{array}
\qquad
\begin{array}{c}
\Gamma, [P] \\
\vdots \\
Q \\
\hline
P \to Q
\end{array}
\qquad
\begin{array}{cc}
\Gamma & \Delta, [P(a)] \\
\vdots & \vdots \\
(\exists x)P(x) & R \\
\hline
\multicolumn{2}{c}{R}
\end{array}
$$

with the usual restrictions on a. In the propositional rules, the sentences in square brackets represent *virtual* information that may not be (and typically is not) contained in the information that is actually "given" in the premises. In the existential quantifier rule, the sentence in the square brackets may represent information on an individual that is not actually "given" in the set of individuals associated with the quantifiers that occur in the premises.

After making this fundamental distinction between inferences that make use only of actual information and those that require the use of virtual information, we can ask ourselves the following question: can we fix the meaning of the logical operators in terms of the *information that is actually possessed* by an agent, that is, without appealing to virtual information?

The *informational semantics* of the logical operators is based on the following principle:

> The meaning of an *n*-ary logical operator $\star$ is determined by specifying the sufficient (necessary) conditions for an agent x to actually hold the information that a sentence of the form $\star(P_1, \ldots, P_n)$ is true, respectively false, in terms of the information that x actually holds about the truth or falsity of $P_1, \ldots, P_n$.

Here by saying that x *actually holds* the information that P is true (respectively false) we mean that this is information *practically available* to x and with which x can operate (e.g., in decision-making).

In [10, 8, 7, 6] a suitable set of introduction and elimination (*intelim*) rules for the Boolean operators were presented that comply with the basic principle of informational semantics. These rules characterize a subsystem of classical propositional logic that is a logic in Tarski's sense and is *tractable*. Interestingly enough, this logical system is sound and complete w.r.t. to a *non-deterministic matrix*, in the sense of [1], that complies with the basic principle of informational semantics and was first proposed by W.V.O. Quine in [25] to capture the "primitive" meaning of the logical operators.[5] The full deductive power of classical propositional logic is retrieved by adding a single discharge rule that governs the use of virtual information and consists in a form of *classical dilemma* rule. The maximum number of nested applications of this single discharge rule that are needed to obtain a conclusion from a given set of premises provides a natural measure of the *propositional depth* of the associated inference. For each given k, k-depth validity can also be decided in polynomial time, so providing an infinite hierarchy of *tractable* approximations to classical propositional logic.[6] In the next section we shall propose a way of extending these rules to first-order logic, to provide a similar measure of the quantificational depth of an inference and a hierarchy of *decidable* approximations to full classical logic.

To summarize, we have examined the main ideas of two theories that reject the logical empiricists' tenet that logic is analytic and tautological. Hintikka focuses on the tension between this tenet and the undecidability of first-order logic. In his conceptual framework, an inference is analytic if it does not introduce new individuals into the argument beyond those that one needs to consider in order to grasp the premises and the conclusion. The approach of DBBLs focuses on the similar tension between the (probable) intractability of Boolean logic and the claim, shared

---

[5]See [8] for a discussion.

[6]Recently in [2] the DBBL approach has been adopted as the logical foundation of a depth-bounded approach to belief functions.

by Hintikka and the logical empiricists, that it is informationally trivial. According to this perspective, an inference is analytic if and only if the informational meaning of the logical operators is sufficient to derive the conclusion from the premises; an inference is synthetic (at different *degrees*) when virtual information (up to a certain depth) is needed to derive the conclusion from the premises.

Although each of these two theories suggests a compelling reason for which logic is informative, neither of them, if taken in isolation, is sufficient to provide a *complete* vindication of the thesis that first-order logical inferences are synthetic and informative. On the one hand, Hintikka's work classifies as analytic all propositional and monadic inferences. On the other, DBBLs are restricted to propositional logic and do not capture the dimension of quantificational depth that in Hintikka's work is related to the introduction of new individuals that were not "given" in a surface understanding of the premises.

The main contribution of this paper consists in merging the two approaches by introducing a new family of logical systems, that we call *Depth-Bounded First-Order Logics* (DBFOLs), which extends DBBLs to first-order languages by exploiting Hintikka's insight (in particular, appropriate rules for quantifiers are added to the introduction and elimination rules for DBBLs). The structure of DBFOLs, which resembles that of DBBLs, is given by an infinite hierarchy of logics representing increasing levels of syntheticity or informativeness of classical first-order logic. The logic $\vdash_0$, which is the basic element of the hierarchy, validates only analytic inferences; for every $k > 0$, the logic $\vdash_k$ validates synthetic inferences in such a way that the greater k is, the more synthetic and informative are the inferences that are valid in it. Here, the terms "analytic" and "synthetic" are given a new meaning that conciliates the intuitions of Hintikka's work and of the DBBL-approach.

5   An intuitive informational semantics for quantifiers

In order to define inference rules that comply with the basic ideas of informational semantics outlined in the previous section, we need to ask ourselves a fundamental question. What do we mean when we say that we hold the information that a sentence of the form $\forall x F$ or $\exists x F$ is true, respectively false?

Let's start with the notion of *actually possessing* the information that $\forall x F$ is true. The answer cannot be that we actually possess the information that $F[x/a]$ is true for *all* the infinitely many individuals that may be denoted by a. A more feasible answer is the following: we are in the disposition to assent to *any* sentence of the form $F[x/a]$. A typical analogy widely used in this context is that we have an urn W whose composition is unknown to us, and if we draw an individual at random from this urn, assign to it the name a (or a is its given name, in case it already has one), then we are in the disposition to assent to $F[x/a]$. We may also imagine that, once an individual has been drawn from the W urn, we move it into a box D that represents the known domain of discourse. The composition of the box D, unlike that of W, is fully known and always consists of a finite number of individuals at each stage of the reasoning process. How many draws from W are needed in association with the given universal quantifier in order to grasp the meaning of the sentence? It makes sense to say that, in order to grasp the meaning of $\forall x F$, a minimal sufficient condition consists in envisaging a situation in which an agent is in the disposition to assent to $F[x/a]$ for *any single* random draw of an individual a from W, *as well as* for all individuals a contained in D.

Similar considerations can be made for the case of the falsity of $\exists x F$. That an agent actually possesses the information that $\exists x F$ is false, or equivalently that $\neg \exists x F$ is true, means in essence that for any possible draw of an individual a from the W urn and for all the finitely many individuals in D, the agent is in the disposition to dissent from $F[x/a]$, i.e., to assent to $\neg F[x/a]$, and this explanation is sufficient to grasp the meaning of $\neg \exists x F$.

What about the notion of actually possessing the information that $\exists x F$ is true? A natural answer is that one is informed that a search for an individual that fits the description given by the open sentence F (assuming that F contains x as a free variable) will eventually be successful. The search involves both the urn W and the box D, meaning that the sought individual might be unknown or already known. Similar considerations hold for the notion of actually possessing the information that the sentence $\forall x F$ is false, or equivalently that $\neg \forall x F$ is true. We are guaranteed that the search for an individual that fits the description given by the open sentence $\neg F$ will eventually succeed. We can call this explanation the *surface meaning* of $\forall$ and $\exists$.

6   Perfect PNF and analytic rules for quantifiers

Our aim is to put forward a set of introduction and elimination rules for quantifiers that are in accordance with their surface meaning as fixed by the intuitive informational semantics outlined in the previous section. Their application will therefore be *analytic* as well as informationally trivial.

In order to keep technicalities to minimum and focus on the conceptual analysis, we shall assume that all *premises* of an inference are in prenex normal form.[7] It is well-known that this involves no loss of generality, for it is computationally easy to

---

[7]Since we are restricting our attention to classical logic, this is always possible.

transform every formula in a formula in prenex normal form such that the matrix has exactly the same Boolean structure as the original formula — namely it results from the original formula by removing all quantifiers. We shall not impose, however, that any formula other than the premises is in prenex normal form.

In what follows we shall assume also that the quantified variables in the *set of premises* Γ of an inference are renamed in such a way that (i) the number of existentially quantified variables occurring in Γ is maximal and (ii) the number of universally quantified variables occurring in Γ is minimal, modulo logical equivalence (see Example 1 below). When such a renaming has been performed we shall say that Γ is in *perfect prenex normal form*. This kind of transformation of the premises, albeit non essential, makes for easier formulation of suitable analytic rules for quantifiers. The transformation can be avoided at the price of more contrived restrictions on the quantifier rules or of a new kind of format for proofs, other than the standard Gentzen-style or Fitch-style format. For the purposes of this preliminary investigation we shall therefore adopt this simplification and leave other options for future research.

Recall that a formula is in *prenex normal form* (PNF) if it has the form

$$Q_{x_1} \cdots Q_{x_n} F[x_1, \ldots, x_n]$$

where each Q_i is either an occurrence of $\forall$ or an occurrence of $\exists$, F is quantifier-free and all variables in F are bound by some quantifier in the prefix. A formula is in *minimal PNF* (*min-PNF*) if there is no logically equivalent formula with the same matrix and a lower number of occurrences of quantifiers in the prefix.

Example 1. $\forall x \exists y \forall z (Rxy \wedge Ryz)$ *is in min-PNF.* $\forall x \exists y \forall z \forall w (Rxy \wedge Ryz)$ *is not, for the last occurence of $\forall$ is redundant.* $\forall x \forall y (Px \wedge Qy)$ *is not, for it is equivalent to* $\forall x (Px \wedge Qx)$.

A *set* Γ of formulae is in *perfect prenex normal form* (PPNF) if:

- Every formula in Γ is in min-PNF;

- All occurrences of existential quantifiers in Γ bind variables that are different from each other and from all the universally quantified variables;

- The number of distinct universally quantified variables occurring in Γ is minimal.

Every set Γ of formulae in min-PNF can be easily transformed into a set Γ' in PPNF, by renaming of variables, in such a way that every formula A in Γ is transformed into a logically equivalent formula A' in Γ' such that the matrix of A' is the same as the matrix of A modulo renaming of variables.

Example 2. *The set*

$$\{\forall x \exists y \forall z (Rxy \wedge Rxz), \forall x \exists y \forall w (Sxy \wedge Sxw)\}$$

is not in PPNF. A possible transformation of this set into PPNF is the following:

$$\{\forall x \exists y \forall z (Rxy \wedge Rxz), \forall x \exists w \forall z (Sxw \wedge Sxz)\}.$$

The main motivation behind the requirement that the set of premises of an inference is in PPNF is given by the following:

Proposition 1. *Let Γ be a set of formulae in PPNF. Then, the number of distinct bound variables in Γ is the same as the number of distinct bound variables in any min-PNF of $\bigwedge \Gamma$.*

So, we can argue that when a set of premises is in PPNF the number of distinct individuals that are considered in these premises and are required to grasp their surface meaning is mirrored by the number of distinct bound variables. Moreover, the situation does not change if we take any min-PNF of the conjunction of the premises.

The preliminary transformation of any set of premises into PPNF makes the explanation of the inference rules quite transparent. All inferences that can be justified by means of these rules will be "analytic" in the following (informal) sense:

> No more individuals need to be considered in proving the conclusion   (PQA)
> than those that were already considered in grasping the surface
> meaning of the premises.

PQA stands for "Principle of Quantificational Analyticity". This sense of analyticity is not strictly equivalent to any of the senses discussed by Hintikka in [19], except perhaps the sense IIIc (p. 181).

Recall that when a set of premises is in PPNF, all occurrences of the existential quantifier bind different variables, while different occurrences of the universal quantifier may bind the same variable. When a set of premises is in PPNF, every distinct universally quantified *variable* involves the consideration of a distinct "arbitrary" individual, and every distinct existentially quantified variable involves the consideration of a distinct "specific" individual. These are all the individuals that may be regarded as been "thought of" in the premises. To take a simplest example, in order to grasp the surface meaning of the set of premises $\{\exists x \forall y Rxy, \forall y \exists z Syz\}$, we need to consider three distinct individuals. The first is a specific one that results from the search, associated with the existentially quantified variable x, of an individual that fits the description given by the open formula $\forall y Rxy$. The second is an unknown one, say b, which is drawn at random from the W urn and is associated with the universally quantified variable y. The same "drawn" individual b can be taken to instantiate the universally quantified variable y in the second formula. Finally we need to consider a third specific individual, say c, which results from the search associated with the existentially quantified z and fits the description given by the open formula Sbz.

Note that, according to our explanation of the informational meaning of $\forall$, holding the information that $\forall x \exists y Rxy$ means that we are in the disposition to assent to $\exists y Ray$ for *any* single unknown individual drawn from the W urn, as well as for *all* already known individuals taken from the D box (when it is not empty). Being in the disposition to assent to $\exists y Ray$ means that we hold the information that the search for a suitable individual y that fits Ray will eventually be successful. This does not immediately imply that we hold information about a specific individual that does fit the description. However, when $\forall x \exists y Rxy$ is used as premise of an inference, we may consider as part of the surface meaning of $\forall x \exists y Rxy$ that the result of this search can somehow be "given" to us. This allows us to choose a new name for this individual, say b, and infer Rab. These are the two individuals that are thought of in grasping the meaning of the premises, as witnessed by the fact that any "concrete" (e.g. graphical) explanation of what counts as a model of this sentence would need to involve two individuals and no more. In essence, what $\forall x \exists y Rxy$ says is: "let a be an arbitrary unknown individual drawn from W and let b any result of the search for an individual that fits the description Ray, we are in the disposition to assent to Rab".

Now, it makes sense to claim that the meaning of $\forall$ implies that we are also in the disposition to assent to $\exists y Rby$. However, it would not be equally natural to assume that the new search for an individual that fits the description Rby and the result of this search has already been thought of in the premise $\forall x \exists y Rxy$ and therefore required to grasp its surface meaning.

For the sake of simplicity we assume that our first-order language contains *no constants* and is equipped with a *set of parameters* (as in [28]) $a, b, c, \ldots$ possibly with subscripts, that may occur in the proof, but neither in the premises, nor in the conclusion. Given that formulae are in PPNF, let the *Q-complexity* of a finite set Γ of premises be the number of distinct variables that occur in Γ (we assume all variables are bound by some quantifier in the prefix). Then our notion of analytic proof in the sense of (PQA) above, that is in a sense that is restricted to the informational meaning of the quantifiers, can be simply rephrased as follows:

A proof is analytic *only if* the number of distinct parameters that (PQA*) occur in it never exceeds the Q-complexity of its initial premises.

The use of "only if" stems from the fact that a proof may be analytic in the sense of (PQA*) above and yet be synthetic in that it may still make use of virtual information at the propositional level. In the sequel we shall see how the structural DBBL rule that governs the introduction of virtual information may be used to mark the transition to the next degree of depth both at the propositional and the quantificational level, so as to provide a unified approach to the notion of synthetic proof for full first-order logic.

Our next problem is then: can we define a set of intelim rules for quantifiers that comply with the surface informational meaning of the quantifiers outlined in Section 5 and deliver *only* analytic inferences in the sense of (PQA*)? Can we construe the transition from analytic to synthetic inferences and from one degree of depth to the next *only* in terms of the nested use of virtual information?

As far as no virtual information is used, we shall display our proofs in the format of sequences of *signed formulae*, of the form $\mathsf{T}\,A$ or $\mathsf{F}\,A$. In accordance with the informational approach to classical logic our interpretation of signed formulae will be non-standard. We take "$\mathsf{T}\,A$" to mean "we actually possess the information that A is true" and "$\mathsf{F}\,A$" to mean "we actually possess the information that A is false". So the signs T and F do not refer to classical truth and falsity (as in Smullyan's semantic tableaux [28]), but to "informational truth" and "informational falsity". As mentioned above, a natural way of thinking of these notions is in terms of an agent's disposition to *assent* to a sentence or *dissent* from it depending on the available information. A straightforward consequence of this epistemic interpretation of the signs is that one cannot assume, in general, that for every sentence A an agent is either in the disposition of assenting to A or in the disposition of dissenting from A. When neither is the case, the agent may abstain for lack of sufficient information.

Although the use of signed formulae is appropriate for conceptual clarity, it is by no means essential in our approach. For all practical purposes one can always revert to standard formulae simply by removing all the T signs and replacing all the F signs with the negation operator.

In the DBBL approach, for each logical operator, there are intelim rules for a signed formula containing it as main operator as well as for its conjugate (the conjugate of "$\mathsf{T}\,A$" is "$\mathsf{F}\,A$" and viceversa). This feature is shared by the tableau method (which, however, is restricted to refutations of sets of formulae via elimination rules only) and other bilateral systems of deduction, such as Bendall's [3] or Rumfit's [26]. The first-order version of the propositional DBBL rules, that allows for the presence of parameters in formulae, is given in Figures 2 and 3. In the sequel we shall make use of the following notation:

- F_a^x denotes the result of replacing every occurrence of the variable x in F with

$$\frac{\begin{array}{c}\mathsf{T}\,A \to B \\ \mathsf{T}\,A\end{array}}{\mathsf{T}\,B}\,\mathsf{T}\!\to\! E_1 \qquad \frac{\begin{array}{c}\mathsf{T}\,A \to B \\ \mathsf{F}\,B\end{array}}{\mathsf{F}\,A}\,\mathsf{T}\!\to\! E_2$$

$$\frac{\mathsf{F}\,A \to B}{\mathsf{T}\,A}\,\mathsf{F}\!\to\! E_1 \qquad \frac{\mathsf{F}\,A \to B}{\mathsf{F}\,B}\,\mathsf{F}\!\to\! E_2$$

$$\frac{\begin{array}{c}\mathsf{T}\,A \vee B \\ \mathsf{F}\,A\end{array}}{\mathsf{T}\,B}\,\mathsf{T}\!\vee\! E_1 \qquad \frac{\begin{array}{c}\mathsf{T}\,A \vee B \\ \mathsf{F}\,B\end{array}}{\mathsf{T}\,A}\,\mathsf{T}\!\vee\! E_2$$

$$\frac{\mathsf{F}\,A \vee B}{\mathsf{F}\,A}\,\mathsf{F}\!\vee\! E_1 \qquad \frac{\mathsf{F}\,A \vee B}{\mathsf{F}\,B}\,\mathsf{F}\!\vee\! E_2$$

$$\frac{\mathsf{T}\,A \wedge B}{\mathsf{T}\,A}\,\mathsf{T}\!\wedge\! E_1 \qquad \frac{\mathsf{T}\,A \wedge B}{\mathsf{T}\,B}\,\mathsf{T}\!\wedge\! E_2$$

$$\frac{\begin{array}{c}\mathsf{F}\,A \wedge B \\ \mathsf{T}\,A\end{array}}{\mathsf{F}\,B}\,\mathsf{F}\!\wedge\! E_1 \qquad \frac{\begin{array}{c}\mathsf{F}\,A \wedge B \\ \mathsf{T}\,B\end{array}}{\mathsf{F}\,A}\,\mathsf{F}\!\wedge\! E_2$$

$$\frac{\mathsf{F}\,\neg A}{\mathsf{T}\,A}\,\mathsf{F}\neg E \qquad \frac{\mathsf{T}\,\neg A}{\mathsf{F}\,A}\,\mathsf{T}\neg E$$

Figure 2: Elimination rules for the propositional operators.

the parameter a;

- F_x^a denotes the result of replacing every occurrence of the parameter a with the variable x;

- $F[a/x]$ denotes the result of replacing some or all occurrences of a with x.

$\mathsf{T}\forall$-elimination and $\mathsf{F}\exists$-elimination.

$$\frac{\mathsf{T}\,\forall x F}{\mathsf{T}\,F_a^x}\,\mathsf{T}\forall E \qquad\qquad \frac{\mathsf{F}\,\exists x F}{\mathsf{F}\,F_a^x}\,\mathsf{F}\exists E$$

where a is any parameter that already occurs above in the proof; or else a is a *new* parameter, provided that no other parameter has been already introduced above by an application of the same rule to a formula of the form $\mathsf{T}\forall x G$, respectively $\mathsf{F}\exists x G$.

$$\frac{\mathsf{F}\,A}{\mathsf{T}\,A \to B}\,\mathsf{T}\to I_1 \qquad \frac{\mathsf{T}\,A}{\mathsf{T}\,A \vee B}\,\mathsf{T}\vee I_1 \qquad \frac{\mathsf{F}\,A}{\mathsf{F}\,A \wedge B}\,\mathsf{F}\wedge I_1$$

provided that every parameter occurring in B already occurs above.

$$\frac{\mathsf{T}\,B}{\mathsf{T}\,A \to B}\,\mathsf{T}\to I_2 \qquad \frac{\mathsf{T}\,B}{\mathsf{T}\,A \vee B}\,\mathsf{T}\vee I_2 \qquad \frac{\mathsf{F}\,B}{\mathsf{F}\,A \wedge B}\,\mathsf{F}\wedge I_2$$

provided that every parameter occurring in A already occurs above.

$$\frac{\begin{array}{c}\mathsf{T}\,A \\ \mathsf{F}\,B\end{array}}{\mathsf{F}\,A \to B}\,\mathsf{F}\to I \qquad \frac{\begin{array}{c}\mathsf{F}\,A \\ \mathsf{F}\,B\end{array}}{\mathsf{F}\,A \vee B}\,\mathsf{F}\vee I \qquad \frac{\begin{array}{c}\mathsf{T}\,A \\ \mathsf{T}\,B\end{array}}{\mathsf{T}\,A \wedge B}\,\mathsf{T}\wedge I$$

$$\frac{\mathsf{T}\,A}{\mathsf{F}\,\neg A}\,\mathsf{F}\neg I \qquad \frac{\mathsf{F}\,A}{\mathsf{T}\,\neg A}\,\mathsf{T}\neg I$$

Figure 3: Introduction rules for the propositional operators.

In essence, each bounded variable x in the premise of these rules can be instantiated at most once by a new parameter, denoting an unknown individual drawn from the W urn, although it can be instantiated by all the old parameters denoting known individuals in the box D.

Example 3. *Example of a* wrong *application of the* $\mathsf{T}\forall E$ *rule (the quantifier $\forall y$ has been used at step 3 to introduce the new parameter b and again at step 5 to introduce* *the new parameter c).*

1	$\mathsf{T}\,\forall x\forall y Rxy$	*Premise*
2	$\mathsf{T}\,\forall y Ray$	$\mathsf{T}\,\forall E, 1$
3	$\mathsf{T}\,Rab$	$\mathsf{T}\,\forall E, 2$
4	$\mathsf{T}\,\forall y Rby$	$\mathsf{T}\,\forall E, 1$
5	**$\mathsf{T}\,Rbc$**	$\mathsf{T}\,\forall E, 4$
6	$\mathsf{T}\,Rab \wedge Rbc$	$\mathsf{T}\,\wedge I, 3, 5$
7	$\mathsf{T}\,\forall z(Rab \wedge Rbz)$	$\mathsf{T}\,\forall I, 6$
8	$\mathsf{T}\,\forall y\forall z(Ray \wedge Ryz)$	$\mathsf{T}\,\forall I, 7$
9	$\mathsf{T}\,\exists x\forall y\forall z(Rxy \wedge Ryz)$	$\mathsf{T}\,\exists I, 8$

$\mathsf{T}\,\exists$-elimination and $\mathsf{F}\,\forall$-elimination.

$$\frac{\mathsf{T}\,\exists x F}{\mathsf{T}\,F_a^x}\;\mathsf{T}\,\exists E \qquad\qquad \frac{\mathsf{F}\,\forall x F}{\mathsf{F}\,F_a^x}\;\mathsf{F}\,\forall E$$

provided that a is a *new* parameter and no other parameter has been already introduced above by an application of the same rule to a formula of the form $\mathsf{T}\,\exists x G$, respectively $\mathsf{F}\,\forall x G$.

In essence, each bounded variable x in the premise of these rules can be instantiated at most once by a new parameter, denoting the result of a search for an individual that fits the description in F.

When these rules are applied, we say that the new parameter a in the conclusion of the rule is *critical* and *depends* on all the other parameters occurring in F.

Example 4. *Example of a wrong application of the $\mathsf{T}\,\exists E$ rule (the quantifier $\exists y$ has been used at step 3 to introduce the new parameter b and again at step 5 to introduce* *the new parameter c).*

1	$\mathsf{T}\,\forall x \exists y Rxy$	*Premise*
2	$\mathsf{T}\,\exists y Ray$	$\mathsf{T}\,\forall E, 1$
3	$\mathsf{T}\,Rab$	$\mathsf{T}\,\exists E, 2$
4	$\mathsf{T}\,\exists y Rby$	$\mathsf{T}\,\forall E, 1$
5	$\mathsf{T}\,\boldsymbol{Rbc}$	$\mathsf{T}\,\exists E, 4$
6	$\mathsf{T}\,Rab \wedge Rbc$	$\mathsf{T}\,\wedge I, 3, 5$
7	$\mathsf{T}\,\exists z(Rab \wedge Rbz)$	$\mathsf{T}\,\exists I, 6$
8	$\mathsf{T}\,\exists y \exists z(Ray \wedge Ryz)$	$\mathsf{T}\,\exists I, 7$
9	$\mathsf{T}\,\forall x \exists y \exists z(Rxy \wedge Ryz)$	$\mathsf{T}\,\forall I, 8$

$\mathsf{T}\,\forall$-introduction and $\mathsf{F}\,\exists$-introduction.

$$\frac{\mathsf{T}\,F}{\mathsf{T}\,\forall x F_x^a}\ \mathsf{T}\,\forall I \qquad\qquad \frac{\mathsf{F}\,F}{\mathsf{F}\,\exists x F_x^a}\ \mathsf{F}\,\exists I$$

Provided that a is not critical and F does not contain critical parameters depending on a. Moreover, x is not bound in F.

Example 5. *Example of a* wrong *application of the* $\mathsf{T}\forall I$ *rule (b is a critical parameter depending on a).*

1	$\mathsf{T}\,\forall x \exists y Rxy$	*Premise*
2	$\mathsf{T}\,\exists y Ray$	$\mathsf{T}\,\forall E, 1$
3	$\mathsf{T}\,Rab$	$\mathsf{T}\,\exists E, 2$
4	$\mathsf{T}\,\boldsymbol{\forall x Rxb}$	$\mathsf{T}\,\forall I, 3$
5	$\mathsf{T}\,\exists y \forall x Rxy$	$\mathsf{T}\,\exists I, 4$

$\mathsf{T}\,\exists$-introduction and $\mathsf{F}\,\forall$-introduction.

$$\frac{\mathsf{T}\,F}{\mathsf{T}\,\exists x F[a/x]}\ \mathsf{T}\,\exists I \qquad\qquad \frac{\mathsf{F}\,F}{\mathsf{F}\,\forall x F[a/x]}\ \mathsf{F}\,\forall I$$

provided x is not bound in F.

General restriction on quantifier eliminations. In order to guarantee that the Principle of Quantificational Analyticity (PQA*) is always satisfied, we also need the following *general restriction* on the application of the quantifier elimination rules: their application is allowed only if *their premise is not the conclusion of an introduction*. It is easy to see how the violation of this general restriction may lead to violations of (PQA*). Consider, for example, the following proof:

1	$\mathsf{T}\,\forall x \exists y Rxy$	premise
2	$\mathsf{T}\,\exists y Ray$	from 1
3	$\mathsf{T}\,Rab$	from 2
4	$\mathsf{T}\,\exists z Raz$	from 3
5	$\mathsf{T}\,\forall x \exists z Rxz$	from 4
6	$\mathsf{T}\,\exists z Rbz$	from 5
7	$\mathsf{T}\,Rbc$	from 6
8	$\mathsf{T}\,Rab \wedge Rbc$	from 3 and 7
·9	$\mathsf{T}\,\exists z (Rab \wedge Rbz)$	from 8
10	$\mathsf{T}\,\exists y \exists z (Ray \wedge Ryz)$	from 9
11	$\mathsf{T}\,\forall x \exists y \exists z (Rxy \wedge Ryz)$	from 10

Here, the number of parameters occurring in the proof exceeds the Q-complexity of the premise. The proof is not analytic.

0-depth inferences (analytic sequences). An *analytic sequence based on* Γ, where Γ is in PPNF, is any sequence of signed formulae starting from the formulae in $\mathsf{T}\,\Gamma = \{\mathsf{T}\,B \mid B \in \Gamma\}$ and such that each subsequent signed formula results from signed formulae previously occurring in the sequence by means of an application of the intelim rules. An *analytic proof of A from* Γ is an analytic sequence based on $\mathsf{T}\,\Gamma$ that ends with $\mathsf{T}\,A$. We say that A is *deducible from* Γ *at depth* 0 when there is an analytic proof of A from Γ.

Note that $A, \neg A \vdash_0 B$ for any B, as shown by the following analytic sequence:

1	$\mathsf{T}\,A$	premise
2	$\mathsf{T}\,\neg A$	premise
3	$\mathsf{F}\,A$	from 2 by $\mathsf{T}\neg E$
4	$\mathsf{T}\,A \vee B$	from 1 by $\mathsf{T}\vee I$
5	$\mathsf{T}\,B$	from 4 and 3 by $\mathsf{T}\vee E$

However, this sequence is not "analytic" in one of the widespread senses of this word, in that it does not enjoy the subformula property (and indeed there is no 0-depth proof of B from $\{A, \neg A\}$ with the subformula property). If we want the subformula property to hold in general, we need to modify our definition of 0-depth deducibility as follows. An analytic sequence based on Γ is *closed* if it contains both $\mathsf{T} B$ and $\mathsf{F} B$ for some formula B. Otherwise we say that it is *open*. A 0-*depth refutation* of Γ is a closed analytic sequence based on Γ. Then we say that A is *deducible from* Γ *at depth* 0 when either there is 0-depth proof of A from Γ or a 0-depth refutation of Γ. On the other hand, if our notion of analytic proof is restricted to sequences with the subformula property, then the previous notion of analytic proof delivers a paraconsistent notion of 0-depth deducibility. Since our aim in this paper is to outline a depth-bounded approach to classical first-order logic, we shall adopt the amended definition of 0-depth deducibility. Note, however, that according to this definition, *not all classically inconsistent set of formulae are explosive*, but only those whose inconsistency can be detected at depth 0, i.e., by virtue of the surface informational meaning of the logical operators.[8]

The notion of 0-depth inference intends to capture the idea of an inference that is performed by virtue of the surface informational meaning of the quantifiers and makes no use of virtual information (no discharge of temporary hypothetical assumptions). The following example illustrates the restriction on the Boolean introduction rules that are needed to preserve the quantificational analyticity of proofs.

Example 6. *Example of a* wrong *application of the* $\mathsf{T} \vee I$ *rule (at step 4, a new parameter, namely c, occurs in the second disjunct of* $\mathsf{T} Rab \vee Rbc$*).*

1	$\mathsf{T} \forall x \exists y Rxy$	*Premise*
2	$\mathsf{T} \exists y Ray$	$\mathsf{T} \forall E, 1$
3	$\mathsf{T} Rab$	$\mathsf{T} \exists E, 2$
4	$\mathbf{T} \, \boldsymbol{Rab} \vee \boldsymbol{Rbc}$	$\mathsf{T} \vee I, 3$
5	$\mathsf{T} \forall z (Rab \vee Rbz)$	$\mathsf{T} \forall I, 4$
6	$\mathsf{T} \exists y \forall z (Ray \vee Ryz)$	$\mathsf{T} \exists I, 5$
7	$\mathsf{T} \exists x \exists y \forall z (Rxy \vee Ryz)$	$\mathsf{T} \exists I, 6$

---

[8]For a further discussion of this point see [6, Section 8].

Example 7. $\mathsf{T}\forall x\exists y Rxy,\ \mathsf{T}\forall x\forall w(Rxw \to Rwx) \vdash_0 \mathsf{T}\forall x\exists y(Rxy \wedge Ryx)$

1	$\mathsf{T}\,\forall x\exists y Rxy$	*Premise*
2	$\mathsf{T}\,\forall x\forall w(Rxw \to Rwx)$	*Premise*
3	$\mathsf{T}\,\exists y Ray$	$\mathsf{T}\forall E, 1$
4	$\mathsf{T}\,Rab$	$\mathsf{T}\exists E, 3$
5	$\mathsf{T}\,\forall w(Raw \to Rwa)$	$\mathsf{T}\forall E, 2$
6	$\mathsf{T}\,Rab \to Rba$	$\mathsf{T}\forall E, 5$
7	$\mathsf{T}\,Rba$	$\mathsf{T}\to E, 4, 6$
8	$\mathsf{T}\,Rab \wedge Rba$	$\mathsf{T}\wedge I, 4, 7$
9	$\mathsf{T}\,\exists y(Ray \wedge Rya)$	$\mathsf{T}\exists I, 8$
10	$\mathsf{T}\,\forall x\exists y(Rxy \wedge Ryx)$	$\mathsf{T}\forall I, 9$

Let us write $\Gamma \vdash_0 A$ whenever A is 0-depth deducible from Γ, and let $\vdash_C$ denote the relation of deducibility in classical first-order logic. The soundness of $\vdash_0$ with respect to $\vdash_C$ is trivial.

Proposition 2. $\Gamma \vdash_0 A \Longrightarrow \Gamma \vdash_C A.$

It can also be shown that:

Proposition 3. *If $\Gamma \vdash_0 A$, then there exists an analytic proof of A from Γ with the subformula property.*

For the propositional part, the proof can be found in [6]. Its extension to the first-order case is immediate given the general restriction on quantifier eliminations, according to which the premise of a quantifier elimination cannot be the conclusion of an introduction. In principle one could impose a similar restriction on the propositional elimination rules, so as to obtain only proofs with the subformula property.

Note that, as for its propositional counterpart, $\vdash_0$ is a Tarskian logic, i.e. it satisfies reflexivity, monotonicity, transitivity and substitution invariance. Moreover, it is not difficult to show that 0-depth inferences satisfy (PQA*), and that this fact implies the following:

Proposition 4. *The logic $\vdash_0$ is decidable.*

Given that $\vdash_0$ is tractable for its propositional fragment, we conjecture that it is *tractable* also in the first-order case, but a proof of this conjecture will be the topic of future research.

To summarize, the 0-depth first-order logic captures a notion of analytic inference that makes no use of virtual information and satisfies the (PQA*) principle of quantificational analyticity.

7 Depth-bounded natural deduction for full first-order logic

Virtual information. The role of virtual information in the DBBL approach has been briefly illustrated in Section 4 and is discussed at length in [11, 10, 8, 7, 6]. In the context of this section it will be convenient to work with the relation of k-*depth derivability* between finite sets of *signed formulae* and *signed formulae* defined in the obvious way. Accordingly we can say that A is k-depth deducibile from Γ whenever $\mathsf{T}\,A$ is k-depth derivable from the set of signed formulae in $\mathsf{T}\,\Gamma$. However, the notion of k-depth derivability is defined for arbitrary finite sets of signed formulae (not necessarily in PPNF) and arbitrary formulae. We shall use X, Y, Z, etc. as variables ranging over finite sets of signed formulae and φ, ψ, χ, etc. as variables ranging over signed formulae.

Starting from 0-depth derivability, the transition from one degree of depth to the next is associated with the use of a *structural* rule that governs the use of virtual information in a proof. This is *the only* discharge rule of the system and takes the following form:

> If φ is k-depth derivable from $X \cup \{\mathsf{T}\,A\}$ and from $Y \cup \{\mathsf{F}\,A\}$, then   (RB)
> φ is $k+1$-depth derivable from $X \cup Y$.

This rule simulates the transition from an information state in which we do not possess any information about the truth or falsity of A, to a richer one, in which the formula A is decided, that is, either we actually possess the information that it is true or we actually possess the information that it is false. It can be seen as a principle of *potential omniscience* and it is the informational version of the classical *principle of bivalence*. Accordingly we call this rule "Rule of Bivalence" (RB).

Given that the virtual assumptions $\mathsf{T}\,A$ and $\mathsf{F}\,A$ introduced by an application of this rule may contain parameters, its use suggests the need for a further restriction on the $\mathsf{T}\forall I$ and $\mathsf{F}\exists I$ rules, namely that *the parameter a does not occur in any undischarged virtual assumption on which the premise of the rule application depends*.

In the following examples we shall use boxes to represent the subproofs to which the RB rule is applied. The depth of a derivation is nothing but the maximum number of nested boxes occurring in it. We write $\Gamma \vdash_k A$ to mean that A is k-depth deducible from Γ (T A is k-depth derivable from T Γ).

Example 8. T$\forall x(Qx \to Rx)$, T$\forall x(Rx \to Sx) \vdash_1$ T$\forall x(Qx \to Sx)$

1	T $\forall x(Qx \to Rx)$	*Premise*
2	T $\forall x(Rx \to Sx)$	*Premise*
3	T $Qa \to Ra$	T $\forall E, 1$
4	T $Ra \to Sa$	T $\forall E, 2$

5	T Qa		F Qa		
6	T Ra	T $\to E, 3, 5$	T $Qa \to Sa$		T $\to I, 5$
7	T Sa	T $\to E, 4, 6$			
8	T $Qa \to Sa$	T $\to I, 7$			

9	T $Qa \to Sa$	
10	T $\forall x(Qx \to Sx)$	T $\forall I, 9$

Note that in the above proof the application of the rule T $\forall I$ at step 10 is allowed because the virtual assumptions containing the parameter a have already been discharged.

Example 9. *Consider again the example discussed in Section 3. We transform the premises of the argument in PPNF and obtain the following set $\Gamma = \{\forall x \forall y \exists z(Rxy \to (Gxz \wedge Gzy)), \forall x \forall y \exists w(Gxy \to (Bxw \wedge Bwy)), \forall x \forall y((Bxy \wedge Cx) \to Cy)\}$. Figure 4 illustrates the configuration of the premises in Γ. The derivation of the conclusion T $\forall x \forall y((Rxy \wedge Cx) \to Cy)$ from the premises T Γ, which is shown in Figure 5, has depth 2. Notice that, unlike the former, the latter application of the RB introduces a new quantifier $\exists v$, the elimination of which permits the introduction into the argument of a new individual e. As a result, this derivation violates (PQA) and vindicates Hintikka's insights.*

Liberalized introduction rules. In [6], D'Agostino, Gabbay and Modgil have shown that, towards the normalization result, it is convenient to prove that every derivation can be transformed into its RB-canonical form, i.e. into a derivation in which there is no application of a rule below the conclusion of an application of RB. In that paper, the authors have shown that this outcome can be achieved by applying the transformation depicted in Figure 6, where χ is the conclusion of a rule

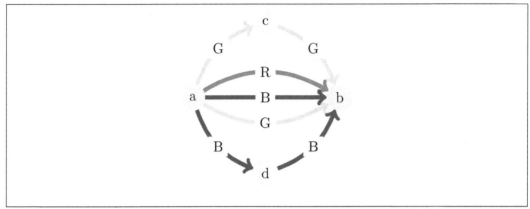

Figure 4: This figure illustrates the configuration of the premises in PPNF of the example shown in Section 3.

having as its premise(s) ψ (and φ). The iterated application of this transformation results in pushing downwards all the applications of RB so that, eventually, the conclusion of an application of RB is never used as a premise of a rule and must be identical to the conclusion of the whole derivation.

This theorem concerning RB-canonical derivations, that D'Agostino, Gabbay and Modgil proved for *propositional* logic, might be useful also in DBFOLs. However, the result of the transformation shown in Figure 7 is not sound, because, as we have seen above, the $\mathsf{T}\forall I$ rule might be applied to a formula such as $\mathsf{T}B(c)$ whenever c does not occur in any undischarged assumption on which the premise of the rule application depends. Therefore, if we want to prove that every derivation of DBFOLs can be transformed into an equivalent one in RB-canonical form, we are required to liberalize the use of the rule $\mathsf{T}\forall I$ (and of the rule $\mathsf{F}\exists I$), in such a way that the transformation shown in Figure 7 turns out to be sound.

We say that an individual denoted by a is *arbitrary* for a certain property $F(x)$ expressed by an open formula with a free variable x, if *either* F_a^x is false, *or* F_a^x is true for *every* individual a in the domain. Note that we can always assume, with no loss of generality, that a *new* parameter introduced in a virtual assumption via an application of the RB denotes an individual that is arbitrary for a certain property $F(x)$. The crucial point is that if a is arbitrary for $F(x)$, then it might not be arbitrary for a syntactically distinct property $G(x)$. The idea behind the liberalized versions of $\mathsf{T}\forall I$ and $\mathsf{F}\exists I$ is that if an individual denoted by the parameter a is arbitrary for a certain property, the same individual cannot be arbitrary for a different property:

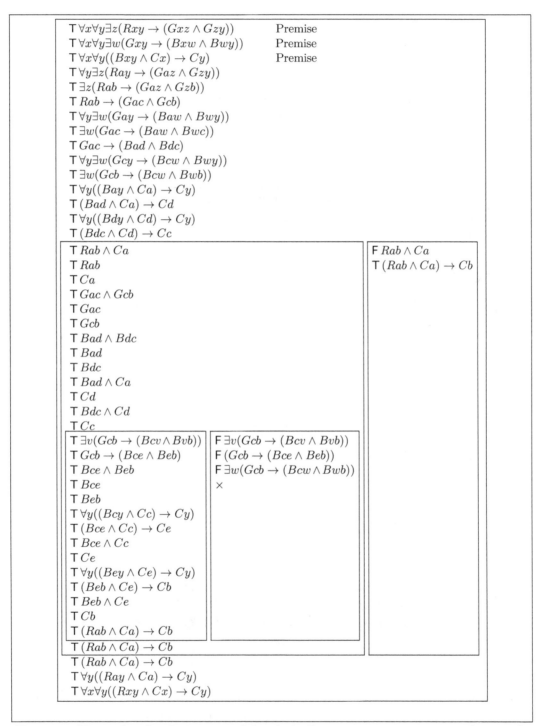

Figure 5: Example of a derivation of depth 2. Due to space restrictions the justifications of the steps are omitted.   447

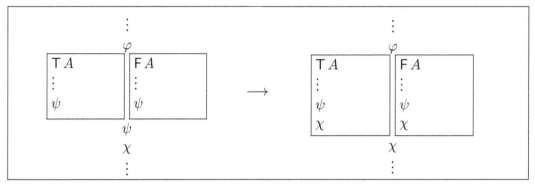

Figure 6: Iterated applications of this transformation turn any derivation into an RB-canonical one: χ is the conclusion of an *intelim*-rule having as its premise(s) ψ (and φ) [6].

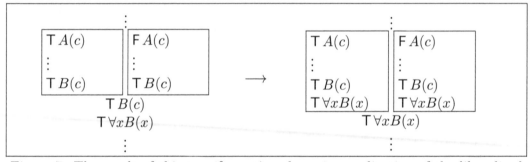

Figure 7: The result of this transformation shows an application of the liberalized $\mathsf{T}\forall I$ rule.

$$\frac{\mathsf{T}\,F}{\mathsf{T}\,\forall x F_x^a}\ \mathsf{T}\forall I \qquad\qquad \frac{\mathsf{F}\,F}{\mathsf{F}\,\exists x F_x^a}\ \mathsf{F}\exists I$$

Provided that a is not critical, F does not contain critical parameters depending on a and x is not bound in F. Moreover, the rule $\mathsf{T}\forall I$ (respectively $\mathsf{F}\exists I$) is not applied to $\mathsf{T}\,G(a)$ (respectively, $\mathsf{F}\,G(a)$) obtaining $\mathsf{T}\,\forall x G_x^a$ (respectively, $\mathsf{F}\,\exists x G_x^a$) for any formula G syntactically distinct from F.

Example 10. *Example of a wrong application of the liberalized I-rule. The $\mathsf{T}\forall I$ rule cannot be applied both at step 2 left and at step 3 right, because the individual*

448

denoted by c cannot be arbitrary for both $\mathsf{T}\,Ac$ *and* $\mathsf{T}\,\neg Ac$.

1	$\mathsf{T}\,Ac$		$\mathsf{F}\,Ac$	
2	$\mathsf{T}\,\forall x Ax$	$\mathsf{T}\,\forall I, 1$	$\mathsf{T}\,\neg Ac$	$\mathsf{T}\,\neg I, 1$
3	$\mathsf{T}\,\forall x Ax \vee \forall x \neg Ax$	$\mathsf{T}\,\vee I, 2$	$\mathsf{T}\,\forall x \neg Ax$	$\mathsf{T}\,\forall I, 2$
4			$\mathsf{T}\,\forall x Ax \vee \forall x \neg Ax$	$\mathsf{T}\,\vee I, 3$

5    $\mathsf{T}\,\forall x Ax \vee \forall x \neg Ax$

Example 11. *Example of a correct application of the liberalized I-rule:*
$\vdash_1 \mathsf{T}\,\forall x Ax \vee \exists x \neg Ax$

1	$\mathsf{T}\,Ac$		$\mathsf{F}\,Ac$	
2	$\mathsf{T}\,\forall x Ax$	$\mathsf{T}\,\forall I, 1$	$\mathsf{T}\,\neg Ac$	$\mathsf{T}\,\neg I, 1$
3	$\mathsf{T}\,\forall x Ax \vee \exists x \neg Ax$	$\mathsf{T}\,\vee I, 2$	$\mathsf{T}\,\exists x \neg Ax$	$\mathsf{T}\,\exists I, 2$
4			$\mathsf{T}\,\forall x Ax \vee \exists x \neg Ax$	$\mathsf{T}\,\vee I, 3$

5    $\mathsf{T}\,\forall x Ax \vee \exists x \neg Ax$

Example 12. *Example of a correct application of the liberalized I-rule:*
$\mathsf{T}\,\forall x \exists y (\neg Ay \vee Ax) \vdash_1 \mathsf{T}\,\exists y \neg Ay \vee \forall x Ax$

1	$\mathsf{T}\,\forall x \exists y (\neg Ay \vee Ax)$	*Premise*		
2	$\mathsf{T}\,\exists y (\neg Ay \vee Aa)$	$\mathsf{T}\,\forall E, 1$		
3	$\mathsf{T}\,\neg Ab \vee Aa$	$\mathsf{T}\,\exists E, 2$		
4	$\mathsf{T}\,Aa$		$\mathsf{F}\,Aa$	
5	$\mathsf{T}\,\forall x Ax$	$\mathsf{T}\,\forall I, 4$	$\mathsf{T}\,\neg Ab$	$\mathsf{T}\,\vee E, 3, 4$
6	$\mathsf{T}\,\exists y \neg Ay \vee \forall x Ax$	$\mathsf{T}\,\vee I, 5$	$\mathsf{T}\,\exists y \neg Ay$	$\mathsf{T}\,\exists I, 5$
7			$\mathsf{T}\,\exists y \neg Ay \vee \forall x Ax$	$\mathsf{T}\,\vee I, 6$

8    $\mathsf{T}\,\exists y \neg Ay \vee \forall x Ax$

The propositions stated above, at the end of Section 6 for the notion of 0-depth deducibility, can be extended to the general case. Here we state them without proof.

Proposition 5. *A formula A is a classical consequence of* Γ *if and only if there is a k-depth proof of A from* Γ *for some* $k \in \mathbb{N}$.

We just observe that completeness can be proven via simulation of a classical natural deduction system (with or without the liberalized introduction rules). Applications of the standard quantifier eliminations that increase the number of distinct parameters beyond the Q-complexity of the premises can be simulated via suitable applications of the RB rule introducing virtual information and so increasing the depth of the proof.

Proposition 6. *If A is a classical consequence of Γ, there is a k-depth proof of A from Γ with the subformula property for some $k \in \mathbb{N}$.*

A detailed proof-theoretical investigation of the normalization problem will be the topic of a future paper.

Proposition 7. *The notion of k-depth inference is decidable for every fixed k.*

Decidability follows from depth-boundedness: only a finite number of new parameters can be introduced by increasing the depth of the proof. It is open whether k-depth inference (for normal proofs with the subformula property) is tractable, i.e., if there exists a polynomial time decision procedure. This problem will also be a crucial topic of further research.

References

[1] A. Avron and A. Zamansky, "Non-Deterministic Semantics for Logical Systems", in Gabbay D., Guenthner F., eds, *Handbook of Philosophical Logic*, vol 16. Springer, Dordrecht, 2011.

[2] P. Baldi and H. Hosni, "Depth-bounded Belief functions", *International Journal of Approximate Reasoning* 123: pp. 26-40, 2020.

[3] K. Bendall, "Natural Deduction, Separation and the Meaning of Logical Operators", *Journal of Philosophical Logic* 7: pp. 245-276, 1978.

[4] G. Boolos, "Don't Eliminate Cut", *Journal of Philosophical Logic* 13: pp. 373-378, 1984.

[5] R. Carnap, H. Hahn and O. Neurath, "The Scientific Conception of the World: The Vienna Circle", in O. Neurath, *Empiricism and Sociology*, Dordrecht: D. Reidel, 1973, pp. 299-328.

[6] M. D'Agostino, D.M. Gabbay and S. Modgil, "Normality, Non-Contamination and Logical Depth in Classical Natural Deduction", *Studia Logica* 108: pp. 291-357, 2020.

[7] M. D'Agostino, "An Informational View of Classical Logic", *Theoretical Computer Science* 606: pp. 79-97, 2015.

[8] M. D'Agostino, "Analytic Inference and the Informational Meaning of the Logical Operators", *Logique et Analyse* 57 (227): pp. 407-437, 2014.

[9] M. D'Agostino, "Semantic Information and the Trivialization of Logic: Floridi on the Scandal of Deduction", *Information* 4 (1): pp. 33-59, 2013.

[10] M. D'Agostino, M. Finger and D. Gabbay, "Semantics and Proof-Theory of Depth Bounded Boolean Logics", *Theoretical Computer Science* 480: pp. 43-68, 2013.

[11] M. D'Agostino and L. Floridi, "The Enduring Scandal of Deduction. Is Propositional Logic Really Uninformative?", *Synthese* 167 (2): pp. 271-315, 2009.

[12] M. Dummett, *The Logical Basis of Metaphysics*, London: Duckworth, 1991.

[13] G. Frege, *The Foundations of Arithmetic*, J. L. Austin, trans., Evanston, Illinois: Northwestern University Press, 1960^2.

[14] M.R. Garey and D.S. Johnson, *Computers and Intractability: A Guide to the Theory of NP-Completeness*, New York: W.H. Freeman and Company, 1979.

[15] G. Gentzen, *Collected Papers*, M. E. Szabo, ed., Amsterdam: North-Holland, 1969.

[16] H. Hahn, "Logic, Mathematics and Knowledge of Nature", in A.J. Ayer, ed., *Logical Positivism*, Glencoe, Illinois: The Free Press, 1959.

[17] A.P. Hazen, "Logic and Analyticity", in A.C. Varzi, ed., *The Nature of Logic*, Stanford: Center for the Study of Language and Information, 1999, pp. 79-110.

[18] J. Hintikka, *Logic, Language-Games and Information. Kantian Themes in the Philosophy of Logic*, Oxford: Clarendon Press, 1973.

[19] J. Hintikka, "Are logical truths analytic?", *The Philosophical Review*, 74: pp. 178-203, 1965.

[20] I. Kant, *Critique of Pure Reason*, P. Guyer and A. Wood, trans. and eds., Cambridge: Cambridge University Press, 1998.

[21] C.H. Langford, "The Notion of Analysis in Moore's Philosophy", in P.A. Schilpp, ed., *The Philosophy of G.E. Moore*, La Salle: Open Court, 1992, pp. 321-342.

[22] C. Larese, *The Principle of Analyticity of Logic*, PhD Thesis, Scuola Normale Superiore, Pisa 2019.

[23] S. Negri and J. von Plato, *Structural Proof Theory*, Cambridge: Cambridge University Press, 2001.

[24] W.V.O. Quine, "Two Dogmas of Empiricism", *The Philosophical Review* 60: pp. 20-43, 1951.

[25] W.V.O. Quine, The Roots of Reference, Open Court, 1973.

[26] I. Rumfit, "'Yes' and 'No' ", *Mind* 109: pp. 781-823, 2000.

[27] R.M. Smullyan, *What is the Name of this Book? The Riddle of Dracula and Other Logical Puzzles*, Englewod Cliffs: Prentice-Hall, 1978.

[28] R.M. Smullyan, *First-Order Logic*, New York: Dover Publications, Inc., 1968.

Received 8 June 2020

Reflections on Logics
for Assertion and Denial

Ciro De Florio[*]

Department of Philosophy, Università Cattolica, Milan
ciro.deflorio@unicatt.it

Abstract

The aim of this paper is to develop and extend a framework of pragmatic logic for assertions. The base system has been developed by Dalla Pozza and Garola [7]; here, we want to improve the analysis of the justification conditions of the act of assertion. Furthermore, we explore the possibility of a pragmatic logic of denial. We analyse the relationships between denial, negation, and assertion.

1 A Logic for Illocutionary Acts

Consider the following short story:

> Emma *asserts* that she left her car in the garage; Tom, however, *disagrees*: Emma didn't leave her car in the garage. Maybe, Tom *hypothesizes*, the car is still at Francis' house.

Emma and Tom assert, deny, and conjecture in their short interaction. Classically, these acts are considered *illocutionary acts*: asserting, denying, hypothesising, conjecturing or doubting that p are acts that can (though need not) be performed by saying that one is doing so.[1]

In general, the logical characterization of these acts has to take into account at least two phases: the *characterization* of the logical structure of illocutionary acts and the *analysis* of the admissibility (or eligibility) of those acts. Analysis of admissibility is the investigation of the conditions on the basis of which one can

---

[*]I would like to thank Massimiliano Carrara and Daniele Chiffi for their friendship and our many discussions on pragmatics.

[1]See, for instance, Searle and Vanderveken [12].

consider an illocutionary act as *justified* or *unjustified*. Thus, we can define a relation of consequence among pragmatic sentences which describe illocutionary acts.

Consider the following structure of an illocutionary act:

$$Act(Content) \tag{1}$$

'*Act*' indicates a speech act intended to be complete at a certain level of idealization; in fact, it can be noticed that in (1) there is no reference to the subject (Emma or Tom, in the example) who actually performs the illocutionary act. As we will see in the following, treatment of the justification conditions of the acts of assertion and denial actually refers to the epistemic subject which, accordingly, asserts and denies. However, the level of idealization at which the investigation is conducted is deliberately left vague.

The first phase, then, consists of introducing a specific language in order to describe the illocutionary acts; analogously, another language describes the content of those acts. Dalla Pozza and Garola [7] has seminal findings regarding such train of thought; the authors emphasize that a logic for pragmatics must keep distinct the two dimensions (*viz.* act and content) that constitute the structure of an illocutionary act (hereafter, we indicate Dalla Pozza and Garola's system by LP). The content of an illocutionary act is thus a proposition (which has a determinate truth-value according to the Tarskian semantics); the illocutionary acts, on the other hand, are described by a series of pragmatic operators. So, this general framework could include operators for the act of asserting, conjecturing, doubting and so on. Obviously, there are many options about the logic through which we describe the contents of acts; Dalla Pozza and Garola chose propositional logic but it is also possible to make more complex the language of contents by exploiting, for instance, the expressive resources of FOL (First-Order Logic) or HOL (Higher-Order Logics) or their modal extensions.

The second phase of the logical characterisation of illocutionary acts concerns the setting of the eligibility conditions of such acts. The following returns to the story of Emma and Tom:

> Emma: – Vaccines cause tremendous pathologies; among them, there is autism!
> Tom: – That's ridiculous! How could you say such a silly thing?
> Emma: – What? How dare you deny it!?

In this rather heated discussion, Emma and Tom are not merely performing some illocutionary acts, but they are wondering about the permissibility of these acts. The other grounding intuition of Logic for Pragmatics is the following: the eligibility condition of an illocutionary act lies in its *justification*. Emma is allowed to assert something when her assertion is justified; in the same vein, Tom is allowed to deny p if and only if his act is somehow justified.

But what is the justification of an illocutionary act? How do you formally characterize it? Moreover, it seems that different illocutionary acts require different justification conditions. In that case, one may wonder if there are connections between those acts and if these alleged relationships can be analyzed through a specification of their justification conditions.

This paper adopts this train of thought. It develops Dalla Pozza's and Garola's framework providing a *Pragmatic Meta-Language*: this allows characterization of intuitive notion of proof which is at the basis of the justification conditions of assertions. Moreover, we extend our framework to the pragmatic act of *denial*, focusing on the relations between denial and negation. In the end, we cast some light on the interaction between the logic of assertion and the logic of denial, emphasizing *the structural asymmetry* between the justification conditions of those two acts.

2   Pragmatic Logic for Assertions

Dalla Pozza and Garola settled a logical framework in order to characterize the logic of assertions. Their aim was to provide a pragmatic interpretation to intuitionistic logic:

> We propose a pragmatic interpretation of intuitionistic logic that is based on a translation of an intuitionistic propositional calculus [...] and of a classical propositional calculus [...] into a formalized pragmatic language. [...] The purpose of our interpretation is mainly philosophical. Indeed we aim to settle the conflicts between classical and intuitionistic logic. [7, p. 81]

The key ideas of their work are the following: there are two kinds of signs, *descriptive* signs and *pragmatic* signs. The former constitutes the language which describes the content of the assertions; it is a stock of propositional letters $(p, q, r, ...)$ with the usual logical connectives $(\neg, \wedge, \vee, \rightarrow)$. We can define, then, the *radical* formulas (**RF**) as follows:

FR_1 Every propositional letter is an **RF**

FR_2 If A is an **RF**, then $\neg A$ is an **RF**

FR_3 If A_1 is an **RF** and A_2 is an **RF**, then $A_1 \wedge A_2$, $A_1 \vee A_2$, $A_1 \rightarrow A_2$ are **RF**s.

The pragmatic signs are constituted by the primitive sign of assertion (⊢) and by the pragmatic connectives ($\sim, \cap, \cup, \supset$), which allow construction of complex pragmatic formulas (here, we follow the original label, **AF**, viz. Assertion Formulas):

AF_1 If A is an **RF**, then ⊢ A is an **AF**,

AF_2 If Δ is an **AF**, then $\sim \Delta$ is an **AF**,

AF_3 If Δ_1 and Δ_2 are **AF**, then $\Delta_1 \cap \Delta_2$, $\Delta_1 \cup \Delta_2$, $\Delta_1 \supset \Delta_2$ are **AF**.

Therefore, Dalla Pozza and Garola provide both a semantic and a pragmatic interpretation – for the radical and the pragmatic formulas respectively. The semantic interpretation is the classical Tarskian semantics for the propositional language. On the other hand, the pragmatic interpretation is defined on an intuitive concept of proof. In summary, the idea is that the assertion of p is justified if and only if there exists a proof of p.

The reference to the concept of proof triggers a further question: Are we referring to logico-mathematical proofs or empirical proofs? Dalla Pozza and Garola are quite neutral on this point:

> [W]e intend to introduce a purely formal pragmatics, in order to establish general semantic properties of the (metalinguistic) concept of justification that are independent of the specific empirical and logical procedures of proof that can be selected, so that our pragmatics can be considered neutral with respect to the choice between different procedures. In addition we note that our approach is also neutral with respect to the interpretation of proof as *actual* or *potential* [...] so that we consider the expressions "proven" and "provable", "justified" and "justifiable", "asserted" and "assertable", as equivalent, hence interchangeable in our framework. [7, p. 87]

What distinguishes the concept of proof is its *factivity*: the justification of the assertion of p is grounded in the existence of a proof that p is true. It is, fair enough, a very strong requirement: a certain fallibilist attitude imposes some cautions in considering as infallible the empirical proofs. Maybe this *status* could be conceded to the mathematical proofs (given the complete epistemic reliability of the axioms; but 150-year debate on the foundations of mathematics again seems to suggest caution).

Dalla Pozza and Garola introduce, then, an evaluation pragmatic function π which maps pragmatic formulas into justification values: *Justified* and *Unjustified* (J, U).

$$(JR_1) \quad \pi(\vdash A) = J \qquad \text{iff} \quad \text{there is proof that } A \text{ is true}$$

$$(JR_2) \quad \pi(\sim \Delta) = J \qquad \text{iff} \quad \text{there is proof that } \Delta \text{ is unjustified}$$

$$(JR_3) \quad \pi(\Delta_1 \cap \Delta_2) = J \quad \text{iff} \quad \text{there is proof that } \Delta_1 \text{ is justified}$$
$$\text{and there is a proof that } \Delta_2 \text{ is justified}$$

$$(JR_4) \quad \pi(\Delta_1 \supset \Delta_2) = J \quad \text{iff} \quad \text{there is proof that } \Delta_2 \text{ is justified}$$
$$\text{whenever } \Delta_1 \text{ is justified}$$

Some remarks. The most interesting clauses are those concerning pragmatic negation and pragmatic implication. A pragmatic negation of an assertion is very strong: it is the proof that the assertion is unjustified. So, analogously, the relation of pragmatic entailment $(\Delta_1 \supset \Delta_2)$ is justified if there is a proof that every proof of Δ_2 is also a proof of Δ_1.

These features clearly exhibit the intuitionistic flavor of the logic for pragmatic framework. Notice that within the syntax of the logic for pragmatics the iteration of pragmatic signs is 'forbidden' (that is, a formula as $\vdash (\vdash A)$ is ill-formed). This fits well with the intuition according to which pragmatic logic is a logic of acts and it is not possible to 'pack' an act into the content of another act. At the most, the content of assertion can *describe* another assertion, but it cannot *assert* anything.

However, within the justification conditions of the pragmatic interpretation, the 'proof operator' is actually *iterated*. This fact does not seem devoid of consequences. Although Dalla Pozza and Garola emphasize the neutrality of their notion of proof concerning the actuality or potentiality of proofs, it is clear that – in the case of iterations – the question becomes relevant. Consider the case of $\sim\vdash A$; according to the standard interpretation, this formula is justified if and only if there exists a proof that A is not justified. But here, the ambiguity comes up: according to one construal, the *actual* conception, $\sim\vdash A$ is justified if there is a proof that it is not justified. But this leaves open the possibility that, one day, it will be proved. According to the other approach, the *possible* conception, $\sim\vdash A$ is justified if there is a proof that A is not provable; in other words, $\sim\vdash A$ is justified only if a kind of impossibility proof of A is available.[2]

The next section characterizes the pragmatic interpretation of the language of assertions and then extends the framework to the illocutionary act of denial.

3 Meta-Language for Pragmatics

In this section the idea that the justification conditions of an assertion are *always* grounded in the existence of a proof is maintained (thus justifying, at least partially,

---

[2]See [9].

the logical behavior of the pragmatic negation and implication). However, we introduce some new elements which serve both to clarify the original framework and to prepare the extension of our framework to the illocutionary act of denial.[3]

The first element is the introduction of a parameter which indicates the *epistemic subject* which makes the assertion. Notice that it is crucial to consider the level of idealization at which we are working. The epistemic subject can be idealized with reference to the contingent limitations which any empirical individuals experience; but, moreover, the epistemic subject can (ideally) represent a community of epistemic subjects (think about the assertion that human activities are causing global warming is intended as an assertion of the *scientific community* – or at least of a large majority of it).[4]

Let s be an epistemic subject; $\mathcal{F}(s)$ is the subject's *epistemic framework*. The epistemic framework of a subject is constituted by a series of *conceptual resources* (which can be considered beliefs and evidences) which can be used to prove the contents of the assertions.

The relation of proof is characterized by a two-place operator; $\mathbb{P}(\varphi, \psi)$ means that φ is a proof of ψ. But what is a proof? Our suggestion is probably much wider than Dalla Pozza's and Garola's. An example can clarify.

At the base level, we have propositions which describe the state of affairs; for instance, the proposition <Emma is at the movies> can describe an actual, obtaining, state of affairs. The pragmatic level, in that case, has to do with the assertion that Emma is at the movies, and the justification of this illocutionary act lies in the existence of a proof of the content of assertion.

But what can a proof of <Emma is at the movies> be? Assuming that Tom is making the assertion, there are many options:

- Tom, the epistemic subject, saw her.

- Luke said that he saw her.

- Emma is either at the movies or at the restaurant and Emma is not at the restaurant.

- ...

It is important to notice that the relation of proof is many-many. The conceptual resource <Tom saw Emma at the movies> is a proof that Emma is at the movies

---

[3]Extensions and applications of Dalla Pozza's and Garola's work are [4], [5], [6].

[4]Within the original framework of logic for pragmatics, there is no mention of epistemic subject (and related notions); however, we think it allows a better characterization of *dynamics* of assertions and denials.

and that the cinema is open. So, analogously, <Emma is at the movies> can be proved both by the conceptual resource according to which the epistemic subject saw her and by a witness.[5]

The pragmatic meta-language (MP) is then constituted as follows:

- A set of well-formed closed formulas of any complexity: $A, B, C, ...$[6]

- A set of higher order variables for sets of well-formed closed formulas: $X, Y, ...$

- Higher order quantifiers: $\forall, \exists$

- Three logical signs: $\rightarrow, \cup, \perp$

The formation rules are straightforward:

MP$_1$ If A is a well-formed closed formula of propositional logic, or first-order predicate logic, $A \in MP$

MP$_2$ If $A \in MP$ then $\exists/\forall X \mathbb{P}(X, A) \in MP$

MP$_3$ If $A, B \in MP$ then $A \rightarrow B, A \cup B \in MP$

MP$_4$ $\perp \in MP$

Now, consider the justification conditions; define, then, a pragmatic *evaluation function* π with respect to the epistemic subject s:

$$(4) \quad \pi_s(\vdash A) = J \quad \text{iff} \quad \exists X \subseteq \mathcal{F}(s), \mathbb{P}(X, A)$$
$$(5) \quad \pi_s(\vdash A) = U \quad \text{iff} \quad \neg \exists X \subseteq \mathcal{F}(s), \mathbb{P}(X, A), \text{ that is}$$
$$\forall X \subseteq \mathcal{F}(s), \mathbb{P}(X, A) \rightarrow \perp$$

(4) says that the subject s is justified to assert A if and only if within the epistemic framework of s there is a proof of A; analogously, the subject s is not justified to assert A if and only if any conceptual resource aiming to be a proof of A leads to absurdity.

$$(6) \quad \pi_s(\sim \Delta) = J \quad \text{iff} \quad \exists X \subseteq \mathcal{F}(s), \mathbb{P}(X, \forall Y (\mathbb{P}(Y, \Delta) \rightarrow \perp))$$
$$(7) \quad \pi_s(\Delta_1 \cup \Delta_2) = J \quad \text{iff} \quad \exists X, Y \subseteq \mathcal{F}(s), \mathbb{P}(X, \Delta_1) \text{ and } \mathbb{P}(Y, \Delta_2)$$
$$(8) \quad \pi_s(\Delta_1 \supset \Delta_2) = J \quad \text{iff} \quad \exists X \subseteq \mathcal{F}(s), \mathbb{P}(X, \forall Y (\mathbb{P}(Y, \Delta_1) \rightarrow \mathbb{P}(Y, \Delta_2)))$$

---

[5]Of course, it could trigger a sort of regress of justification and proofs.

[6]However, for the sake of simplicity, we consider the case in which the object language is formalized in propositional logic.

(6) states that the subject s is justified to negate the assertion of A if (and only if) the subject has a proof such that all alleged proofs of A lead to contradiction. Notice that in clauses (6) and (8), the membership to $\mathcal{F}(s)$ condition is dropped. It ensures that the subject needs simply to have a proof about *other* proofs. In other words, a subject is authorized, so to speak, to $\sim\vdash A$ if and only if she has at disposal a proof that *every proof* of A leads to contradictions.

An important principle in Dalla Pozza's and Garola's framework is the exportation of negation principle. Within LP, it holds that

$$\vdash \neg A \supset \sim\vdash A \tag{2}$$

but we have that

$$\sim\vdash A \not\supset \vdash \neg A \tag{3}$$

Let us check if our meta-theoretical framework accounts for (6). Assume that $\pi_s(\vdash A) = J$. Therefore, $\exists X \subseteq \mathcal{F}(s), \mathbb{P}(X, \neg A)$. If the subject has among her conceptual resources an X which proves $\neg A$, then she cannot have a Y which proves A, that is $\forall Y(\mathbb{P}(Y, A) \rightarrow \bot)$. But this very argument is part of the conceptual resources of the subject; therefore $\exists X \subseteq \mathcal{F}(s), \mathbb{P}(X, \forall Y(\mathbb{P}(Y, A) \rightarrow \bot))$, that is $\sim\vdash A$. $\square$

The converse does not hold. Having a conceptual resource X, which proves that all the (alleged) proofs of A fail, does not mean that there is *another* conceptual resource (say, Z) which proves $\neg A$. The point is that the required proof for the soundness of (6) is a proof about the very structure of the (possible) proofs of A. The subject, thus, reflecting on the structure of her epistemic framework, $\mathcal{F}$, gets this proof . But clearly the converse does not hold; in order to justify the assertion of $\neg A$ ($\vdash \neg A$) we need a proof which has as object $\neg A$ and not the (failed) proofs of A.

4   A Pre-Theoretical Introduction to Denial

As an example, suppose Emma disagrees with Tom as to whether Padua is North of Venice. Tom asserts that it is, and Emma disagrees. Emma may express her disagreement with Tom by denying what Tom said: Padua is *not* North of Venice. Whatever the exact details of one's account of disagreement are, a way for a speaker — Emma in this case — to disagree with another, Tom in this case, is *to deny, or to negate, what the other is asserting*.[7] In general one can argue that two speakers

---

[7]For a general background, see Asenjo [2], Asenjo & Tamburino [1]; Beall [3]; and Priest [10].

disagree if they have *incompatible* beliefs or perform speech acts that cannot be jointly correct. A standard way to express that one has an *incompatible* belief with another is to *negate* the other's *assertion*, as in the case of Emma and Tom.

In daily discussions, it is very common to disagree with the other speaker. When we do that, we are, at least *prima facie*, performing a specific illocutionary act: the act of *denial*. One may wonder whether the denial is only apparently an autonomous illocutionary act and if it could be *reduced* to (some forms of) assertion. Accordingly, we call *Frege's thesis* the equivalence between the denial and the assertion of a negation:

Frege's Thesis to deny $A \equiv$ to assert $\neg A$

It is a perfectly plausible view: that rejecting thesis A is merely asserting $\neg A$. Looking back to our example, Emma could disagree with Tom by asserting that Padua is *not* North of Venice. Moreover, Frege's thesis has an undoubted economical advantage: within the project of a logic for pragmatics, the framework for the assertions is perfectly able to include the denial as a particular case of assertion, governed, thus, by the same justification rules of acts of assertion.[8]

Assume, for the sake of discussion, that Frege's thesis holds and that LP for assertions is able to catch the logic of denial. What are the permissibility conditions of Emma's act? That is, when is Emma justified in denying Tom's view? Following what has been previously said, we can claim that Emma is justified if and only if there is a proof that A is false or, better, within her epistemic framework there are the conceptual resources for proving $\neg A$.

Now, the justified assertion of $\neg A$ is a denial of A; actually, it is the strongest form of denial of A. Not only is A rejected but it is given a proof of $\neg A$. However, the act of denial has less demanding justification conditions than the act of assertion; this is our grounding intuition about denial. There are a large amount of cases in which an epistemic subject is intuitively justified to deny A, even if she is not able to provide a proof of $\neg A$.

Greg Restall underlines the two possible negations of Frege's thesis:

> Friends of truth-value **gaps** and truth-value **gluts** both must distinguish the *assertion of a negation* (asserting $\neg A$) and *denial* (denying A). If you take there to be a truth-value glut at A the appropriate claim to make (when asked) is to assert $\neg A$ without thereby denying A. If you

---

[8]As we will see in the following, in LP for assertions, we have at disposal two forms of negation of an assertion: $\vdash \neg A$ and $\sim\vdash A$. The advocates of Frege's thesis are supposed to choose what form of assertion corresponds to the denial.

take there to be a truth-value gap at A the appropriate claim to make (when asked) is to deny A without thereby asserting $\neg A$. [11, p. 1]

We assume a position of *justification gap*: it is possible that the denial of A is justified, and the assertion of $\neg A$ is not justified. How is this possible? In the following we take into account two case studies which should help explain the point.

CASE 1. Suppose that Emma is a naturalistically minded philosopher. Emma is not concerned with metaphysics; she is a philosopher of language. Karl is a close friend of Emma and he is a philosopher of religion. Karl is convinced that immaterial entities, souls, exist. Emma, rather likely, disagrees that these exist. Her denial is grounded on her background on metaphysical assumptions. Nevertheless, Emma is not able to exhibit a *proof* of the non-existence of the souls.

CASE 2. You hear that your friend Pat has been arrested for violence and drug dealing. You are shocked: Pat is a kind, gentle person, who seemed far from being involved in the criminal activity. Based on your long friendship, you are authorized to deny the thesis according to which Pat is guilty. However, this does not constitute at all *a proof* of Pat's innocence in the court.

The two cases are quite different regarding the content of the belief on which there is disagreement; nevertheless, they share a crucial structural feature. In both cases, it is emphasized how the act of denial can satisfy the justification conditions which have nothing to do with the concept of proof. It seems that in many cases we are justified in denying a certain thesis on the basis of contextual considerations.

This is the grounding intuition of the extension of LPD (logic for pragmatic denial). The act of denial is, thus, structurally different from the act of assertion and, consequently, it has different justification conditions.

5 A Pragmatic Logic for Denial

Here we sketch the language for a pragmatic logic for denial. The radical part is identical to LP; the pragmatic part is, obviously, different.

The pragmatic signs are constituted by the primitive sign of denial ($\dashv$) and by the pragmatic connectives ($\cap, \cup, \supset$) which allow to construct complex pragmatic formulas (here, we introduce the label **DF**, viz. Denial Formulas). As it can be noticed, it lacks a sign of negation. The reason is straightforward: within $\dashv$ the idea of negation is already packed and it would be quite complicated to ascribe an intuitive meaning to something like $\sim\dashv A$.[9]

---

[9] Alternatively, to keep the pragmatic connectives of LPA in LPD, could establish the following meta-theoretical equivalence: $\pi_s(\sim\dashv A) = J$ iff $\pi_s(\dashv A) = U$.

AF_1 If A is an **RF**, then $\dashv A$ is a **DF**

AF_2 If Δ_1 and Δ_2 are **DF**, then $\Delta_1 \cap \Delta_2$, $\Delta_1 \cup \Delta_2$, $\Delta_1 \supset \Delta_2$ are **DF**

Regarding the pragmatic interpretation of LPD, the base concept is no more the concept of proof but the concept of *incompatibility* with the epistemic framework. The concept of incompatibility has a long history (see [8] for an illuminating overview on that idea). In what follows, we will not provide an in-depth analysis of the concept of incompatibility. We would like to remark that the notion of incompatibility at play here is a typically semantic relation which has to do with the content of the involved propositions. Paradigmatic examples are the well-known cases:

(8) a is a square and a is a circle

(9) a is red and a is green

But there are also more interesting cases:

(10) a is green and a is a bureaucratic procedure

(10) expresses a relation of incompatibility since it seems to be a categorial mistake. It is not possibile that a is green and a bureaucratic procedure *because* the bureaucratic procedures cannot have a color.

Thus, in the same way we did for the meta-theory of LP, we shall introduce a two-place predicate **Inc**. The intended meaning of $\mathbf{Inc}(A, X)$ is that proposition A is incompatible with the class (of propositions) X. Let us see now the justification conditions; as before, we have a pragmatic *evaluation function* π with respect to the epistemic subject s:

$$
\begin{array}{lll}
(\mathbf{DR1}) & \pi_s(\dashv A) = J & \text{iff} \quad \exists X \subseteq \mathcal{F}(s), \mathbf{Inc}(A, X) \\
(\mathbf{DR2}) & \pi_s(\Delta_1 \cup \Delta_2) = J & \text{iff} \quad \pi_s(\Delta_1) = J \text{ and } \pi_s(\Delta_2) = J \\
(\mathbf{DR3}) & \pi_s(\Delta_1 \supset \Delta_2) = J & \text{iff} \quad \pi_s(\Delta_1) = U \text{ or } \pi_s(\Delta_2) = J
\end{array}
$$

Principle (**DR3**) is a simple material conditional: if A is incompatible, then B is incompatible means that either A is compatible or B is incompatible.

6 Denial and Negation

Let us go back to the first characterization of illocutionary acts:

$$Act(Content) \tag{4}$$

We established that a logic of the illocutionary acts must provide the structural conditions for the justification of these acts; thereby, we defined a function of pragmatic interpretation:

$$\pi : \{\text{Pragmatic Formulas}\} \rightarrow \{J, U\}$$

That said, in LP we have three levels of negation:

i. One can deny the content

ii. One can deny the act

iii. One can deny the justification value

(i) does not exhibit particular problems since the language used to describe our contents has a sign for negation; the negation of the pragmatic act – (ii) – is more puzzling: as we have previously stated, the intended meaning of $\sim\vdash A$ is not the negation of the act of assertion but, on the contrary, a sort of meta-assertion about the impossibility of asserting A. In case of denial, then, the very idea of denying a denial seems too unusual. At the end – (iii) – the negation can operate – so to speak – at the metalanguage level, on the value of the justification function. In the following, we explore the relationships between the act of denial and the negation of the denied content.

(1) $\pi_s(\dashv A) = J \nRightarrow \pi_s(\dashv A) = U$

(2) $\pi_s(\dashv A) = U \nRightarrow \pi_s(\dashv A) = J$

These two conditions simply state that our logic of denial is consistent. Much more interesting are the following relations:

(3) $\pi_s(\dashv A) = J \Rightarrow \pi_s(\dashv \neg A) = U$

Suppose that A is incompatible with our framework and therefore we are justified to deny it. From that, does it follow the compatibility of $\neg A$? In some cases, this seems to be straightforward. Consider Emma, who denies that Padua is North of Venice since this is incompatible with a series of geographical information Emma has about Veneto. Therefore, Emma would consider totally compatible with her framework that Padua is not North of Venice. In some other cases, these conclusions are not so easy to make. Let us imagine that Emma believes in horoscopes and claims that Jane is shy since she is a Virgo and it is well-known that Virgo people are shy. Ann does not believe in horoscopes and therefore she denies that Virgo people are shy. And she is

464

justified since that proposition is incompatible with her epistemic framework. But if (3) holds, Ann should, in some way, accept the negation of A. The problem is that the negation of A could be ambiguous: it could mean that Virgo people are not shy but, on the contrary, they are very exuberant. But if this is the intended meaning of $\neg A$ it is clear that Ann would be justified to deny it as well.

If this is plausible, there could be counter-examples to (3), that is, cases in which it is justified both the denial and the denial of the negation. Since the relation of incompatibility is semantic, this happens because the contents at play presuppose a series of background hypotheses which can be incompatible with the epistemic framework.

We call this property *negative undeniability*; thus, B is negatively undeniable if there is at least a case in which:

(NU) $\pi_s(\dashv B) = J$ and $\pi_s(\dashv \neg B) = J$

The other interesting principle to discuss is the following:

(4) $\pi_s(\dashv A) = U \Rightarrow \pi_s(\dashv \neg A) = J$

The intuitive justification for (4) follows the schema of (3); however, there could be cases in which the denial is always unjustified. Consider, for instance, a very abstract proposition about the inner structure of energy and matter, say A. Then consider its negation, $\neg A$. It is possible that both A and $\neg A$ are compatible with the referring framework constituted by the standard model in fundamental physics and, thus, is not justified neither the denial of A nor the denial of $\neg A$. In this case, we have cases of *positive undeniability*:

(PU) $\pi_s(\dashv A) = U$ and $\pi_s(\dashv \neg A) = U$

7 Denial and Assertion

In this last section, we will take into account some relationships between the logics of assertions and denial. The asymmetry between the conditions of justification is due to the fact that the existence of a proof is a stronger requirement than incompatibility. We have the following result:

(5) $\pi_s(\dashv A) = J \not\Rightarrow \pi_s(\vdash \neg A) = J$

A subject can be justified in denying A without any proof of $\neg A$. Analogously:

(6) $\pi_s(\dashv A) = U \neq \pi_s(\vdash A) = J$

The mere compatibility with subject's epistemic framework does not guarantee the existence of a proof. However, it holds that

(7) $\pi_s(\dashv A) = J \Rightarrow \pi_s(\vdash A) = U$

The subject cannot prove something which is incompatible with her epistemic framework; logically

(8) $\pi_s(\vdash A) = J \Rightarrow \pi_s(\dashv A) = U$

The existence of a proof is sufficient to assure the compatibility with the epistemic framework.

(7) and (8) are clearly consistency conditions; (5) and (6) claim that the conditions according to which it is rational to deny A are not sufficient to prove $\neg A$ and, by consequence, A not being rational to deny does not mean that there is a proof of A. On the contrary

(5*) $\pi_s(\vdash \neg A) = J \Rightarrow \pi_s(\dashv A) = J$

(6*) $\pi_s(\vdash A) = J \Rightarrow \pi_s(\dashv A) = U$

hold and they show the asymmetric feature of denial and assertion. Let us see (5*). $\pi_s(\vdash \neg A) = J$, that is, $\exists X \subseteq \mathcal{F}(s), \mathbb{P}(X, \neg A)$; now, $\neg A$ is proved and therefore it belongs to the conceptual framework of s. But clearly, $\mathbf{Inc}(A, \neg A)$ therefore $\exists Y \subseteq \mathcal{F}(s), \mathbf{Inc}(A, Y)$. That is $\pi_s(\dashv A) = J$. ∎

The converse does not hold since from $\exists Y \subseteq \mathcal{F}(s), \mathbf{Inc}(A, Y)$ does not follow that the set Y is a proof of $\neg A$.

8   Conclusion

In this paper we developed a system of logic for pragmatics following Dalla Pozza's and Garola's seminal paper. We provided a pragmatic interpretation for the conditions of justification of the acts of assertion and denial. The key concepts involved are, respectively, the concept of *proof* and the concept of *incompatibility*. We then emphasized a deep asymmetry between the conditions to assert a proposition and the conditions to deny it. Further developments can concern the extension of the framework to other illocutionary acts (conjecturing, doubting, hypothesising and so on) and a more fine-grained analysis of the pragmatic models.

References

[1] F. G. Asenjo and J. Tamburino. Logic of antinomies. *Notre Dame Journal of Formal Logic*, 16(1):17–44, 1975.

[2] F.G. Asenjo. A calculus of antinomies. *Notre Dame Journal of Formal Logic*, 7(1):103–105, 1966.

[3] J.C. Beall and J. Beall. *Spandrels of truth*. 2009.

[4] Massimiliano Carrara, Daniele Chiffi, and Ciro De Florio. Assertion and hypothesis: a logical framework for their opposition relations. *Logic Journal of the IGPL*, 25(2):131–144, 2017.

[5] Massimiliano Carrara, Daniele Chiffi, and Ciro De Florio. On assertion and denial in the logic for pragmatics. *Journal of Applied Logic*, 25:S97–S107, 2017.

[6] Massimiliano Carrara, Daniele Chiffi, and Ciro De Florio. Pragmatic logics for hypotheses and evidence. *Logic Journal of the IGPL*, 2019.

[7] Carlo Dalla Pozza and Claudio Garola. A pragmatic interpretation of intuitionistic propositional logic. *Erkenntnis*, 43:81–109, 1995.

[8] Nils Kürbis. An argument for minimal logic. *Dialectica*, 73(1-2):31–63, 2019.

[9] Enrico Martino and Gabriele Usberti. Temporal and atemporal truth in intuitionistic mathematics. *Topoi*, 13(2):83–92, 1994.

[10] G. Priest. *Doubt Truth to be a Liar*. Oxford University Press, 2006.

[11] G. Restall, Assertion, denial and non-classical theories, in: K. Tanaka, F. Berto, E. Mares, F. Paoli (Eds.), *Paraconsistency: Logic and Applications*, Springer, 2013, pp.81–99.

[12] John R Searle and Daniel Vanderveken. *Foundations of illocutionary logic*. CUP Archive, 1985.

Received 19 December 2019

HTLC:
Hyperintensional Typed Lambda Calculus

Michal Fait
Department of Computer Science
VSB-Technical University of Ostrava,
Czech Republic
michal.fait@vsb.cz

Giuseppe Primiero
Department of Philosophy
University of Milan, Italy
giuseppe.primiero@unimi.it

Abstract

In this paper we introduce the logic HTLC, for Hyperintensional Typed Lambda Calculus. The system extends the typed λ-calculus with hyperintensions and related rules. The polymorphic nature of the system allows to reason with expressions for extensional, intensional and hyperintentsional entities. We inspect meta-theoretical properties and show that HTLC is complete in Henkin's sense under a weakening of the cardinality constraint for the domain of hyperintensions.

This collaboration was possible thanks to the EU project "Science without borders" No. CZ.02.2.69/0.0/0.0/16_027/0008463. The first author was funded by the Grant Agency of the Czech Republic (GACR) project GA18-23891S "Hyperintensional Reasoning over Natural Language Text", by the internal grant agency of VSB-Technical University of Ostrava, project No. SP2019/40, "Application of Formal Methods in Knowledge Modelling and Software Engineering I", and by the Moravian-Silesian regional program No. RRC/10/2017 "Support of science and research in Moravian-Silesian region 2017". The second author was supported by the Department of Philosophy "Piero Martinetti" of the University of Milan under the project "Departments of Excellence 2018-2022" awarded by the Ministry of Education, University and Research (MIUR). The authors wish to thank two anonymous reviewers for valuable comments on previous versions of this paper which have helped improving its presentation.

1 Introduction

The literature in philosophical [11] and computational logic [18] has increasingly been paying attention to the crucial distinction between reasoning about extensional (functional values, like individuals or truth-values); intensional (functions); and hyperintensional entities (abstract procedures, [6]; linguistic expressions, [18, 19, 10, 13]; proofs, [9, 8, 17, 21, 22]; or computations [3, 2]), including the dynamics of hyperintensions, [1, 15].

The encapsulation of extensional, intensional and hyperintensional layers of reasoning in one logical system has been offered by Transparent Intensional Logic (TIL), a hyperintensional, partial typed λ-calculus [6]: hyperintensional, because the meaning of TIL λ-terms are procedures producing functions rather than the denoted functions themselves; partial, because TIL is a logic of partial functions; and typed, because all the entities of TIL ontology receive a type. TIL is endowed with a procedural semantics which explicates the meanings of language expressions as abstract procedures encoded by the expressions. TIL is powerful in accounting for contexts and their relations, especially for some natural language phenomena like partial denotations and modal modification, see e.g. [12]. But although it is technically an extension of typed λ-calculus, it still misses a well-defined and agreed upon proof theory. Because of this, it is not possible to reflect its semantics in properties of derivations. A full system providing a proof-theoretic semantics to reason about all these types of entities (extensional, intensional, hyperintensional) seems still to be missing in the literature. In this paper, our goal is to provide an inferential engine common to extensions, intensions and hyperintensions, able to express their relations as well. Previous attempts in this direction are: either impure λ-calculi, because they attempt to capture all of TIL, [16]; or ND-systems for TIL, which do not offer a rule based semantics, see e.g. [7]. Our approach is limited compared to the semantics of TIL, because we only aim at expressing what can be formulated in standard proof-theoretic terms.

The system HTLC presented in this paper is an extension of a typed λ-calculus with hyperintensions. Expressions in this language are of the form $\Gamma \vdash t : T$ where, as usual in typed λ-calculus: Γ is the context of assumptions; t is a term and T is its type. Types express the extensional, intensional or hyperintensional nature of terms. Hence, terms of HTLC denote, as usual in λ-calculus, functions from set to set and their values, with the added hyperintensional terms denoting procedures or computations which we call *constructions*; the output produced by a construction is called its *product*. Hyperintensional terms are defined proof-theoretically by introduction and elimination rules, thereby extending a standard typed λ-calculus, see Figure 2. The Trivialization rule works as an introduction rule: given an extensional or intensional term t, Trivialization returns a hyperintensional term t^*, whose denotation is a construction. The Execution rule works as an elimination: given a hyperintensional term t^*, Execution returns the term t denoting the product of the corresponding construction.[1] Execution eventually provides a non-hyperintensional term as an output, in which case we say that the construction denoted by the hyperintensional term *produces* the object denoted by the non-hyperintensional one. We also obtain higher-order hyperintensional terms by Trivialization on a term denoting a construction; Execution applied on a term denoting a higher-order construction results in a term denoting a lower-order construction, until it is applied to a term denoting a construction, producing an extensional or intensional object.

To offer a very basic example, consider the number 2, which in our system is a term of type $\mathbb{N}$. Many different functional expressions may denote this number, for example $[+\ 1\ 1]$ and $[-\ 5\ 3]$ are two of those. Each of those terms denoting the number 2 may have different constructions, or hyperintensions. For example, hyperintensional terms having $[+\ 1\ 1]$ as their product are: $Integer.sum(1, 1)$ where the operation at hand is the Java command for addition, or $S(S(0) + 0)$ which is the recursive equation presenting the number 2 with addition as successor. When moving from the term $[+\ 1\ 1]$ to the corresponding hyperintensional level, any of those two terms could be obtained; we will use in our language the expression $[+\ 1\ 1]^*$ to denote *any* hyperintensional term producing $[+\ 1\ 1]$. When moving back to the functional level, the term $[+\ 1\ 1]$ should always be produced, together with its denotation the number 2. Hyperintensional terms having $[-\ 5\ 3]$ as their products could be $Integer.minus(5, 3)$ and $S(S(S(S(S(0))))) - S(S(S(0)))$, where we assume subtraction can be redefined as a predecessor function. Each of those terms produces $[-\ 5\ 3]$ and this in turn denotes 2. We will use in our language the expression $[-\ 5\ 3]^*$ to stand for any linguistic expression (written for instance in some programming language) denoting the operation $[-\ 5\ 3]$. The former is thus an hyperintension for the latter. Again, when moving from the hyperintensional level to the functional level, a unique functional term should be obtained. This functional term denotes only one entity, its hyperintensional counterpart denotes instead different constructions.

Although partly inspired by TIL, and reflecting some of its terminology, our approach differs in several aspects. First, we use a bottom-up approach: we start from well-typed extensional and intensional terms and define hyperintensional ones from them. Because of this strategy, there is always a term obtained by an instance of the Execution rule. In other words, our system does not allow the derivation of improper constructions, i.e. hyperintensions that do not produce any object. TIL cases of improperness caused by execution of non-constructions are avoided in our

---

[1] We borrow the names for these rules from TIL.

system by requiring that only trivialized canonical terms can be executed.[2] Second, another source of improperness in TIL is composition applied to partial functions, i.e. when application may fail for some specific argument: we deal only with total functions, thus function application always returns an output. Third, Composition in TIL can fail also if the types of arguments do not match with the type required by the bound variables in the body of the expression: in our system, type checking will forbid the rule application in such cases. Fourth, in our calculus the product of a construction is obtained by explicit application of Execution, i.e. it is never implicitly denoted. Fifth, our semantics does not use quantification over worlds and times. Finally, to show the logical and analytical (i.e. under term decomposition) equivalence of terms denoting constructions, our only means is to perform Execution and check identity by reduction on the terms denoting the corresponding products. In the following, to aid readability, we will sometimes avoid referring to the terms of the language as denoting constructions, functions, numbers etc., and we will conventionally refer to their denotations: hence we might say that in a derivation both intensions and hyperintensions occur, or that a function of constructions occurs, while technically we intend that terms denoting such objects occur in the derivation.[3]

The rest of this paper is structured as follows. In Section 2 we present the language of HTLC . We define first the polymorphic set of rules which technically reduces to a typing system à la Curry, i.e. where types are assigned to pure λ-terms; for HTLC, this means that the same rules set can be instantiated not only by extensional and intensional terms, but also by hyperintensional ones. We then formulate those rules for each of the relevant types, offering thereby the extensional, intensional and hyperintensional fragments. We offer examples of derivations with terms of different types, and in particular in Section 3 we formulate an example where the same expression is treated first at the extensional/intensional level, and then at the hyperintensional level. In Section 4 we provide the meta-theoretical results, including the definition of term occurrence, normalization across contexts and completeness with respect to a Henkin's style of general model. Finally we provide some conclusions and ideas for possible further extensions of our work.

---

[2]The restriction on the canonicity of trivialized terms to be executed is close in spirit to the constraint imposed by Martin-Löf on the application of β-reduction w.r.t the terms of his theory of types, according to which a λ-term has to be β-reduced only "from without" and not "from within", i.e. guaranteeing that the way in which β-reduction is performed eventually coincides with the evaluation of closed λ-terms, [14, p.160]. We owe this observation to one of the reviewers.

[3]For clarity: we use the expression "function of constructions" (or of functions, or of anything else) to indicate a function that takes constructions (or functions, or anything else) as input and is allowed to be heterogeneous, thus having something else as codomain.

2   The System HTLC

The syntax of HTLC is a typed λ-calculus extended with a type for hyperintensional terms.

Definition 1 (Grammar).

$$T ::= \alpha \mid (T\ T_1 \ldots T_n) \mid *^T$$

$$\alpha ::= o \mid \iota \mid \tau \mid \ldots$$

$$t ::= x_i \mid [\lambda x_1 \ldots x_n.t] \mid [t\ t_1 \ldots t_n] \mid t^*$$

The type syntax for T is inductively defined by three kinds of entities:

- Extensional entities of type α;

- Intensional entities of type $(T\ T_1 \ldots T_n)$;

- Hyperintensional entities of type $*^T$.

The set of atomic types can depend on the application, including o (set of truth values: T, F), ι (infinite set of individuals, members of the universe), τ (as a meta-variable type for numbers: $\mathbb{N}, \mathbb{R}$ - e.g. sets of natural and real numbers respectively which might be added and should be defined through appropriate rules), and so on. Functions will be defined accordingly, as mappings from the Cartesian product of types $(T_1 \times \cdots \times T_n)$ into the type T, for any arbitrary type (hence involving sets of individuals, of truth values, of numbers and so on). We constraint these to total functions. We simplify the arrow notation of multi-argument functions $(T_1 \to \cdots \to T_n \to T)$ with the pair notation $(((T\ T_n)\ldots)\ T_1)$. As in standard typed λ-calculi we use association to the left when dealing with function types, so the curried version can be rewritten as $(T\ T_n\ \ldots\ T_1)$.[4] We can build higher order functions that take functions as arguments and return a function as value. Terms typed as hyperintensions denote abstract procedures whose outputs can be of any type (including hyperintensions as we admit higher-order constructions); these entities are constructions, computations or different "senses" in which lower order constructions, extensional or intensional entities can be produced.

---

[4]The notation $(T\ T_1\ \ldots\ T_n)$ from Definition 1 and the notation $(T\ T_n\ \ldots\ T_1)$ are equivalent: the second one can be obtained from the first one by a simple and harmless renumbering of arbitrary types T_x, and vice versa.

Terms of the language have the following form: variables x_i; abstraction terms $[\lambda x_1 \ldots x_n.t]$ denoting functions; application terms $[t\ t_1 \ldots t_n]$, denoting values of functions on given arguments; and finally hyperintensional terms t^*, recursively constructed and denoting constructions. We call a formula of our language an expression $t : T$, which declares a term t to belong to a given type T; as usual in the literature, a closed formula is one whose term does not contain any free variables, i.e. only λ-bound ones; a list of assumptions of formulae $\{x_1 : T_1, \ldots, x_n : T_n\}$ is called a context; a judgement is the assertion of a formula under a given context, denoted as $x_1 : T_1, \ldots, x_n : T_n \vdash t : T$.

Rules of the system HTLC are given below in four parts. First, we describe the polymorphic fragment of our system with rules applicable to arbitrary terms of any type. Then we focus on the extensional fragment, where types are extensional values, especially truth values and functions defined on them; then we move up to the intensional fragment, where objects of interest are functions of basic types, and functions of higher order; finally, we present the hyperintensional fragment with constructions and functions of constructions.

2.1 The Polymorphic Fragment

Rules for terms of arbitrary types are inference rules of the standard typed λ-calculus (see Figure 1) extended with rules to define the meaning of hyperintensional terms (see Figure 2): the latter are called intra-context rules as they allow to move reasoning from the extensional and intensional contexts to the hyperintensional one, and back. Rules are applied on formulas of the form $t : T$, but we often say that a rule is applied on a term t of type T, or that a rule returns such a term.

The Assumption rule allows for the derivation of a typed variable from its own assumption. The Abstraction rule allows to construct a λ-term for a function type $(T\ T_1 \ldots T_n)$ from the corresponding judgement inferring a term $t : T$ from variables $x_1, \ldots, x_n$ having types $T_1, \ldots, T_n$. The Application rule creates a term of type T denoting the value of a function, expressed by a term t on the n-tuple of arguments expressed by terms $t_1, \ldots, t_n$.

While inferring terms denoting intensional entities is guaranteed by Abstraction, and their elimination is the result of Application, appropriate rules are given in Figure 2 to shift to and from terms denoting hyperintensional entities. In line with proof-theoretic semantics, we provide the meaning of hyperintensional terms by defining an introduction and an elimination rule. The former establishes the necessary conditions for the formulation of a construction; the latter provides the minimal consequences of its use. The rules must be considered in pairs: a detour Trivialization/Execution is well-behaving (i.e. harmonious) if given a term t of type

$$\frac{}{x_i : T \vdash x_i : T} \text{ Assumption}$$

$$\frac{\Gamma, x_1 : T_1, \ldots, x_n : T_n \vdash t : T}{\Gamma \vdash [\lambda x_1 \ldots x_n . t] : (T \ T_1 \ldots T_n)} \text{ Abstraction}$$

$$\frac{\Gamma \vdash t : (T \ T_1 \ldots T_n) \quad \Gamma_1 \vdash t_1 : T_1 \ldots \Gamma_n \vdash t_n : T_n}{\Gamma, \Gamma_1, \ldots \Gamma_n \vdash [t \ t_1 \ldots t_n] : T} \text{ Application}$$

Figure 1: HTLC: Polymorphic Rules System

$$\frac{\Gamma \vdash t : T}{\Gamma \vdash t^* : *^T} \text{ Trivialization}$$

$$\frac{\Gamma \vdash t^* : *^T}{\Gamma \vdash t : T} \text{ Execution}$$

Figure 2: HTLC: Intra-context Rules

T inducing a term t^* of type $*^T$ by Trivialization, an instance of the Execution rule applied to the latter will return the former. The rules are formulated for a general term, and their version for complex terms is explained below.

The Trivialization rule defines the process of going from a given term t to the hyperintensional term t^* which denotes a construction of the object denoted by t. When the trivialised term t is of type α, Trivialization allows to shift from a term denoting an extensional entity to a hyperintensional one producing it. When the trivialised term t is of type $(T \ T_1 \ldots T_n)$, Trivialization allows to shift from a term denoting an intensional entity to a hyperintensional one producing it. When the trivialised term is of type $*^T$, Trivialization results in a higher-order hyperintensional entity denoted by t^{**}, which produces still a (lower-order) hyperintensional entity. In this latter case we will iterate on the type: Trivialization on a term t^* of type $*^T$ returns a formula $t^{**} : *^{*^T}$. By convention and to simplify notation, in the following we do not iterate $*$ on terms that are trivialized and were already of type $*^T$, but we agree just to rename the term; the relevant information on the iterated Trivialization is carried by the type. This allows to introduce the notion of order of the hyperintensional type:

$$
\begin{array}{c}
\vdots \qquad\qquad \vdots \qquad\qquad \vdots \\
\dfrac{\vdash Plus : (\tau\tau\tau) \quad \vdash 5 : \tau \quad \vdash 1 : \tau}{\vdash [Plus\,5\,1] : \tau} \ \text{Application} \\
\end{array}
$$

Figure 3: HTLC: Trivialization example

Definition 2 (Hyperintensional type of order n). *We say that a term t^* is of a hyperintensional type of order n, if and only if t^* results from n instances of the Trivialization rule. We denote the formula containing such term as*

$$
t^* : *^{*^{\cdot^{\cdot^{*^T_{n-1}}}}_1}
$$

where T is either an extensional or intensional type.

In all the cases above, the necessary condition in order for a term of type $*^T$ to be properly typed is that the term to be trivialized be a properly typed term of type T. This, in turn, means that Trivialization is always defined in its argument, and therefore we do not allow improper constructions. As a result, the trivialized term can always be returned (by Execution, see below). An example of the use of the Trivialization rule is illustrated in Figure 3. To aid readability, in this example we keep empty the contexts on the left-hand side of all formulas: the *Compute* relation takes by Application an individual as first argument, and the construction of a function to add numbers 5 and 1 as second argument; the latter is obtained by Trivialization on the functional term whose denotation is the object "6"; the Application returns a truth-value; the proposition "Michal computes the sum of 5 and 1" is then Trivialized in the last step of the derivation.

Execution is the opposite process of going from a hyperintensional type (eventually of higher order) to an extensional or intensional one (or to a hyperintensional type of lower order). Given a canonical term t^* of type $*^\alpha$, Execution returns the term t denoting the product of this construction, that is a term of type α. When the term t^* to be executed is of type $*^{(T\ T_1 \ldots T_n)}$, Execution returns the term t that denotes an intensional entity of type $(T\ T_1 \ldots T_n)$. Given a term t^* of hyperintensional type of order $n + 1$, Execution returns a term t^* of hyperintensional type of order n.[5] The condition on the canonicity of the term which is executed allows a

---

[5]It might be noted that our Execution rule has a stronger requirement than what is typical for

general formulation of the rule under a non-empty context, required not to include variables occurring in that term. On the other hand, this also means that a term to be executed might require first additional steps according to the computational rules of the system (see Figure 6) when not in canonical form. Hence, by closure under Abstraction of trivialized terms, there can be also a term of the form $[\lambda x^*.t^*]$ and type $(*^T *^{T_1})$: in this case, Execution is obtained by the following detour:

$$\frac{\vdash [\lambda x_1^*.t^*] : (*^T *^{T_1})}{\vdash [\lambda x_1.t] : (T\ T_1)} \text{ Execution} \qquad \rightsquigarrow$$

$$\frac{\dfrac{\dfrac{\dfrac{x_1:T_1 \vdash x_1:T_1}{x_1:T_1 \vdash x_1^* : *^{T_1}} \text{Triv.} \quad \vdash [\lambda x_1^*.t^*] : (*^T *^{T_1})}{x_1:T_1 \vdash [[\lambda x_1^*.t^*]\,x_1^*] : *^T} \text{App.} \quad [[\lambda x_1^*.t^*]\,x_1^*] \rightarrow_\beta t^*[x_1^*/x_1^*] \equiv t^* : *^T}{\dfrac{x_1:T_1 \vdash t^* : *^T}{\dfrac{x_1:T_1 \vdash t : T}{\vdash [\lambda x_1.t] : (T\ T_1)}} \text{Execution}} \text{Sub.Red.}}{} \text{Abstraction}$$

In this tree we start from trivializing an assumed variable to type $*^{T_1}$, to which we apply our λ-term. We then have a subject reduction step where $t^*[x_1^*/x_1^*]$ is syntactically equivalent to t^*, execute this closed term,[6] and abstract to obtain our desired (now executed) λ-term. In the following we always abbreviate this detour by direct Execution on each trivialized component of a λ-term and assume the computational step to obtain a closed term is always performed before execution.

By closure under Application of trivialized terms, there can be a term of the form $[[\lambda x_1^*.t^*]\,t_1^*]$ of type $*^T$: in this case, Execution returns products for each of the composing terms, combining the previous reduction with one additional available premise:

$$\frac{\dfrac{\vdash [\lambda x_1^*.t^*] : *^T *^{T_1} \qquad \vdash t_1^* : *^{T_1}}{\vdash [[\lambda x_1^*.t^*]\,t_1^*] : *^T} \text{Application}}{\vdash [[\lambda x_1.t]\,t_1] : T} \text{Execution}$$

$$\rightsquigarrow$$

---

proof-theoretic semantics, namely that it applies to canonical terms. Typically, an elimination rule is applicable to arbitrary terms of the required type and, as a result, a selector s is applied to this term. Then, if the term obtained is a redex (i.e. the selector is applied on a constructor), we can apply reduction. Our rule requires the term already in canonical form (namely to be built with constructor $*$), and we consider the reduction step as already performed and hidden, in order to avoid executing improper constructions.

[6]Note that t^* in this expression does not actually depend on x_1 in the context. Despite the fact that the variable x_1^* appearing in t^* has been obtained from x_1 by the application of a Trivialization rule, these two variables have to be taken as different, since in the derivation they are associated with two different types: the variable x_1 is associated with the type T_1, while the variable x_1^* is associated with the type $*^{T_1}$.

$$\cfrac{\cfrac{\vdots}{\Gamma \vdash Loves^* : *^{(o\iota\iota)}}}{\Gamma \vdash Loves : (o\iota\iota)}\text{Exec.} \quad \cfrac{\cfrac{\vdots}{\Gamma_1 \vdash John^* : *^\iota}}{\Gamma_1 \vdash John : \iota}\text{Exec.} \quad \cfrac{\cfrac{\vdots}{\Gamma_2 \vdash Mary^* : *^\iota}}{\Gamma_2 \vdash Mary : \iota}\text{Exec.}$$

$$\cfrac{\cfrac{\Gamma, \Gamma_1, \Gamma_2 \vdash [Loves\ John\ Mary] : o}{\Gamma, \Gamma_1, \Gamma_2 \vdash [Loves\ John\ Mary]^* : *^o}\text{Triv.}}{}\text{App.}$$

Figure 4: HTLC: Execution example

$$\cfrac{\cfrac{\Gamma, x^* : *^\iota, y^* : *^\iota \vdash Loves^* : *^o}{\Gamma \vdash [\lambda x^* y^*.Loves^*] : (*^o *^\iota *^\iota)}\text{Abs.} \quad \cfrac{\Gamma_1 \vdash John : \iota}{\Gamma_1 \vdash John^* : *^\iota}\text{Triv.} \quad \cfrac{\Gamma_2 \vdash Mary : \iota}{\Gamma_2 \vdash Mary^* : *^\iota}\text{Triv.}}{}\text{App.}$$

$$\cfrac{\Gamma, \Gamma_1, \Gamma_2 \vdash [[\lambda x^* y^*.Loves^*]\ John^*\ Mary^*] : *^o}{\Gamma, \Gamma_1, \Gamma_2 \vdash [[\lambda x\, y.Loves]\ John\ Mary] : o}\text{Exec.}$$

Figure 5: HTLC: Second Execution example

$$\cfrac{\cfrac{\vdash [\lambda x_1^*.t^*] : *^T *^{T_1}}{\vdash [\lambda x_1.t] : T\ T_1}\text{Execution} \quad \cfrac{\vdash t_1^* : *^{T_1}}{\vdash t_1 : T_1}\text{Execution}}{\vdash [[\lambda x_1.t]\ t_1] : T}\text{Application}$$

Note that also in this case we require the application to be done on a closed term. Again, we always abbreviate this derivation by direct Execution on each trivialized component of an applied term.

An example of the use of Execution is depicted in Figure 4: we first derive a construction of the relation "loves" between two individuals (e.g. its linguistic sense which appears on this very page between the written names of those individuals); the hyperintensional term denoting the function "loves" as well as its arguments are all executed for the Application to be well-typed and to obtain the propositional content "John loves Mary", whose type is a truth value; finally we can apply Trivialization back to obtain a construction of such propositional content, of type $*^o$. In the second example from Figure 5, the Execution of a construction is obtained by Application of trivialised terms, where each component of this application is a construction. In this case, Execution is required to act on all subterms according to the detour presented above, bringing each term to its product: the formula $[\lambda x^* y^*.Loves^*] : (*^o *^\iota *^\iota)$ is executed to obtain the formula $[\lambda x\, y.Loves] : (o\ \iota\ \iota)$.

As an extension of the lambda calculus, in HTLC β-reduction is present in the

$$\Gamma \vdash [[\lambda x_1 \ldots x_n .t]\ t_1 \ldots t_n] \rightarrow_\beta t[x_1/t_1 \ldots x_n/t_n] : T$$

$$\frac{\Gamma, x_1 : T_1, \ldots, x_n : T_n \vdash t \rightarrow_\beta t' : T}{\Gamma \vdash [\lambda x_1 \ldots x_n .t] \rightarrow_\beta [\lambda x_1 \ldots x_n .t'] : (T\ T_1 \ldots T_n)}\ \beta\text{-Abstr}$$

$$\frac{\Gamma \vdash t \rightarrow_\beta t' : (T\ T_1 \ldots T_n) \quad \Gamma_1 \vdash t_1 : T_1 \ldots \Gamma_n \vdash t_n : T_n}{\Gamma, \Gamma_1 \ldots \Gamma_n \vdash [t\ t_1 \ldots t_n] \rightarrow_\beta [t'\ t_1 \ldots t_n] : T}\ \beta\text{-App}$$

$$\frac{\Gamma \vdash t : (T\ T_1 \ldots T_i \ldots T_n) \quad \Gamma_1 \vdash t_1 : T_1 \ldots \Gamma_n \vdash t_n : T_n \quad \Gamma_i \vdash t_i \rightarrow_\beta t_i' : T_i}{\Gamma, \Gamma_1 \ldots \Gamma_n \vdash [t\ t_1 \ldots t_i \ldots t_n] \rightarrow_\beta [t\ t_1 \ldots t_i' \ldots t_n] : T}\ \beta\text{-App}$$

Figure 6: HTLC: β-rules

$$\frac{\Gamma \vdash t : T \quad t \twoheadrightarrow_\beta t'}{\Gamma \vdash t' : T}$$

Figure 7: HTLC: Subject reduction

form of a computational step, i.e. it expresses a purely syntactic term transformation to go from a syntactically more complex to a reduced term. We present such computation step applied to each of the possible rules, see Figure 6: β-reduction on a lambda term $[[\lambda x_1 \ldots x_n.t]\ t_1 \ldots t_n]$ corresponds to the substitution of the terms $t_1 \ldots t_n$ for variables $x_1 \ldots x_n$ occurring inside the term t; it is closed under the rules for Abstraction and Application; it is moreover a type-preserving operation when we take its transitive and reflexive closure $\twoheadrightarrow_\beta$, a fact which can be formulated as a simple rule known as Subject reduction and illustrated in Figure 7.

The last set of rules for the polymorphic fragment reflects the structural behaviour of the system, see Figure 8: Weakening expresses the usual principle that context extension is a monotonic operation in view of derivable terms; Exchange reflects the unstructured nature of contexts; Contraction allows to merge two variables of the same type occurring in the same context (this latter operation is expressed in terms of variable substitution inside the derivable term).

$$\frac{\Gamma \vdash t : T}{\Gamma, x : T_1 \vdash t : T} \text{ Weakening}$$

$$\frac{\Gamma, x_1 : T_1, x_2 : T_2 \vdash t : T}{\Gamma, x_2 : T_2, x_1 : T_1 \vdash t : T} \text{ Exchange}$$

$$\frac{\Gamma, x_1 : T_1, x_2 : T_1 \vdash t : T}{\Gamma, x_3 : T_1 \vdash t[x_1/x_3; x_2/x_3] : T} \text{ Contraction}$$

Figure 8: HTLC: Structural rules

2.2 The Extensional Fragment

In the extensional implementation of the HTLC rules, we reason with atomic types and functions defined over them. By creation of a function from atomic types, we move from a term occurring extensionally to a term occurring intensionally; and viceversa, by application of a function on an argument, we move from a term occurring intensionally to a term occurring extensionally.[7]

A first obvious interpretation for atomic types is by propositional terms with truth-values o as types, and functions defined on them. Rules of this fragment are illustrated in Figure 9. The system can be extended with connectives for conjunction and disjunction by adding pairs of propositions and projection on pairs respectively for appropriate introduction and elimination rules. A second possible interpretation of the extensional fragment is given by considering computational terms (programs) and the type of their outputs.

2.3 The Intensional Fragment

In the intensional fragment, we are able to reason about functions and higher order functions, see Figure 10. Beginning with functions of atomic types (i.e. functions $(T\ T_1 \dots T_n)$, where T and the T_i are all atomic types and hence are considered of order one), we can create functions defined over them (functions of higher order). Functions of hyperintensions can also be obtained by the Abstraction rule in the hyperintensional fragment (see Section 2.4). When reasoning with functions, we work with terms occurring intensionally; and when applying functions, we generate terms that occur either extensionally (if one deals with a function of atomic types), or intensionally themselves (if one deals with a higher-order function), or hyperinten-

---

[7]See Section 4.1 for an appropriate definition of term occurrence.

Assumption

$$\frac{}{x_i : o \vdash x_i : o}$$

Conditional proof (CP)

$$\frac{\Gamma, x_i : o \vdash t : o}{\Gamma \vdash \lambda x_i . t : (o \, o)}$$

Modus ponendo ponens (MPP)

$$\frac{\Gamma \vdash \lambda x_i . t : (o \, o) \qquad \Gamma_1 \vdash t_i : o}{\Gamma, \Gamma_1 \vdash t : o}$$

Figure 9: HTLC: Extensional fragment - propositional version

sionally (if one deals with a function of constructions). Function evaluation occurs therefore within the extensional fragment when we are reasoning with functions of atomic types; it occurs within the intensional fragment when working with higher-order functions; and it occurs within the hyperintensional fragment when working with functions of constructions.

An example of a higher order function is $Map : ((o \, T_2) \, (T_2 \, T_1) \, (o \, T_1))$, which takes two arguments of the function type, and it returns an object of a function type. In functional programming languages Map is used to apply a function to a list and return another list. In order to replace lists, whose type we do not have explicitly in our language, we can give up on ordering and define a set of type T by using a characteristic function of type $(o \, T)$. For the Map function, consider a set expressed by its characteristic function $(o \, T_1)$, and a function $(T_2 \, T_1)$ applied to every element of the input set, to obtain an output set of type $(o \, T_2)$, again as the characteristic function of a set. For example, consider a term $Square$ of type $(\mathbb{N} \, \mathbb{N})$ denoting the function that transforms any natural number into its square; and consider the term $Primes$ of type $(o \, \mathbb{N})$ denoting the characteristic function that selects the subset of prime numbers from $\mathbb{N}$. Then we can think of $[Map \; Square \; Primes] : (o \, \mathbb{N})$ as the application of $Square$ on all members of $Primes$. The result is a new set containing the squares of prime numbers. In this particular example, the typing of our map function is $Map : ((o \, \mathbb{N}) \, (\mathbb{N} \, \mathbb{N}) \, (o \, \mathbb{N}))$.

481

$$\frac{}{x_i : (T\ T_1 \dots T_n) \vdash x_i : (T\ T_1 \dots T_n)} \text{ Assumption}$$

$$\frac{\Gamma, x_1 : F_1, \dots, x_n : F_n \vdash t : F}{\Gamma \vdash [\lambda x_1 \dots x_n . t] : (F\ F_1 \dots F_n)} \text{ Abstraction}$$

$$\frac{\Gamma \vdash t : (F\ F_1 \dots F_n) \qquad \Gamma_1 \vdash t_1 : F_1 \dots \Gamma_n \vdash t_n : F_n}{\Gamma, \Gamma_1, \dots, \Gamma_n \vdash [t\ t_1 \dots t_n] : F} \text{ Application}$$

where $F_i = (T_i\ T_{i_1} \dots T_{i_n})$

Figure 10: HTLC: Intensional fragment

2.4 The Hyperintensional Fragment

At the highest level, we introduce the hyperintensional fragment, where our objects of interest are procedures whose products are objects of either an extensional, or an intensional type, or procedures of lower order. Procedures are obtained by Trivialization on a term t of a given type T. Here, we assume that the term t is well-typed. Given a relation from the domain of basic types and function types to their constructions as co-domain, in our calculus this relation is one-to-many.

Execution works as an elimination rule for the type $*^T$; if the rule is applied to a higher-order construction, it lowers its degree. Note, however, that by Abstraction on constructions, we move from a term occurring hyperintensionally to a term occurring intensionally; and viceversa, by Application on constructions, we move from a term occurring intensionally to a term occurring hyperintensionally.[8] Therefore, rules of the hyperintensional level allow us to reason about constructions, the creation of functions of constructions and their evaluation; to reason about functions of constructions, we move down to the intensional fragment. Given a relation from the set of constructions as domain to the set of their products as co-domain, in our calculus such relation is many-to-one. The construction rules of this fragment are shown in Figure 11. Note that by the explicit requirement that the term t^* in the syntax is defined recursively, we admit variables x_i^*. While for terms appearing on the right-hand side of the symbol $\vdash$, the operator $*$ is obtained by Trivialization, in the case of the Assumption rule for the hyperintensional fragment, it is possible instead to let appear the $*$ operator also on terms appearing on the left-hand side of $\vdash$, namely when these terms are variables. This is required to avoid improper constructions at the level of variables, i.e. obtaining hyperintensional terms on which no

---

[8]Again, in Section 4.1 we provide proper definitions of term occurrence.

$$\frac{}{x_i^* : *^T \vdash x_i^* : *^T} \text{ Assumption}$$

$$\frac{\Gamma, x_1^* : *^{T_1}, \ldots, x_n^* : *^{T_n} \vdash t^* : *^T}{\Gamma \vdash [\lambda x_1^* \ldots x_n^* . t^*] : (*^T \ *^{T_1} \ \ldots \ *^{T_n})} \text{ Abstraction}$$

$$\frac{\Gamma \vdash t : (*^T \ *^{T_1} \ \ldots \ *^{T_n}) \qquad \Gamma_1 \vdash t_1^* : *^{T_1} \ldots \Gamma_n \vdash t_n^* : *^{T_n}}{\Gamma, \Gamma_1, \ldots, \Gamma_n \vdash [t \ t_1^* \ldots t_n^*] : *^T} \text{ Application}$$

Figure 11: HTLC: Hyperintensional fragment

Execution rule can be applied, and hence for which no product can be associated. To avoid this, we consider variables for hyperintensional types of the form x_i^*, then the Execution rule can always be applied on them, producing a variable of extensional or intensional type T.

In order to exemplify a derivation in which both terms of the hyperintensional and intensional types occur, we show a tree where we move from a construction $t^* : *^{(T \ T_1)}$ whose product is a function of type $(T \ T_1)$, to a function of type $(*^T *^{T_1})$: for this, we require first that the function at the *intensional* level be obtained by Execution and Application, followed by Trivialization and finally Abstraction:

$$\frac{\dfrac{\dfrac{\Gamma, x_1^* : *^{T_1} \vdash t^* : *^{(TT_1)}}{\Gamma, x_1^* : *^{T_1} \vdash t : (TT_1)} \text{ Execution} \qquad \dfrac{\vdots}{\vdash t_1 : T_1}}{\dfrac{\Gamma, x_1^* : *^{T_1} \vdash [t \ t_1] : T}{\quad}} \text{ Application}}{\dfrac{\Gamma, x_1^* : *^{T_1} \vdash [t \ t_1]^* : *^T}{\Gamma \vdash [\lambda x_1^* . [t \ t_1]^*] : (*^T *^{T_1})} \text{ Abstraction}} \text{ Trivialization}$$

In the opposite direction, we can easily proceed from a function of type $(*^T *^{T_1})$ whose product is a function of type $(T \ T_1)$ obtained by the detour illustrated in Section 2.1 for non-canonical terms, to a construction of type $*^{(T \ T_1)}$:

$$\frac{\dfrac{\Gamma \vdash [\lambda x_1^* . t^*] : (*^T *^{T_1})}{\Gamma \vdash [\lambda x_1 . t] : (T \ T_1)} \text{ Execution}}{\Gamma \vdash [\lambda x_1 . t]^* : *^{(T \ T_1)}} \text{ Trivialization}$$

$$\vdash =: (o\tau\tau) \qquad \dfrac{\overline{x : \tau \vdash x : \tau}\ \text{Assum} \qquad \vdash 1 : \tau}{\dfrac{x : \tau \vdash [= x\, 1] : o}{\dfrac{\vdash [\lambda x.[= x\, 1]] : (o\tau)}{}}\ \text{Abstraction}}\ \text{Application} \qquad \dfrac{\vdash [Succ\, 0]^* : *^{\tau}}{\vdash [Succ\, 0] : \tau}\ \text{Execution}$$

$$\dfrac{}{\vdash [[\lambda x.[= x\, 1]]\, [Succ\, 0]] : o}\ \text{Application}$$

Then we can perform β-reduction:

$$[[\lambda x.[= x\, 1]]\, [Succ\, 0]] \rightarrow_{\beta} [= [Succ\, 0]\, 1]$$

Figure 12: HTLC: Functional identity between numbers

3 A Comparative Example

In a language like TIL, it is possible to compute with non-executed constructions and their products. For example, one could construct the set of constructions producing number one, and then take one element in such a set, e.g. "Successor of 0". The process of checking whether this element belongs to that set eventually results in checking the equality of the product of this construction with number 1. HTLC allows similar expressions, although it is more strict in terms of type matching, so that the type a function requires for its argument must always be properly met: i.e. either the functional term is of type $(T_1 \ *^{T_2})$ and its argument of type $*^{T_2}$ (the output type of this function is not relevant, and it can be $*^{T_1}$ as well); or respectively $(T_1\ T_2)$ and T_2.

The formulation of such a function at the intensional level between numbers is reflected in the tree in Figure 12. In this example, the function $=$ takes two numbers as arguments $(\tau\tau)$ and it returns a truth-value (o). Given a variable in the type of numbers, and number 1, by Application and Abstraction we build the λ-term that takes a number to be substituted for a variable and it compares for equality with 1, in order to return a truth-value. Consider then a construction to produce the successor function of the number 0, i.e a term of type $*^{\tau}$ which denotes one of the possible ways of expressing the successor of 0, for example by stroke notation $0'$: by Execution we obtain the term denoting the actual product of the construction; by Application we pass the term denoting this number to the function of type $(o\tau)$, to obtain a truth value. In this case the identity is at the extensional level, between the product of a procedure (of order 1) and a number.

On the other hand, it is possible to express the same content at the hyperintensional level as the equality between procedures, see Figure 13. In this case we

$$\dfrac{\vdots}{\vdash \approx : (o *^\tau *^\tau)} \quad \dfrac{\overline{x : *^\tau \vdash x : *^\tau}\ \text{Assum}}{\dfrac{x^* : *^\tau \vdash [\approx\ x^*\ 1^*] : o}{\vdash [\lambda x^*.[\approx\ x^*\ 1^*]] : (o\ *^\tau)}\ \text{Abstraction}} \quad \dfrac{\vdots}{\vdash 1^* : *^\tau}\ \text{App.} \qquad \dfrac{\vdots}{\dfrac{\vdash (Succ\,0) : \tau}{\vdash [Succ\,0]^* : *^\tau}\ \text{Trivializ.}}$$

$$\dfrac{\vdash [[\lambda x^*.[\approx\ x^*\ 1^*]]\,[Succ\,0]^*] : o}{}\ \text{Application}$$

Then we can perform β-reduction:

$$[[\lambda x^*.[\approx,\,x^*\ 1^*]]\,[Succ\,0]^*] \to_\beta [\approx\ [Succ\,0]^*\ 1^*]$$

Figure 13: HTLC: Functional Identity between Procedures

consider equality between a construction for a number and a construction for the number 1, returning a truth value. In this example, the function $\approx$ takes two constructions as arguments ($*^\tau *^\tau$) and it returns a truth-value (o). Given a variable for the construction of numbers, and a construction for number 1, by Application and Abstraction we build the λ-term that takes a construction for the successor of 0 to be substituted for a variable, and it compares for identity with a construction for 1, in order to return a truth-value. Note that we derive here the argument by Trivialization. This term β-reduces to the identity between a construction for the successor of 0 and a construction for 1, with the identity being false or true, depending from which construction is chosen for number 1, i.e whether the same construction is selected. Note that this corresponds to the function between constructions and their products being many-to-one.

4 Meta-theory

4.1 Term occurrence

HTLC allows to identify extensional, intensional and hyperintensional terms by inspecting the derivation tree under consideration and looking at the rule applied at the relevant step. In the following, we provide appropriate definitions of the extensional, intensional or hyperintensional occurrence of a term at a given step of a derivation.[9]

---

[9] For the same properties TIL relies on the inductive definition of the structure of the relevant construction. For details, see [5].

Definition 3 (Extensional Occurrence). *A term t occurs extensionally at step n of a tree if and only if it results from:*

1. *an Assumption rule, and it is of type α;*

2. *an Application rule, and it is of type α;*

3. *an Execution rule on a term of type $*^{\alpha}$.*

Consider as an example the following derivation:

$$
\cfrac{
  \cfrac{
    \cfrac{
      \cfrac{
        \cfrac{
          \cfrac{\Gamma \vdash [\lambda x^*.t^*] : (*^o*^o) \qquad \vdash t_1^* : *^o}{\Gamma \vdash [[\lambda x^*.t^*]\ t_1^*] : *^o} \text{ Application}
        }{\Gamma \vdash [[\lambda x.t]\ t_1] : o} \text{ Execution}
      }{\Gamma, x_1 : o \vdash [[\lambda x.t]\ t_1] : o} \text{ Weakening by } x_1
    }{\Gamma \vdash [\lambda x_1.[[\lambda x.t]\ t_1]] : (o\ o)} \text{ Abstraction}
  }{\Gamma \vdash [\lambda x_1.[[\lambda x.t]\ t_1]]^* : *^{(o\ o)}} \text{ Trivialization}
}
$$

The term $[[\lambda x.t]\ t_1]$ of type o resulting by Execution from the hyperintensional term $[[\lambda x^*.t^*]\ t_1^*]$ of type $*^o$ occurs extensionally at the third step of the derivation.

Definition 4 (Intensional Occurrence). *A term t occurs intensionally at step n of a tree if and only if it results from:*

1. *an Assumption rule and it is of type $(T\ T_1 \ldots T_n)$;*

2. *an Abstraction rule;*

3. *an Application rule, and it is of type $(T\ T_1 \ldots T_n)$;*

4. *an Execution rule on a term of type $*^{(T\ T_1 \ldots T_n)}$.*

Consider as an example the term $[\lambda x_1.[[\lambda x.t]\ t_1]]$ of type $(o\ o)$ in the above derivation: it results from Abstraction, and it occurs therefore intensionally at the fifth step of the derivation.

Definition 5 (Hyperintensional Occurrence). *A term t occurs hyperintensionally at step n of a tree if and only if it results from:*

1. *a Trivialization rule;*

2. *an Assumption rule, and it is of type $*^T$;*

3. an Application rule, and it is of type $*^T$;

4. an Execution rule on a term of type $*^{*^T}$.

Consider as an example the term $[\lambda x_1.[[\lambda x.\, t]\ t_1]]^*$ of type $*^{(o\ o)}$ in the above derivation: it results from Trivialization, and it occurs therefore hyperintensionally at the last step of the derivation.

Finally, we are also able to define the occurrence of a term within a hyperintensional term by inspecting the result of an Execution rule.

Definition 6. *A term t occurs extensionally, respectively intensionally, or hyperintensionally in a hyperintensional term t^* at step n of a tree if and only if at step $n+1$ the term t occurs:*

1. *extensionally according to Definition 3, case 3;*

2. *intensionally according to Definition 4, case 4;*

3. *hyperintensionally according to Definition 5, case 4.*

Consider as an example again the term $[\lambda x_1.[[\lambda x.\, t]\ t_1]]^*$ of type $*^{(o\ o)}$ obtained by Trivialization in the above derivation from this section: it is a hyperintensional term in which a term occurs intensionally, i.e. an application of the Execution rule in a next step would return an intensional term.

4.2 Normalization

Execution is a converging rule, i.e. it is possible that distinct constructions can be generated from β-equivalent terms (i.e. terms for which the symmetric closure of $\twoheadrightarrow_\beta$ holds) of a base type or of a function type by Trivialization, and hence they return the same entity when executed. This is the classical example of failing identity for hyperintensions, where the expressions "bachelor" and "unmarried man" might fail to be identified as equal, but would eventually be applied truthfully to the same individual. Accordingly, an application of Trivialization on distinct but reducible terms t_1, t_2 may induce distinct hyperintensional terms t_1^*, t_2^*, each denoting a distinct construction of the same product. This relation in its general formulation is an instance of the so-called Diamond Lemma for terms related by Trivialization:

Lemma 1 (Trivialized Diamond). *Let $\Gamma \vdash t_1 : T$, $\Gamma \vdash t_2 : T$ and $t_1 \twoheadrightarrow_\beta t_2$. Let, moreover, $t_1^* : *^T$ be obtained from $t_1 : T$ by Trivialization, and $t_2^* : *^T$ be obtained from $t_2 : T$ by Trivialization. Then $t_1^* \twoheadrightarrow_\beta t_2^*$.*

Proof. By induction on t_1^*, we only reason on the base atomic case. First reduce t_1^* to t_1 by Execution; now obtain t_2^* from t_2 by Trivialization. Let us now denote with $\twoheadrightarrow_{Exec}$ the transitive closure of Execution and $\twoheadrightarrow_\beta$; with $\rightarrow_{Triv}$ the term transformation resulting from Trivialization and with $\rightarrow_{Exec}$ the term transformation resulting from Execution. Then $t_1^* \twoheadrightarrow_{Exec} t_2 \rightarrow_{Triv} t_2^*$. If t_1^* and t_2^* are syntactically identical terms, the Lemma is trivially satisfied. Else: if it is not the case that $t_1^* \twoheadrightarrow_\beta t_2^*$ but $t_1^* \rightarrow_{Exec} t_1$ and $t_2^* \rightarrow_{Exec} t_2$, then because of failure of subject reduction t_1 and t_2 cannot have the same type, against the assumption that $t_1 \twoheadrightarrow_\beta t_2$. □

Note that subject reduction implies only β-reduction of constructions, which might not be guaranteed and is a weaker requirement than the notoriously problematic identity of hyperintensions. We can also show convergence for the transitive and reflexive closure of β-reduction (for the general case, i.e. also considering hyperintensions, not used in the above last step of the Diamond Lemma):

Theorem 1 (Church-Rosser). *If $\Gamma \vdash t : T$, $t \twoheadrightarrow_\beta t'$ and $t \twoheadrightarrow_\beta t''$ then there is a term u such that $t' \twoheadrightarrow_\beta u$ and $t'' \twoheadrightarrow_\beta u$ and $\Gamma \vdash u : T$.*

Proof. By induction on t, t', t'', and u using subject reduction, and the Diamond Lemma if the term u is of the form t'^* (and thus t''^*). □

4.3 Completeness

A recent conjecture presented in [4] states that the non-hyperintensional fragment of total functions, without modalities (quantifying over possible worlds and times) of TIL is Henkin complete. HTLC only expresses total functions and proper constructions, without modalities quantifying over possible worlds and times. On this basis, we show the following version of completeness:

Theorem 2. *For any consistent set of closed well-formed formulas Λ of the form $t : o$ from HTLC there is a general model, in which*

- *the domain of basic and function types is denumerable,*

- *and the domain for hyperintensional types is non-denumerable but strongly reducible to a model with a denumerable domain for the lower types,*

with respect to which Λ is satisfiable.

Proof. We first consider standard properties of any consistent set of closed formulas Γ in HTLC.[10] We also use normalization by β-equivalence across contexts as a

---

[10]Consistency in a typed λ-calculus is notoriously guaranteed by the impossibility of typing a term $\lambda f.[\lambda x.[f[xx]]\lambda x.[f[xx]]]$. We assume here therefore that recursion can only be externally added to the language in order to preserve consistency of any set of formulas Γ.

congruence relation on terms. We now build the standard model:

$$M := \{D_\alpha, D_{(T\ T_1...T_n)}, D_{(*^T)}\}$$

containing a family of domains, one for each type:

- D_α stands for a meta-variable for each of the domains of basic types, i.e. $D_o, D_\iota, \ldots$ truth-values $\{\textit{True, False}\}$, individuals, and any other required basic type;

- $D_{(T\ T_1...T_n)}$ is the domain of all total functions, with input of types $(T_1, \ldots, T_n)$ and values of type T, i.e. terms which after reduction by Application and possibly Execution are in the domain D_α;

- $D_{(*^T)}$ is the domain of all proper constructions, i.e. the set of elements of type $*^T$ generated from elements in the types α or $(T\ T_1 \ldots T_n)$.

Note that we define the standard model, and subsequently adding an evaluation on the equivalence class of formulas for the general model, by considering only the domain of constructions of order 1. This is required because the full evaluation of such domain can only be given by obtaining the terminal product of the construction, i.e. for proper constructions. This also means that when in the presence of higher order constructions, completeness can be guaranteed only by multiple Execution.

By an assignment ϕ with respect to the standard model we mean a relation from the set of variables into the domain of the appropriate type, i.e. the value of $\phi(x : T)$ is an element of D_T. We now define a relation V_ϕ associated with an assignment ϕ with respect to the standard model such that it assigns every formula to elements of a domain:

- for a formula $x_i : T$, the evaluation $V_\phi(x_i : T) = \phi(x_i : T)$;

- for a formula $[\lambda x_1, \ldots, x_n.t] : (T\ T_1 \ldots T_n)$, the evaluation $V_\phi([\lambda x_1, \ldots, x_n.t] : (T\ T_1 \ldots T_n))$ assigns an element in the domain $D_{(T\ T_1...T_n)}$, i.e. a function whose value for arguments $d_i \in D_{T_i}$ is $V_\psi(t : T)$, where ψ is an assignment with the same values as ϕ for all variables in t except for x_i, while $\psi(x_i : T_i)$ is d_i; and, $V_\phi([\lambda x_1.t] : (o\ o))$ has overall value $\textit{False}$ iff $V_\phi(x_1 : o) = \textit{True}$ and $V_\phi(t : o) = \textit{False}$, otherwise $\textit{True}$;

- for a formula $[t\ t_1 \ldots t_1] : T$, the evaluation $V_\phi([t\ t_1 \ldots t_n] : T)$ assigns the value of the function $V_\phi(t : (T\ T_1 \ldots T_n))$ in the domain D_T for the values of the arguments $V_\phi(t_i : T_i)$ in the domain D_{T_i};

- for a formula $t^* : *^T$, the evaluation $V_\phi(t^* : *^T)$ assigns elements in the domain D_{*T} that produce elements in the domain D_T for each evaluation $V_\phi(t : T)$.

Note that V_ϕ is one-to-many because $V_\phi(t^* : *^T)$ assigns many elements in the domain D_{*T}. With V_ϕ defined, let us define the standard notion of valid formula:

Definition 7 (Validity in the standard sense). *A wff $t : o$ is valid in the standard sense if $V_\phi(t : o) = True$ for every assignment ϕ wrt the standard model.*

We now define a frame F by induction on T containing a family of domains, one for each type as defined above:

$$F := \{D_\alpha, D_{(o\ T_1...T_n)}, D_{(*^T)}\}$$

Recall that we use β-equivalence as a congruence relation, thus two terms t and t' are equivalent iff $t =_\beta t'$ (i.e. the symmetric closure of $\twoheadrightarrow_\beta$). Given a formula $t : T$, we denote with $\lceil t : T \rceil$ the equivalence class of formulas containing terms congruent with t. Then we can define the following:

Definition 8 (General Model). *A frame F is called a general model if a one-to-many relation $f(\lceil t : T \rceil)$ of equivalence classes of closed formulas $t : T$ is such that it assigns elements in the domain D_T.*

We now build the frame which is a model of a maximal consistent set of closed formulas Γ, which is clearly a superset of Λ, as follows:

- $f(\lceil t : o \rceil)$ is the value true or false depending on term t being in the set Γ or not, and hence D_o is the set of truth values $\{True, False\}$;

- $f(\lceil t : \iota \rceil)$ is the equivalence class of individuals $\lceil t : \iota \rceil$, hence D_ι is the set of equivalence classes of all terms of the type of individuals;

- and accordingly so for any other type in α;

- Assuming that D_o and D_{T_i} have been defined, as well as the value of f for all equivalence classes of terms of types o, and T_i and that every element of $D_o, D_{T_1}, \ldots, D_{T_n}$ is the value of f for some $\lceil t : o \rceil$, $\lceil t_1 : T_1 \rceil, \ldots, \lceil t_n : T_n \rceil$, respectively; then $f(\lceil t : (o\ T_1 \ldots T_n) \rceil)$ is the function whose value in domain $D_{(o\ T_1,...,T_1)}$ is given by the value for the element $f(\lceil t_1 : T_1 \rceil)$ of D_{T_1}, up to $f(\lceil t_n : T_n \rceil)$ of D_{T_n} and returns the value of $f(\lceil t : o \rceil)$ of D_o;[11]

---

[11] Note that here we consider functions which have arguments of every possible type, including higher-order functions and hyperintensions, but only outputs of type o, i.e. truth-values. This allows us to define formulas in frames as those for which satisfiability and validity are defined.

- Assuming that D_α and $D_{(o\ T_1...T_n)}$ have been defined, as well as the value of f for all equivalence classes of terms of types α and of type $(o\ T_1...T_n)$ and that every element of D_α is the value of f for some equivalence class $\lceil t : o \rceil$ or $\lceil t : \iota \rceil$ and that every element of $D_{(o\ T_1...T_n)}$ is the value of f for some equivalence class of corresponding terms; then $f(\lceil t^* : *^T \rceil)$ is a construction with value $f(\lceil t^* : *^T \rceil)$ in D_{*T} for some element $f(\lceil t : T \rceil)$ of D_T.

- Assuming that D_{*T} has been defined, as well as the value of f for all equivalence classes of terms of types $*^T$ and that every element of D_{*T} is one of the values of f for some $t^* : *^T$; then $f^{-1}(\lceil t^* : *^T \rceil)$ is the execution of construction whose unique value for the element $f(\lceil t^* : *^T \rceil)$ of D_{*T} is $f(\lceil t : T \rceil)$ of D_T.

Note that f is one-to-many as well because the domain D_{*T} includes many values for $f(\lceil t^* : *^T \rceil)$, while the function $f^{-1}(\lceil t^* : *^T \rceil)$ returns the only input producing all such outputs. We now define formula validity and satisfiability of a set of formulas for the general type $T = \{\alpha, (o\ T_1 \dots T_n), *^T\}$:

Definition 9 (Validity in the general sense). *A wff $t : o$ is valid in the general sense if $V_\phi(t : o) = True$ for every assignment ϕ wrt the general model.*

Definition 10 (Satisfiable set of formulas). *A set of formulas Λ is satisfiable with respect to the frame $\{D_T\}$ for any type T, if there exists a valuation ϕ such that $V_\phi(t : o) = True$ for every formula $t : o$ in Λ.*

Lemma 2. *For every ϕ and $t : T$, $V_\phi(t : T) = f(\lceil t : T \rceil)$*

The proof of this intermediary Lemma is by induction on $t : T$.

- If $t : \alpha$ is of the form $x_i : \alpha$ and $\phi(x : \alpha)$ is the element $f(\lceil t : T \rceil)$ in the domain D_α, then $\phi(x : \alpha)$ is a consistent formula $t : \alpha$ such that $V_\phi(x : \alpha) = \phi(x : \alpha) = f(\lceil t : \alpha \rceil) = V_\phi(t : \alpha)$.

- If $t : T$ is of the form $[\lambda x_1 \dots x_n.t] : (o\ T_1 \dots T_n)$ and $V_\phi([\lambda x_1 \dots x_n.t])$; then the element $f(\lceil t : o \rceil)$ is a consistent formula in the domain D_o when each $\phi(x_i : T_i)$ is a closed formula $t_i : T_i$ in the domain D_{T_i}, and if $V_\phi(x_i : T_i) = \phi(x_i : T_i) = f(\lceil t_i : T_i \rceil) = V_\phi(t_i : T_i)$ then $V_\phi(t : o) = \phi(t : o) = f(\lceil t : o \rceil) = V_\phi(t : o)$.

- If $t : T$ is of the form $[t\ t_1 \dots t_n] : o$ and $V_\phi([t\ t_1 \dots t_n])$ is the element $f(\lceil [t\ t_1 \dots t_n] : o \rceil)$ in the domain D_o, then $\phi([t\ t_1 \dots t_n])$ is a closed formula such that every $t_i : T_i$ is one element in the corresponding domain D_{T_i}, in which case $[t\ t_1 \dots t_1] : o$ is a closed formula interpreted in the domain $D_{(o\ T_1,\dots T_n)}$, such that $V_\phi([t\ t_1 \dots t_n]) = \phi([t\ t_1 \dots t_n] : o) = f(\lceil [t\ t_1 \dots t_n] \rceil : o) = V_\phi([t\ t_1 \dots t_n])$.

- Let us consider here the novel case $t^* : *^T$. We assume that $f(\lceil t : T \rceil)$ has already been defined for $T = \alpha$ or $T = (o\ T_1 \ldots T_n)$, and $V_\phi(t : T) = f(\lceil t : T \rceil)$ and $V_\phi(t : (T\ T_1 \ldots T_n)) = f(\lceil t : (T\ T_1 \ldots T_n) \rceil)$ respectively. Now the value of $f(t^* : *^T)$ is defined as one of the elements in the Domain D_{*T} as the relation is one-to-many. Consider any two terms $\{t_1^*, t_2^*\}$ such that $\{t_1^*, t_2^*\} \in D_{*T}$, then $t_1^* \to_\beta t_2^*$ must hold by Lemma 1, assuming it holds that $t_1 \to_\beta t_2$ for $\{t_1, t_2\} \in D_T$. Hence, if $V_\phi(t_1 : T) = f(t_1 : T) = f(t_2 : T) = V_\phi(t_2 : T)$, then $f(t : T) = f^{-1}(t^* : *^T)$, for t any of t_1, t_2 and t^* any of t_1^*, t_2^*. Hence $V_\phi(t : T) = f^{-1}(\lceil t^* : *^T \rceil)$.

The frame $F = \{D_\alpha, D_{(o\ T_1 \ldots T_n)}, D_{(*T)}\}$ is a general model because for every formula $t : T$ and assignment ϕ, $V_\phi(t : T)$ is an element of the domain D_T for each element of $f(\lceil t : T \rceil)$. The elements of D_α and $D_{(o\ T_1 \ldots T_n)}$ are in one-to-one correspondence with the normalised set (equivalence class) of formulas, hence the domains are infinitely enumerable (and possibly finite). For the domain D_{*T} though, this is not the case as the relation between the values of $f(t : T)$ and $f(t^* : *^T)$ is one-to-many. Any value of $V_\phi(t^* : *^T)$ in the domain D_{*T} normalizes with respect to the value of $V_\phi(t : T)$ in the domain D_T for which $f^{-1}(\lceil t^* : *^T \rceil)$ holds. Since for every formula $t : T$ its denotation in the domain is given by $V_\phi(t : T)$ for ϕ arbitrary, and for every $t : \alpha$ and $t : (o\ T_1 \ldots T_n)$ of any consistent Γ there is such a term in the appropriate domain D_α and $D_{(o\ T_1 \ldots T_n)}$ respectively; and for every $t^* : *^T$ there is a function which returns an element in D_α or $D_{(o\ T_1 \ldots T_n)}$; it follows that for every $\Lambda \subseteq \Gamma$ a valuation $V_\phi(t : T)$ assigns an element in the domain D_T, i.e. Λ is satisfiable with respect to the model M, and D_T is denumerable for $T = \{\alpha, (o\ T_1 \ldots T_n)\}$, while elements of D_{*T} normalize with respect to elements in D_T. $\square$

Theorem 3. *Any closed wff $t : T$ is derivable in HTLC if and only if $t : T$ is valid in the general sense.*

Proof. We prove by induction from the basic case.

$\leftarrow$ 1. For $t : o$, the formula is valid by Definition 9 iff $V_\phi(x_i : \alpha)$ is valid once every free variable x_i occurring in t has been substituted, and there is an element in the corresponding domain of the general model D_o. In such a case, any formula $t' : o$ with variable x' with evaluation $V_\phi(t' : o) = False$ i.e. such that $V_\phi(x_i' : \alpha) = False$ cannot be consistent with $t : o$; in particular, $V_\phi([\lambda x_i'.t : (o\ \alpha)]) = True$ and $V_\phi([t\ t'] : o) = True$;

2. For $t : T$ of the form $[\lambda x_1 \ldots x_n.t] : (o\ T_1 \ldots T_n)$, the argument generalizes the previous one with several arguments;

3. For $t : T$ of the form $t^* : *^T$, such a formula is valid iff for every value of $V_\phi(t^* : *^T)$ the picked element from the domain D_{*T} (satisfying congruence with any other in the same domain) corresponds to $V_\phi(t : o) = True$. Hence, one reduces first to such a value by an application of Execution and then the argument runs according to the previous step for either $t : \alpha$ or $[\lambda x_1 \ldots x_n . t] : (o\ T_1 \ldots T_n)$

$\rightarrow$ Starting from our Assumption as axiom, both Abstraction and Application preserve validity, using Execution where necessary. Note that Trivialization is not invoked and all intra-context operations are from the domain D_{*T} to the domain D_T.

$\square$

To conclude, we reformulate the last step of the completeness proof to show satisfiability based on compactness; the only constraint is again that in order to preserve denumerability of the domain of reference, sets of formulas including hyperintensions need to be reduce to the corresponding formulas with executed terms:

Theorem 4. *A set Γ of closed well-formed formulas is satisfiable with respect to*

- *some model of denumerable domains $D_o, D_o\ _{T_1,\ldots,T_n}$*

- *and some model of a non-denumerable domain D_{*T}*

if and only if every finite subset Λ of Γ is satisfiable.

Proof. $\rightarrow$ – If Γ is not satisfiable with respect to some model of a denumerable domain $D_o, D_o\ _{T_1,\ldots,T_n}$ then it is inconsistent by Theorem 2, i.e. in particular $\Gamma \vdash [\lambda x_n . t_n] : (o_n\ T_n)$. Then there is a finite $\Lambda \subset \Gamma$ such that $\Lambda = \{x_1 : A_1, \ldots x_{n-1} : A_{n-1}\}$ and $\vdash [\lambda x_1, \ldots, x_{n-1}, x_n . t_n] : (o_n\ T_1, \ldots, T_n)$; but then this formula is valid because derivable for Theorem 3 hence there is also a $V_\phi(x_i : T_i) = False, i = 1 \ldots n - 1$ for every ϕ with respect to any model, hence Λ is not satisfiable. Then, if every $\Lambda \subset \Gamma$ is satisfiable, also Γ is satisfiable with respect to a denumerable domain.

   – if Γ has formulas of the form $t : *^T$, then the domain D_{*T} is non-denumerable; apply Execution to reduce to Γ' with respect to denumerable domains $D_o, D_o\ _{T_1,\ldots,T_n}$. Proceed as above.

$\leftarrow$ Immediate: if every subset of Γ is satisfiable, then Γ is.

$\square$

5 Conclusions

The system HTLC is an extension of typed λ-calculus with hyperintensions. The system is presented with a polymorphic rules set which can be applied to terms of arbitrary types: a triplet of rules, namely Assumption, Abstraction and Application known from λ-calculus are extended by Trivialization and Execution rules for terms denoting hyperintensions, and reason with them. We have provided formal definitions of term occurrence (which corrresponds to a proof inspection for type-checking) and we formulated appropriate versions of the Diamond Lemma and the Church-Rosser Theorem valid with respect to the extension to terms denoting hyperintensions. Finally, the system is shown to be complete in Henkin's sense, with respect to a general model of basic types, functions whose values belong to the set of truth values, and constructions of these types. The important difference to be drawn with standard completeness for Henkin's model concerns the cardinality of the model for hyperintensions, which cannot be denumerable. Nonetheless, our systems guarantees strong reducibility to the denumerable model of the trivialised term for each hyperintensional one.

Further possible investigations of this system concern: a computational interpretation of the extensional fragment, and the appropriate interpretation of both intensional (by higher order computations) and hyperintensional fragments (e.g. in terms of monads); a modal extension of the language, to express more precisely contexts in which lifting to hyperintensional terms is valid, e.g. on the lines formulated in [20]; and an implementation for type-checking purposes.

References

[1] F. Berto. Simple hyperintensional belief revision. *Erkenntnis*, 84:559–575, 2019.

[2] L. Burke. P-HYPE: A monadic situation semantics for hyperintensional side effects. *Proceedings of Sinn und Bedeutung*, 23(1):201–218, 2019.

[3] S. Charlow. *On the semantics of exceptional scope*. PhD thesis, New York University, 2014.

[4] M. Duží. Hyperintensions as abstract procedures. Presented at Congress on Logic Methodology and Philosophy of Science and Technology 2019.

[5] M. Duží, M. Fait, and M. Menšík. Context recognition for a hyperintensional inference machine. In *AIP Conference Proceedings of* ICNAAM 2016, *International Conference of Numerical Analysis and Applied Mathematics*, volume 1863, 2017.

[6] M. Duží, B. Jespersen, and P. Materna. *Procedural Semantics for Hyperintensional Logic - Foundations and Applications of Transparent Intensional Logic*, volume 17 of *Logic, Epistemology, and the Unity of Science*. Springer, 2010.

[7] M. Duží and M. Menšík. Inferring knowledge from textual data by natural deduction. *Computación y Sistemas*, 24(1), 2020.

[8] Ch. Fox and S. Lappin. Type-theoretic logic with an operational account of intensionality. *Synthese*, 192(3):563–584, 2015.

[9] N. Francez. The granularity of meaning in proof-theoretic semantics. In Nicholas Asher and Sergei Soloviev, editors, *Logical Aspects of Computational Linguistics - 8th International Conference, LACL 2014, Toulouse, France, June 18-20, 2014. Proceedings*, volume 8535 of *Lecture Notes in Computer Science*, pages 96–106. Springer, 2014.

[10] M. Jago. Hyperintensional propositions. *Synthese*, 192(3):585–601, 2015.

[11] B. Jespersen and M. Duží. Introduction. *Synthese*, 192(3):525–534, Mar 2015.

[12] B. Jespersen and G. Primiero. Alleged assassins: Realist and constructivist semantics for modal modification. In Guram Bezhanishvili, Sebastian Löbner, Vincenzo Marra, and Frank Richter, editors, *Logic, Language, and Computation - 9th International Tbilisi Symposium on Logic, Language, and Computation, TbiLLC 2011, Kutaisi, Georgia, September 26-30, 2011, Revised Selected Papers*, volume 7758 of *Lecture Notes in Computer Science*, pages 94–114. Springer, 2011.

[13] H. Leitgeb. HYPE: A system of hyperintensional logic (with an application to semantic paradoxes). *J. Philosophical Logic*, 48(2):305–405, 2019.

[14] P. Martin-Löf. Constructive mathematics and computer programming. In *Proc. of a Discussion Meeting of the Royal Society of London on Mathematical Logic and Programming Languages*, page 167–184, USA, 1985. Prentice-Hall, Inc.

[15] A. Özgün and F. Berto. Dynamic hyperintensional belief revision. *Review of Symbolic Logic*, pages 1–46, https://doi.org/10.1017/S1755020319000686.

[16] I. Pezlar. *Investigations into Transparent Intensional Logic: A Rule-based Approach*. PhD thesis, Masaryk University, 2016.

[17] I. Pezlar. Proof-theoretic semantics and hyperintensionality. *Logique et Analyse*, 61(242):151–161, 2018.

[18] C. Pollard. Hyperintensions. *Journal of Logic and Computation*, 18(2):257–282, 2008.

[19] C. Pollard. Agnostic hyperintensional semantics. *Synthese*, 192(3):535–562, 2015.

[20] G. Primiero. A contextual type theory with judgemental modalities for reasoning from open assumptions. *Logique & Analyse*, 55(220):579–600, 2012.

[21] L. Tranchini. Proof-theoretic harmony: towards an intensional account. *Synthese*, 2016, https://doi.org/10.1007/s11229-016-1200-3.

[22] L. Tranchini and A. Naibo. Harmony, stability, and the intensional account of proof-theoretic semantics. Presented at Congress on Logic Methodology and Philosophy of Science and Technology 2019, HaPoC Symposium on Identity in computational formal and applied systems.

Received 6 February 2020

Bilateralism based on Corrective Denial

Nissim Francez

Computer Science Dept., the Technion-IIT, Haifa, Israel

`francez@cs.technion.ac.il`

Abstract

The standard notion of denial in the bilateralism literature is based on *exclusion*, in some sense, of the denied φ.

I present a new variant of bilateralism based on a different, stronger notion of denial, not being *excluding only*, but also *corrective*. A corrective denial, while also excluding, points to an *atomic incompatible alternative* to the denied φ, the latter serving as the *ground for the denial*.

An *atomic incompatibility class* is a finite set of *atomic sentences* with at least two elements, with the following intended interpretation: *exactly one of its members can be asserted, provided all others are denied.*

the paper presents a bilateral natural-deduction proof-system for corrective denial, with connective-independent introduction and elimination rules. Rumfitt's connective-dependent rules are derivable in my system.

1 Introduction

Bilateralism is an approach to meaning taking *denial* (or *rejection*) as a primitive attitude, on par with *assertion* (or *acceptance*) (see, for example, [1], which also contains references to earlier work). It puts forward a claim, that there are good reasons, pace Frege [2, pp. 384–5], and Geach [3], *not* to regard the denial of φ to be adequately represented by an assertion of its (sentential) negation $\neg\varphi$. That is, bilateralism rejects the thesis called in [4] *the denial equivalence* thesis. For arguments to this end, see [5]. Rather, negation is explicated *in terms of denial*: explicated by means of — but *not reduced to*. Negation is still needed as an operator when some content is *embedded* in some other contents; denial, as a speech act, cannot be embedded or iterated. See [4] for a discussion of this issue.

The standard notion of denial in the bilateralism literature is based on *exclusion*, in some sense, of the denied φ. For example, Price [6] identifies denial as originating from *disagreement* (in dialog), where denying φ by a participant in the dialog excludes φ from the beliefs, the latter underlying action plans.

Exactly what φ is excluded from does not matter much for my concern in this paper, so I leave it as not further specified. What does matter is that an explication of negation *in terms of* a denial that only excludes (without offering an atomic incompatible alternative) leads to a (sentential) negation operator that is also viewed as an operator that only excludes.

Consider, for example, Rumfitt's bilateral natural deduction rules in [1], defining[1] classical negation in terms of assertion and denial. The rules use *polarity marked* formulas, where φ^+ means asserting φ and φ^- means denying φ.

$$\frac{\varphi^-}{(\neg\varphi)^+} \, (\neg^+ I) \qquad \frac{\varphi^+}{(\neg\varphi)^-} \, (\neg^- I) \tag{1.1}$$

$$\frac{(\neg\varphi)^+}{\varphi^-} \, (\neg^+ E) \qquad \frac{(\neg\varphi)^-}{\varphi^+} \, (\neg^- E) \tag{1.2}$$

The negation operator emerging from this definition is an operator merely *toggling* between assertion and denial of φ.

My aim in this paper is to present a variant of bilateralism based on a different notion of denial, naturally leading to a different kind of negation, the latter not being *excluding only*, but also *corrective*. Thus, denial needs a more elaborate *ground for denial*. For a detailed discussion of the notion of grounds for denial (as well as grounds for assertion) see [8, Section 4.4]. A family of *contra-classical* logics exhibiting a corrective negation is presented in [9].

2 Corrective denial

The exclusion-only denial can be seen as being reflected by the following simply structured schematic dialog $\mathcal{D}_e$ between two participants 'A' and 'B'. The subscript 'e' on $\mathcal{D}$ stands for 'exclusive (only) denial'.

$$\mathcal{D}_e :: \begin{array}{l} A : \varphi \\ B : \text{No!} \end{array} \tag{2.3}$$

by using No (cf. [1]), participant B expresses his denial of φ, excluding it and thereby disagreeing with A who asserted it.

In order to motivate my intended kind of denial, consider the following schematic dialog between two participants 'A' and 'B' of a somewhat more elaborate form. The subscript 'c' on $\mathcal{D}$ stands for 'corrective denial'

$$\mathcal{D}_c :: \begin{array}{l} A : \varphi \\ B : \text{No!} \, \psi \end{array} \tag{2.4}$$

---

[1]Rumfitt's aim is to provide, by means of those rules, *harmonious* rules [7] for classical logic.

The exact form of ψ and the relationship between φ and ψ within a corrective denial dialog will become clear in the sequel. Here, it suffices to consider the following simple instance of $\mathcal{D}_c$.

$$\mathcal{D}_{\mathsf{apple}} :: \quad \begin{array}{l} A : \text{This apple is green (all over)} \\ B : \text{No! This apple is yellow (all over)} \end{array} \tag{2.5}$$

Here participant B does not use No just for excluding this apple is green, but while excluding it, B also points to the *incompatible* this apple is yellow as the ground for the denial of A's assertion.

The incompatibility of those two *interpreted* sentences results from their meanings, as determined by an underlying semantics for the interpreted (natural) language. I return to incompatibility in the logical, uninterpreted language, abstracting from an interpretive semantics, below (Section 3.1).

An *atomic incompatibility class*, IC, is a finite set of *atomic sentences* with at least two elements, exactly one member of which is[2] true.

To direct the thought, the reader may think of a *generic* atomic incompatibility class as a representation of *colours*, where atomic sentences assign a colour to some *specific* coloured object, say o. Clearly, o has exactly one colour (all over). As another example of an IC, consider atomic sentences assigning to o an *evaluative size*, like o is short, o is long, etc. Disregarding issues of vagueness, an object o has exactly one evaluative size.

Clearly, the set of all atomic sentences is partitioned by the underlying semantics into ICs.

Let p, q range over atomic sentences[3]. Let ic map atomic sentences to their atomic incompatibility classes: for every IC and $p \in IC$, let $ic(p) = IC/\{p\}$, the (non-empty) set of all atomic sentences incompatible with p. The mapping ic satisfies:

-

$$(nref) \quad p \notin ic(p) \tag{2.6}$$

-

$$(sym) \quad p \in ic(q) \text{ iff } q \in ic(p) \tag{2.7}$$

---

[2] At this stage, I am relating to truth simpliciter, as if it is an absolute property of sentences. In the logic resulting from the corrective negation (in Section 3), I will relate to the more usual notion of truth in a model.

[3] By atomic sentences I mean here what is called elsewhere propositional variables. Propositional constants such as '$\perp$' and '$\top$' are not included as atomic sentences.

-

$$(part) \text{ If } p \in ic(q) \text{ then } ic(p) = (ic(q)/\{p\}) \cup \{q\} \qquad (2.8)$$

A conjunction of atoms $\alpha = \wedge_{1 \leq j \leq m} q_j$ is *ic-proper*, denoted by $\pi(\alpha)$, iff no q_{j_1} and q_{j_2} (for $j_1 \neq j_2$) are in the same IC. A conjunction of atoms α which is not proper is *improper*. The role of being proper will become clear in the sequel.

Next, the mapping ic is extended to $\hat{ic}$, mapping also compound formulas to their atomic incompatibility sets, again sets of conjunctions of atoms. This extension is the heart of corrective bilateralism, inducing grounds for denial for arbitrary formulas. Note the relata of the incompatibility relation: it does not relate compound sentences to each other.

Compound formulas are assumed here generated by the standard connectives: '$\neg$' (negation), '$\wedge$' (conjunction), '$\vee$' (disjunction) and '$\rightarrow$' (conditional). I use φ, ψ as meta-variables ranging over formulas, α, β over conjunctions[4] of atomic formulas. The range of $\hat{ic}$ consists of conjunctions of atomic formulas.

$$\hat{ic}(p) = \{q \wedge \alpha \mid q \in ic(p), \ \pi(\alpha), \ p \notin \alpha\} \qquad (2.9)$$

$$\hat{ic}(\neg\varphi) = \cap_{\alpha \in \hat{ic}(\varphi)} \hat{ic}(\alpha) \qquad (2.10)$$

$$\hat{ic}(\varphi \wedge \psi) = \hat{ic}(\varphi) \cup \hat{ic}(\psi) \qquad (2.11)$$

$$\hat{ic}(\varphi \vee \psi) = \hat{ic}(\varphi) \cap \hat{ic}(\psi) \qquad (2.12)$$

$$\hat{ic}(\varphi \rightarrow \psi) = \hat{ic}(\neg\varphi) \cap \hat{ic}(\psi) \qquad (2.13)$$

Note that for every φ, $\hat{ic}(\varphi)$ contains only *ic-proper* atomic conjunctions.

There is a certain redundancy in $\hat{ic}$, as a result of which $\hat{ic}(\varphi)$ can, in general, contain infinitely many conjunctions of atoms. However, since φ may have only finitely many atomic formulas as sub-formulas, only finitely many of the conjunctions in $\hat{ic}(\varphi)$ are relevant to its deductive role. So, I restrict $\hat{ic}(\varphi)$ to its finitely many relevant conjunctions. For every φ, denote by $\mathbf{a}(\varphi)$ the set of all atoms occurring in φ, and by $\alpha/\mathbf{a}(\varphi)$ the restriction of α to such atoms. Let

$$\hat{ic}_r(\varphi) =^{df.} \{\alpha/\mathbf{a}(\varphi) \mid \alpha \in \hat{ic}(\varphi)\} \qquad (2.14)$$

Example 2.1. *For simplicity, assume there are only three colours, and for better readability suppose we have mnemonically-named atomic formulas. Consider*

$$IC_1 = \{red, green, blue\}, IC_2, = \{short, long\}$$

---

[4]Where convenient, a conjunction of atomic formulas is also considered as the *finite* set of those atoms.

Thus

$$ic(red) = \{green, blue\}, \quad ic(green) = \{red, blue\}, \quad ic(blue) = \{red, green\}$$
$$ic(short) = \{long\}, ic(long) = \{short\}$$

To avoid notational clutter, the conjunctions generated by (2.9) are displayed modulo their commutativity.

$$\hat{ic}(red) = \begin{array}{l} \{blue \wedge short, green \wedge short, \\ blue \wedge long, green \wedge long\} \end{array}$$

$$\hat{ic}(green) = \begin{array}{l} \{blue \wedge short, red \wedge short, \\ blue \wedge long, red \wedge long\} \end{array}$$

$$\hat{ic}(blue) = \begin{array}{l} \{red \wedge short, green \wedge short, \\ red \wedge long, green \wedge long\} \end{array}$$

$$\hat{ic}(long) = \{blue \wedge short, green \wedge short, red \wedge short\}$$

$$\hat{ic}(short) = \{blue \wedge long, green \wedge long, red \wedge long\}$$

$$\hat{ic}(red \wedge long) = \hat{ic}(red) \cup \hat{ic}(long) = \begin{array}{l} \{blue \wedge short, green \wedge short, \\ blue \wedge long, green \wedge long, \\ red \wedge short\} \end{array}$$

$$\hat{ic}(red \vee long) = \hat{ic}(red) \cap \hat{ic}(long) = \{blue \wedge short, green \wedge short\}$$

$$\hat{ic}(\neg red) = \{red \wedge short, red \wedge long\}$$

$$\hat{ic}(\neg(red \wedge long)) = \{red \wedge long\}$$

$$\hat{ic}(red \rightarrow long) = \hat{ic}(\neg red) \cap \hat{ic}(long) = \{red \wedge short\}$$

$$\hat{ic}(\neg(red \rightarrow long)) = \{red \wedge long\}$$

$$\hat{ic}(red \vee \neg red) = \emptyset$$

$$\hat{ic}(red \wedge \neg red) = \begin{array}{l} \{blue \wedge short, green \wedge short, \\ blue \wedge long, green \wedge long, \\ red \wedge short, red \wedge long\} \end{array}$$

I now can state the relationship between assertion and denial in the modified conception. I refer to those revised notions as *c-assertion* and *c-denial*.

Corrective assertion and denial:

$$\begin{array}{l} \varphi \text{ is } c - \text{assertible iff every } \alpha \in \hat{ic}_r(\varphi) \text{ is } c - \text{deniable} \\ \varphi \text{ is } c - \text{deniable iff some } \alpha \in \hat{ic}_r(\varphi) \text{ is } c - \text{assertible} \end{array} \qquad (2.15)$$

That is:

- φ can be *correctively denied* (and thereby excluded) only if some *atomic incompatible alternative* to φ is asserted.

- φ can be *correctively asserted* only if every *atomic incompatible alternative* to φ is denied.

Note that:

- A proper α, e.g. in Example 2.1, red∧short, is:

  - c-assertible just in case $\{blue, green, long\}$ are all denied.
  - c-deniable just in case long and one of blue, green is asserted.

- an improper α, e.g. in Example 2.1, red∧blue, cannot be c-asserted, it can be c-denied (in case green is c-asserted).

digression: The term 'incompatible' has been used extensively in the literature, alas in different senses. For example, Restall [10] uses it as a relation between states (points) in an *incompatibility frame*. Brandom [11] uses incompatibility as relation between sets of sentences and subsets of the power-set of sentences; and there are more.

In my use of this term the relata of incompatibility change: these are not associated with points, but are formulas and finite sets of simple formulas (conjunctions of atomic formulas).

(end of digression)

3 A family of corrective bilateral natural deduction systems

In the sequel, I assume a propositional object language freely-generated by a set $P = \{p_i \mid i \geq 0\}$ of *atomic formulas*, ranged over by meta-variables p, q and the standard connectives: '¬' (negation), '∧' (conjunction), '∨' (disjunction and '→' (conditional). I use φ, ψ as meta-variables ranging over formulas, α, β over conjunctions[5] of atomic formulas.

---

[5]Where convenient, a conjunction of atomic formulas is also considered as the *finite* set of those atoms.

3.1 Atomic incompatibility bases

The first issue to be resolved is *how to define incompatibility among atomic formulas*, which are *uninterpreted* in the formal object language. Traditionally, atomic formulas are considered *mutually independent*. In a model-theory based on valuations (assignments of truth-values to formulas), a valuation assigns a truth-value to an atomic formula arbitrarily[6], independently of the truth-value assigned to any other atomic sentences.

Here, I deviate from this tradition, and assume an *imposed* incompatibility relation among atomic formulas, abstracting from some "hidden" interpretation of them. Note that I am referring continuously to atomic formulas, not to propositional variables, as the generators of compound formulas. Indeed, this view does not support uniform substitutions, that would not preserve the imposed incompatibility relationships among atomic formulas.

The definition below of the *atomic incompatibility base* $\mathbf{i}$ is the formal counterpart of the mapping *ic* presented above. I deliberately repeat it, but this time in terms of an uninterpreted language, to emphasise the importance of this definition and the notion of imposed incompatibility on uninterpreted atomic sentences.

Definition 3.1 (Atomic incompatibility base). *An* atomic incompatibility base *is a mapping* $\mathbf{i} : P \Rightarrow \mathcal{P}_f(P)/\emptyset$ *(i.e., the range of* $\mathbf{i}$ *consists of finite, non-empty subsets of atomic formulas), s.t.:*

- $$(nref) \quad p \notin \mathbf{i}(p) \tag{3.16}$$

- $$(sym) \quad p \in \mathbf{i}(q) \text{ iff } q \in \mathbf{i}(p) \tag{3.17}$$

- $$(part) \text{ If } p \in \mathbf{i}(q) \text{ then } \mathbf{i}(p) = (\mathbf{i}(q)/\{p\}) \cup \{q\} \tag{3.18}$$

Remarks:

1. By definition, for every $p \in P$, $\mathbf{i}(p) \neq \emptyset$. Each atomic formula has at least one (other) atomic formula incompatible with it.

2. Incompatibility is irreflexive, symmetric and transitive.

---

[6]Indeed, in the model-theory for the contra-classical logics in [9], my approach is embodied as a restriction on valuations. Truth-values are assigned by valuations in such a way that in every incompatibility class there is exactly one true member.

3. The mapping **i** *partitions* P into *incompatibility classes* s.t. for every incompatibility class IC: if p, $q \in IC$ then $p \in \mathbf{i}(q)$ (and, hence, $q \in \mathbf{i}(p)$). Clearly, incompatibility classes are not equivalence classes, as incompatibility is not reflexive.

Next, the mapping **i** is extended to $\hat{\mathbf{i}}$, mapping also compound formulas, the formal counterpart of $\hat{ic}$ above. The range of $\hat{\mathbf{i}}$ consists of proper conjunctions of atomic formulas.

Definition 3.2 (Proper conjunctions of atoms). *A conjunction of atoms* $\alpha = \bigwedge_{1 \leq j \leq m} q_j$ *is* **i**-*proper, denoted by* $\pi_{\mathbf{i}}(\alpha)$, *iff no* q_{j_1} *and* q_{j_2} *(for* $j_1 \neq j_2$*) are in the same IC of* **i**.

A conjunction of atoms α which is not **i**-proper is **i**-*improper*. Note that properness is relative to an atomic incompatibility base.

Definition 3.3.

$$\hat{\mathbf{i}}(p) = \{q \wedge \alpha \mid q \in \mathbf{i}(p),\ \pi_{\mathbf{i}}(\alpha),\ p \notin \alpha\} \tag{3.19}$$

$$\hat{\mathbf{i}}(\neg \varphi) = \cap_{\alpha \in \hat{\mathbf{i}}(\varphi)} \hat{\mathbf{i}}(\alpha) \tag{3.20}$$

$$\hat{\mathbf{i}}(\varphi \wedge \psi) = \hat{\mathbf{i}}(\varphi) \cup \hat{\mathbf{i}}(\psi) \tag{3.21}$$

$$\hat{\mathbf{i}}(\varphi \vee \psi) = \hat{\mathbf{i}}(\varphi) \cap \hat{\mathbf{i}}(\psi) \tag{3.22}$$

$$\hat{\mathbf{i}}(\varphi \rightarrow \psi) = \hat{\mathbf{i}}(\neg \varphi) \cap \hat{\mathbf{i}}(\psi) \tag{3.23}$$

Note that for every φ, $\hat{\mathbf{i}}(\varphi)$ contains only **i**-proper atomic conjunctions.
From Definition (3.3) we have:

Corollary 1 (non-reflexivity). *For every* α: $\alpha \notin \hat{\mathbf{i}}(\alpha)$.

Proof. Assume, towards a contradiction, that for some $\alpha = q_1 \wedge \cdots \wedge q_l, l \geq 2$

$$\alpha \in \hat{\mathbf{i}}(\alpha) =^{(3.21)} \cup_{1 \leq i \leq l} \hat{\mathbf{i}}(q_i)$$

W.l.o.g, assume $\alpha \in \hat{\mathbf{i}}(q_1) =^{(3.19)} \{p \wedge \beta \mid p \in \hat{\mathbf{i}}(q_1), q_1 \notin \beta\}$.
But this is impossible, as q_1, a conjunct in α, cannot be p because $q_1 \notin \mathbf{i}(q_1)$ and cannot be in β. $\qquad \square$

Corollary 2 (Negation). *For every* φ

$$\hat{\mathbf{i}}(\varphi) \cap \hat{\mathbf{i}}(\neg \varphi) = \emptyset \tag{3.24}$$

Proof. Assume, towards a contradiction, that for some α

$$\alpha \in \hat{\mathbf{i}}(\varphi) \cap \hat{\mathbf{i}}(\neg\varphi)$$

But $\hat{\mathbf{i}}(\neg\varphi) =^{(3.20)} \cap_{\beta \in \hat{\mathbf{i}}(\varphi)} \hat{\mathbf{i}}(\beta)$, and α is one of those βs. Hence, $\alpha \in \hat{\mathbf{i}}(\alpha)$, contradicting the non-reflexivity Corollary 1. $\qquad\square$

Proposition 3.1 (Double negation incompatibility).

$$\hat{\mathbf{i}}(\neg\neg\varphi) = \hat{\mathbf{i}}(\varphi) \tag{3.25}$$

Instead of presenting the tedious computation for the general case, consider the special case of the double negation of a single atom, as in Example 2.1. Consider $\mathbf{i}(\neg\neg red)$. First, recall that $\mathbf{i}(\neg red) = \{red\}$.

$$\mathbf{i}(\neg\neg red) = \cap_{\alpha \in \mathbf{i}(\neg red)} \mathbf{i}(\alpha) = \cap_{\alpha \in \{red\}} \mathbf{i}(\alpha) = \mathbf{i}(red)$$

3.2 Restricting $\hat{\mathbf{i}}$ to finite sets

There is a difficulty of using $\hat{\mathbf{i}}$ in the proof-system, namely the fact that $\hat{\mathbf{i}}(\varphi)$ can, in general, contain infinitely many conjunctions of atoms. However, since φ may have only finitely many atomic formulas as sub-formulas, only finitely many of the conjunctions in $\hat{\mathbf{i}}(\varphi)$ are relevant to its deductive role. So, the first step is to restrict $\hat{\mathbf{i}}(\varphi)$ to its finitely many relevant conjunctions, again as a formal counterpart of the restriction of $\hat{\iota c}$.

Definition 3.4 (Restricting $\hat{\mathbf{i}}$). *Let* $\hat{\mathbf{i}}_r(\varphi) =^{df.} \{\alpha/\mathbf{a}(\varphi) \mid \alpha \in \hat{\mathbf{i}}(\varphi)\}$.

Clearly, $\hat{\mathbf{i}}_r(\varphi)$ is finite for every φ.

3.3 The c-bilateral I/E-rules

In this section, c-bilateral *I/E*-rules are introduced, defining a family $\mathcal{BN}_{\mathbf{i}}$ of ND-systems, one for each atomic compatibility base $\mathbf{i}$. Note that the rules are *independent* of the specific connectives used. I come back to connective-dependent rules below.

Suppose $\hat{\mathbf{i}}_r(\varphi) = \{\alpha_1, \cdots, \alpha_m\}$.

$$\frac{\alpha_j^+}{\varphi^-} \ (I_{j,\mathbf{i}}^-), \ 1 \leq j \leq m \qquad\qquad \frac{\alpha_1^- \quad \cdots \quad \alpha_m^-}{\varphi^+} \ (I_{\mathbf{i}}^+) \tag{3.26}$$

$$\frac{\varphi^- \quad \overset{\displaystyle [\alpha_1^+]_{j_1}}{\underset{\chi}{\vdots}} \quad \cdots \quad \overset{\displaystyle [\alpha_m^+]_{j_m}}{\underset{\chi}{\vdots}}}{\chi} \ (E_{\mathbf{i}}^{-,j_1,\cdots,j_m}) \qquad \frac{\varphi^+}{\alpha_j^-} \ (E_{j,\mathbf{i}}^+) \ , \ 1 \le j \le m \qquad (3.27)$$

In contrast to Rumfitt's denial rules for compound formulas, where every negated connective has its own denial rules, here all negated formulas are c-denied uniformly, by the same rule. Thus, Rumfitt has denial rules (derived below) such as $(\wedge^- I)$, $(\wedge^- E)$ (denial rules for conjunction), $(\vee^- I)$, $(\vee^- E)$ (denial rules for disjunction), etc. In my rules, the dependence on specific connectives is "hidden" in the definition of $\hat{\mathbf{i}}$ for compound formulas.

For negation, I derive the following c-bilateral I/E-rules.

$$\frac{\alpha_j^+}{(\neg\varphi)^+} \ (\neg^+ I_{j,\mathbf{i}}) \ , \ 1 \le j \le m \qquad \frac{(\neg\varphi)^+ \quad \overset{\displaystyle [\alpha_1^+]_{j_1}}{\underset{\chi}{\vdots}} \quad \cdots \quad \overset{\displaystyle [\alpha_m^+]_{j_m}}{\underset{\chi}{\vdots}}}{\chi} \ (\neg^+ E_{\mathbf{i}}^{j_1,\cdots,j_m}) \quad (3.28)$$

$$\frac{\alpha_1^- \quad \cdots \quad \alpha_m^-}{(\neg\varphi)^-} \ (\neg^- I_{\mathbf{i}}) \qquad \frac{(\neg\varphi)^-}{\alpha_j^-} \ (\neg^- E_{j,\mathbf{i}}) \ , 1 \le j \le m \qquad (3.29)$$

For the derivations of these rules, let, for $1 \le j \le m$

$$\hat{\mathbf{i}}(\alpha_j) = \{\gamma_{j_1}, \cdots, \gamma_{j_{m_j}}\} \qquad (3.30)$$

and

$$\hat{\mathbf{i}}(\neg\varphi) =^{(3.20)} \cap_{\alpha \in \hat{\mathbf{i}}(\varphi)} \hat{\mathbf{i}}(\alpha) = \{\beta_1, \cdots, \beta_p\} \qquad (3.31)$$

$(\neg^+ I_{1,\mathbf{i}})$:

$$\frac{\dfrac{\dfrac{\alpha_1^+}{\varphi^-}(I_{1,\mathbf{i}}^-) \quad \dfrac{[\alpha_1^+]_1}{\beta_1^-}(E_{1,\mathbf{i}}^+) \quad \cdots \quad \dfrac{[\alpha_m^+]_m}{\beta_1^-}(E_{m,\mathbf{i}}^+)}{\beta_1^-}(E_{\mathbf{i}}^{-,1,\cdots,m}) \quad \cdots \quad \dfrac{\dfrac{\alpha_1^+}{\varphi^-}(I_{1,\mathbf{i}}^{-,1,\cdots,m}) \quad \dfrac{[\alpha_1^+]_1}{\beta_p^-}(E_{1,\mathbf{i}}^+) \quad \cdots \quad \dfrac{[\alpha_m^+]_m}{\beta_p^-}(E_{m,\mathbf{i}}^+)}{\beta_p^-}(E_{\mathbf{i}}^-)}{(\neg\varphi)^+}(I_{\mathbf{i}}^+)$$

Note that the applications of $E_{\mathbf{i}}^+$ are correct, since for $1 \le k \le p$ $\beta_k \in \hat{\mathbf{i}}(\alpha_j)$, $1 \le j \le m$. The derivations for $(\neg^+ I_{j,\mathbf{i}})$ with $j > 1$ are similar and omitted.

$(\neg^+ E_{1,\mathbf{i}})$:

$$\cfrac{\cfrac{(\neg\varphi)^+}{\gamma_{j_1}^-}\ (E_{1,\mathbf{i}}^+) \quad \cdots \quad \cfrac{(\neg\varphi)^+}{\gamma_{j_{m_1}}^-}\ (E_{m_1,\mathbf{i}}^+)}{\cfrac{\alpha_1^+}{\varphi^-}\ (I_{1,\mathbf{i}}^-)}\ (I_{\mathbf{i}}^+) \qquad [\alpha_1^+]_{j_1} \qquad [\alpha_m^+]_{j_m}}{} $$

$$\cfrac{\cfrac{\cfrac{(\neg\varphi)^+}{\gamma_{j_1}^-}\ (E_{1,\mathbf{i}}^+)\ \cdots\ \cfrac{(\neg\varphi)^+}{\gamma_{j_{m_1}}^-}\ (E_{m_1,\mathbf{i}}^+)}{\cfrac{\alpha_1^+}{\varphi^-}\ (I_{1,\mathbf{i}}^-)}\ (I_{\mathbf{i}}^+) \qquad \overset{\vdots}{\chi} \quad \cdots \quad \overset{\vdots}{\chi}}{\chi}\ (E_{\mathbf{i}}^{-,1,\cdots,m})$$

The derivations for $(\neg^+ E_{j,\mathbf{i}})$ with $j > 1$ are similar and omitted.

3.4 Deriving Rumfitt's bilateral rules for the binary connectives

One can easily realise that Rumfitt's rules in [1], except the I/E-rules for negation, can be carried over to each $\mathcal{BN}_\mathbf{i}$ as derived rules. Below are the derivations for conjunction. The derivation of Rumfitt's disjunction rules is similar and omitted.

Conjunction: Rumfitt's rules are:

$$\cfrac{\varphi_1^+ \quad \varphi_2^+}{(\varphi_1 \wedge \varphi_2)^+}\ (\wedge^+ I) \qquad\qquad \cfrac{(\varphi_1 \wedge \varphi_2)^+}{\varphi_i^+}\ (\wedge^+ E_i), \quad i = 1, 2 \qquad\qquad (3.32)$$

$$\cfrac{\varphi_i^-}{(\varphi_1 \wedge \varphi_2)^-}\ (\wedge^- I_i), \quad i = 1, 2 \qquad\qquad \cfrac{(\varphi_1 \wedge \varphi_2)^- \quad \overset{\vdots}{\chi} \quad \overset{\vdots}{\chi}}{\chi}\ (\wedge^- E^{j,k}) \qquad (3.33)$$

with $[\varphi_1^-]_j \quad [\varphi_2^-]_k$ above.

By (3.21), suppose $\hat{\mathbf{i}}(\varphi_1) = \{\alpha_1, \cdots, \alpha_l\}$ and $\hat{\mathbf{i}}(\varphi_2) = \{\alpha_{l+1}, \cdots, \alpha_m\}$ for some $1 \leq l < m$.

$\wedge^- I_1$:

$$\cfrac{\varphi_1^- \quad \cfrac{[\alpha_1^+]_1}{(\varphi_1 \wedge \varphi_2)^-}\ (I_{\mathbf{i}}^-) \quad \cdots \quad \cfrac{[\alpha_l^+]_l}{(\varphi_1 \wedge \varphi_2)^-}\ (I_{\mathbf{i}}^-)}{(\varphi_1 \wedge \varphi_2)^-}\ (E_{\mathbf{i}}^{-,1,\cdots,l})$$

The derivation for $i = 2$ is similar and omitted.

$\wedge^+ I$:

$$\cfrac{\cfrac{\varphi_1^+}{\alpha_1^-}\ (E_{1,\mathbf{i}}^+) \quad \cdots \quad \cfrac{\varphi_1^+}{\alpha_l^-}\ (E_{l,\mathbf{i}}^+) \quad \cfrac{\varphi_2^+}{\alpha_{l+1}^-}\ (E_{l+1,\mathbf{i}}^+) \quad \cdots \quad \cfrac{\varphi_2^+}{\alpha_m^-}\ (E_{m,\mathbf{i}}^+)}{(\varphi_1 \wedge \varphi_2)^+}\ (I_{\mathbf{i}}^+)$$

$\wedge^- E$:

$$\cfrac{\dfrac{[\alpha_1^+]_{j_1}}{\varphi_1^-}\ (I_{1,\mathbf{i}}^-)}{(\varphi_1\wedge\varphi_2)^-\quad \vdots \atop \chi} \qquad \dfrac{\dfrac{[\alpha_l^+]_{j_l}}{\varphi_1^-}\ (I_{l,\mathbf{i}}^-)}{\vdots \atop \chi} \cdots \dfrac{\dfrac{[\alpha_{l+1}^+]_{j_{l+1}}}{\varphi_2^-}\ (I_{l+1,\mathbf{i}}^-)}{\vdots \atop \chi} \qquad \dfrac{\dfrac{[\alpha_{lm}^+]_{j_m}}{\varphi_2^-}\ (I_{m,\mathbf{i}}^-)}{\vdots \atop \chi}}{\chi}\ (E_{\mathbf{i}}^{-,j_1,\cdots,j_m})$$

$\wedge^+ E_1$:

$$\cfrac{\dfrac{(\varphi_1\wedge\varphi_2)^+}{\alpha_1^-}\ (E_{1,\mathbf{i}}^+) \quad \cdots \quad \dfrac{(\varphi_1\wedge\varphi_2)^+}{\alpha_l^-}\ (E_{l,\mathbf{i}}^+)}{\varphi_1^+}\ (I_{1,\mathbf{i}}^+)$$

The derivation for $i = 2$ is similar and omitted.

The derivations of the denied disjunction and denied implication are similar and omitted.

4 Conclusions

I have introduced corrective bilateralism, a new kind of bilateralism, according to which a ground for denial is an incompatible *atomic* alternative to the denied formula. A bilateral natural-deduction proof-system for corrective denial is presented, in which Rumfitt's I/E-rules are derivable.

It remains as a next step to extend corrective bilateralism to first-order; a major obstacle is the formulation of the atomic incompatible alternatives in a way not depending on the elements of an underlying domain of quantification.

Corrective bilateralism widens the scope of bilateral logics. It indicates that there is still much to be learned from the actual practice of using speech acts in natural language.

References

[1] I. Rumfitt, 'Yes' and 'No', Mind 169 (436) (2000) 781–823.

[2] G. Frege, Negation, in: B. McGuinness (Ed.), Gottlob Frege: Collected papers on mathematics, logic, and philosophy, Blackwell, Oxford, 1984.

[3] P. T. Geach, Assertion, Philosophical Review 74 (1965) 449–465.

[4] D. Ripley, Negation, denial and rejection, Philosophy Compass 69 (2011) 622–629.

[5] G. Restall, Multiple conclusions, in: P. Hàjek, L. Valdès-Villanueva, D. Westerståhl (Eds.), In 12th International Congress on Logic, Methodology and Philosophy of Science, KCL Publications, 2005, pp. 189–205.

[6] H. Price, Why 'not'?, Mind 99 (394) (1990) 221–238.

[7] M. Dummett, The Logical Basis of Metaphysics, Harvard University Press, Cambridge, MA., 1993 (paperback), hard copy 1991.

[8] N. Francez, Proof-theoretic Semantics, College Publications, London, 2015.

[9] N. Francez, Another plan for negation, Australasian Journal of logic 16 (5) (2019) 159–176.

[10] G. Restall, Defining double negation elimination, Logic Journal of the IGPL 8 (6) (2018) 853–860.

[11] R. Brandom, Between saying and doing: towards an analytic pragmatism, Oxford University Press, Oxford, UK., 2001.

    Received 26 September 2019

Two Tales of the Turnstile

Bjørn Jespersen
Utrecht University
Department of Philosophy and Religious Studies
Janskerkhof 13, 3512 BL Utrecht
The Netherlands
VŠB-Technical University of Ostrava
Department of Computer Science
17. listopadu 15,708 33 Ostrava
Czech Republic
`b.t.f.jespersen@uu.nl`

Abstract

I contrast two accounts of assertoric contexts. The Frege-Geach-style 'externalist' account keeps force (judgment) and content (proposition) separate. The act-theoretic 'internalist' inverts the Frege-Geach point by making force integral to content. Assertoric contexts being hyperintensional, act theory cannot assume that extensional logic (such as introduction and elimination rules for the truth-functions) applies to act-theoretic propositions; nor that intensional logic (e.g., distribution axioms) applies. I level an objection against internalism, namely that the internalist is wrong to argue that the assertion of a conjunction entails the assertion of both conjuncts separately. The general insight is that the Frege-Geach point remains intact, but also that the externalist owes an account of the logic of assertoric contexts.

1 Introduction

This paper consists of two parts plus an appendix. The critical Part I explains why it is problematic to weld force and content together. I exemplify the problem by

This research was supported by Grant Academy of the Czech Republic, project no. GA18-23891-S "*Hyperintensional Reasoning over Natural Language Texts*". My participation at the conference *Assertion and Proof* (*WAP 2019*), University of Salento, Lecce, 12-14 September 2019, was made possible by a travel grant from the Department of Philosophy and Religious Studies of Utrecht University. I am indebted to Nils Kürbis, Marie Duží, and an anonymous reviewer for valuable comments, and to Michal Fait for technical assistance.

means of conjunctive propositions. The problem specific to conjunctive propositions is whether the assertion of a conjunction entails the assertion of both conjuncts individually (i.e., whether assertion distributes over conjunction). I also address the problem how to avoid that the assertion of an atomic proposition entails the assertion of a disjunctive proposition. The constructive Part II demonstrates how the broadly Fregean framework of Transparent Intensional Logic (TIL) can block both of the presumed entailments in a principled manner. The discussion is couched in terms of whether force, as represented by the turnstile, should be internal or external to content, which is in this case fine-grained propositions. By implication, the question becomes whether a proposition is embedded within an assertoric context or is itself an assertoric context. Since any particular kind of context is governed by a particular logic, the question therefore becomes which logic should govern assertoric contexts. The appendix contains the central definitions.

2   Force vs content, externalism vs internalism

What is the logic of an assertoric context? In particular, how fine-grained are assertoric contexts? The answer will dictate which substitutions are valid and which sort of closure applies, or fails to apply. Consider this inference schema (A, B are propositional metavariables):

$$A \text{ is asserted}$$
$$A \approx B$$

$$B \text{ is asserted}$$

Where $\approx$ is distinct from self-identity, what constraints apply to $\approx$, so that B may be validly substituted for A within an assertoric context? For instance, if $\approx$ is entailment, then since A entails $A \vee B$ then $A \vee B$ is asserted as soon as A is. By the same token, since $A \wedge B$ entails A, B then A, B are asserted as soon as $A \wedge B$ is. While the former is obviously undesirable, the latter is either trivial or contentious. I will provide arguments against the latter claim and demonstrate how both claims can be invalidated in identical ways.

I adhere to the standard view that an assertoric context is a pragmatic context in which an agent a asserts that a proposition A is true, and that to assert A is to assert that A is true. Hence, if a has asserted A, yet A turns out to be false, then a has a problem. Typically, a will have to retract A, undoing, as it were, the original assertion of A. The general form of an assertoric context would appear to be this:

$$\text{Agent } a \text{ asserts proposition } A$$

i.e. *Assert*(*a*, *A*), which makes an assertion look very much like a psychological attitude context, i.e., a binary relation-in-intension between an agent and a proposition which the agent knows, believes, hopes, etc. to be true. However, if we follow Frege, the particular agent is irrelevant to an assertion and so can be abstracted away.[1] Therefore, instead of "*a* asserts *A*" we would have "*A* is asserted". I consider assertoric contexts to be hyperintensional *par excellence*, and I have two reasons for believing so. First, assertion should be susceptible only to very restrictive principles of logical closure, because real-world agents are very far from being logically impeccable reasoners. For instance, an agent may believe that $A \wedge (B \vee C)$ without believing that $(A \wedge B) \vee (A \wedge C)$, thus missing out on a simple distribution. Furthermore, it is a complicating factor that assertions are *acts*, not *states*, as with attitudes. For this reason, an agent may assert, for instance, that $A \wedge (B \vee C)$ and acknowledge its being equivalent to $(A \wedge B) \vee (A \wedge C)$ and yet coherently abstain from performing the further act of asserting that $(A \wedge B) \vee (A \wedge C)$. While it is perfectly appropriate to point out that such an agent is logically committed to holding that $(A \wedge B) \vee (A \wedge C)$ and so should not, for instance, argue against this disjunction, it hardly follows that the agent is committed by logic to perform an additional act.

Second, an assertion is coloured by the linguistic medium in which the assertion is made, in addition to conveying an attitude to a proposition. An assertoric act may assume the physical form of an utterance (air vibrations) or an inscription (pixels on a screen, ink on paper, oil on canvas), because without a suitable medium a speaker or writer cannot reach out to their audience in the public domain. An attitudinal state does not want to reach out to an audience and needs no intermediate medium. From attitude logic we are familiar with positions such as sententialism and inscriptionalism which demand exceptional fine-graining of attitude contents, because even slight syntactic deviations between any two sentences or other pieces of language get to matter. Therefore, I tend to think that assertoric contexts inherit some of the extreme fine-graining that sententialism and inscriptionalism emphasize and which is less likely to be present in non-syntactic theories of attitude contents which instead relate agents to abstract objects, e.g., truth-conditions or modes of presentation of truth-conditions. If this is on the right track, then assertoric contexts

---

[1]See [15]. [16] argues convincingly that $\vdash A$ does not work like, say, $Bel_a A$, so $\vdash_a A$ would be ill-formed and ill-conceived. An assertion is inherently an act performable by an agent from the agent's own perspective, which is a *first-person perspective*, whereas $Bel_a A$ is a report or attribution from a *third-person perspective* to the effect that *a* believes that *A* is true. The shift in perspective affects whether one deems judgement *factive*. [15] argues that judgement must be factive (a non-factive judgement being a *Scheinurteil*), whereas [16] argues it need not be. When non-factive, the judging agent makes a judgement to the best of their knowledge, and by being entitled to a knowledge claim the agent is right about the judgement, though wrong about the proposition being judged to be true.

will demand a finer individuation, hence a more restrictive logic, than the one which the day-to-day attitude theoreticians standardly impose on their attitude contexts.

To illustrate, first, let a assert that whatever does not kill you makes you stronger, and let b assert that whatever does not make you stronger kills you. Did a and b assert the same thing? Why, no, logical equivalence is insufficient as a principle of individuation. Likewise, we want to distinguish between asserting, say, A and asserting $A \vee (A \wedge B)$.[2] Next, let a assert that chewing marrowbone for fourteen days is healthy and b assert that masticating marrowbone for a fortnight is healthy. Same thing? I would say so, provided we are reporting a's and b's attitudes, for the only variations concern the choice of words and linguistic register and, thus, concern only how the message was conveyed but not the content of the message conveyed. However, the linguistic medium rubs off on the assertion.[3] Syntax-sensitive theories of attitude *complements* will want to say that a and b did not assert the same thing, though they might have come close. They will want to stress that perfect synonymy, as between 'chews' and 'masticates' or 'lasts a fortnight' and 'lasts fourteen days', does not exist in natural language, or if it does, that there is importantly more to an assertion than just the literal message being conveyed. I shall assume, not too controversially, that assertoric contexts enjoy a very fine-grained principle of individuation. Still, whatever particular principle is selected, it must not be too restrictive (hence sterile), for if nothing follows from anything (other than self-substitution and self-implication) then we are simply left without a logic.

I will contrast two conceptions of what an assertoric context is. Since Frege, assertoric force has been indicated by means of the turnstile, '⊢'. As per the Frege-Geach point, force and content are kept separate: a proposition may, or may not, be accompanied by this or that force.[4] I call this position *externalist*, force being external to content.[5] On the other hand, an *internalist* position conceives of propositions as being inherently forceful: those propositions whose satisfaction conditions

---

[2] Hence, where □ represents assertoric acts, the rule of inference (RE) is rendered unsuitable for attitudinal and assertoric contexts: *from $(A \leftrightarrow B)$ infer $(\Box A \leftrightarrow \Box B)$.*

[3] We can generate *mixed* cases where a word, or a string of words, is both used and mentioned – used to point to something beyond itself, and mentioned because the reporter wants to reproduce the attributee's exact choice of word(s); e.g., "Jean-Luc considers *Rochefort* a 'bière de dégustation'". I expect mixed cases to demand exceptional fine-graining, but here I won't explore just how exceptional. Also, there is the complicating factor that their logical analysis must account for the dual occurrence of words as both used and mentioned.

[4] *Vide* [5, §14], [7].

[5] *Vide* [6, p.32]: "Diese Trennung des Urteilens von dem, worüber geurteilt wird, erscheint unumgänglich, weil sonst eine bloße Annahme [...] nicht ausdrückbar wäre. Wir bedürfen also eines besonderen Zeichens, um etwas behaupten zu können [...], so daß wir z. B. mit "⊢ 2 + 3 = 5" behaupten: 2 + 3 ist gleich 5. Es wird also nicht bloß wie in "2 + 3 = 5" ein Wahrheitswert hingeschrieben, sondern zugleich auch gesagt, daß er das Wahre sei." *Cf.* footnote 19.

are truth-conditions are imbued with assertoric force.[6] To perform the act of asserting a proposition is to concretely token the abstract type that is a proposition imbued with assertoric force. Externalism and internalism are the two 'tales of the turnstile' I will unfold here. I side with the former against the latter. While internalism might work for atomic propositions, handling molecular propositions – even conjunctive ones – is problematic. The source of the trouble is that a proposition is now itself an assertoric context, and assertoric contexts are non-extensional, so the truth-functional ('truth-table') logic that standardly applies to conjunctive propositions does not apply, although the internalist assumes it at least occasionally does. Nor should we assume that the logic of possible-world semantics of intermediate granularity, which identifies logical equivalents, applies, as soon as assertoric contexts are acknowledged to obey some hyperintensional logic or other. This dual fact leaves the internalist without a principled and operative logic for assertoric contexts. What we are treated to is instead a mixture of pragmatic maxims, rules of extensional logic, and overruling of this same logic. This is the overall objection I will be levelling against internalism here.[7] Still, the internalist escapade is helpful, as it challenges us externalists to specify exactly how we perceive of assertoric contexts. The positive contribution I will make is a sketch of the hyperintensional logic that an externalist theory is in a position to apply to asserted conjunctive and disjunctive propositions.

I am aware of four ways of understanding the turnstile, '$\vdash$', and I wish to discuss two of them here. The basic taxonomy is this:

(1) $\vdash \alpha$: "There is a proof calculus in which formula α has been proved from its axioms (i.e., α is a theorem) ." This connective can also be binary: $\Gamma \vdash \alpha$ ("Formula α is provable from formulae Γ"). This is the modern-day interpretation.

(2) $\vdash A$: "Proposition A is judged to be true." This is the externalist interpretation.

---

[6] Just to be clear, when speaking of *externalism* and *internalism* above, I have in mind a distinction between two views of how assertoric contexts are structured (i.e., whether or not force is integral to content), and I am not alluding to any sort of conceivable points of contact with *semantic* externalism/internalism.

[7] The present paper is the twin of [12], which levels two objections against act-theoretic internalism. The first is the objection to distribution of assertion over conjunction. The second objection is that the internalist's inversion of the Frege-Geach point requires that force be *suspended* with respect to all the different kinds of molecular propositions, including conjunctive propositions, as soon as the truth-table for conjunction is generated along the lines of, say, Sheffer's stroke (NAND), Quine's dagger (NOR), or De Morgan's $\neg(\neg A \vee \neg B)$. But then what is the dividend of the internalist's inversion of the Frege-Geach point? None, it seems. Forceful propositions with their force suspended behave pretty much like Frege and Geach predict them to behave, anyway. The internalist is back to square one where the externalist has been all along.

(3) $\vdash B$. Proposition $\vdash B$ occurs with assertoric force. B is a sub-propositional fragment of $\vdash B$ in which a property or relation is predicated to apply to one or more objects, or a connective is predicated to apply to one or more propositions. This is the internalist interpretation.

Cases (2) and (3) both treat $\vdash$ as a force indicator. However, (2) comes in two flavours. (2a): while keeping proposition/*Gedanke* and assertion/*Behauptung* separate, Frege still accords logical significance to the turnstile.[8] (2b): while observing a comparable distinction, Tichý, as a modern-day realist, relegates the turnstile entirely to pragmatics.[9] Here, I will be contrasting (2b) with (3). There are two kinds of act theory: one heeding the Frege-Geach point and thus keeping force and content neatly apart; the other inverting the Frege-Geach point such that content is intermingled with force. I will be discussing only the 'inverted', hence more radical, form of act theory. [7, pp.456-457] outlines the inversion, but does not spell out the consequences of going along with it.[10] [8], [9], [10] do just that, without Hanks presenting his theory explicitly as an inversion of the Frege-Geach point, though advertising it as a theory that collapses the wall of separation between force and content.

If assertoric contexts are fine-grained, thus requiring a hyperintensional logic, then the question arises what sort of logic is a or the correct propositional logic to go with act-theoretic propositions. In particular, to what extent can we rely on the truth-functional (i.e., extensional) logic for the connectives $\land, \lor, \neg, \rightarrow, \leftrightarrow$? In [12] I used conjunction as a test case, because conjunction is the least controversial of the truth-functional connectives: its introduction and elimination rules are *stable*, hence also *in harmony*. The root of the trouble with distribution is that, whereas factoring out both conjuncts from a conjunctive proposition is trivial in an extensional context, and the distribution of an operator works just fine if the operator and the context are intensional (*vide* the K axiom) and we remain within normal modal logic, it is far from a foregone conclusion that detachment applies to hyperintensional contexts, too.[11]

---

[8]See [6], [15].

[9]See[17, §33],[19].

[10]See [12] for a brief comparison.

[11]But invoking K – here, in this form: $\Box(A \land B) \rightarrow (\Box A \land \Box B)$ – is actually pointless, for any argument based on it must presuppose the distinction between force (here, the modality embodied by the box) and content (here, the formula/proposition the box is prefixed to). Alternatively, if we turn to $\Diamond(A \land B)$, we work our way from $\Diamond(A \land B)$ to $\Diamond A \land \Diamond B$ via the assumption of $A \land B$, $\land$-elimination, $\Diamond$-introduction applied to A, B, and finally $\land$-introduction. Of course, in Hanks's symbolism nothing matches $\Diamond A$ or $\Box A$ directly, for assertion has already been moved inside A. The proof sketch I just gave presupposes that $\Diamond A$ can be factored out into $\Diamond$ and A.

To be more specific, here is the (highly problematic) key passage I want to discuss:

> It might be argued that in asserting a conjunction a speaker does not assert each conjunct. You might think that this is an illusion generated by the fact that each conjunct is an immediate logical consequence of the whole conjunction. *But we often quite easily distinguish between what gets asserted and what follows immediately and obviously from what we assert.* For any p and q, 'p or q' is a trivial and immediate logical consequence of p. But no one thinks that an assertion that p is also an assertion that p or q. Furthermore, take any account of the necessary and sufficient conditions for assertion – pick whichever account is your favorite. If those conditions hold for the utterance of the conjunction then they will surely also hold for the utterances of each conjunct. [8, p.1395, fn.12] (*My emphasis.*)

See also [9, p.18], [9, p.104], [9, p.105], [10, p.21] where Hanks claims that the act of asserting (i.e., tokening the abstract act type of) a conjunction is tantamount to performing three acts: asserting the conjunction, asserting one conjunct, asserting the other conjunct (and, in some passages, asserting the commuted structure $B \wedge A$, something which requires hyperintensional granularity).

Let us get two interpretations of the above passage out of the way because, appearances notwithstanding, they cannot be what Hanks has in mind. First, we can all agree about this: an agent asserts $A \wedge B$, and from $A \wedge B$ follows A, B; an agent asserts A, and from A follows $A \vee B$. $\wedge$E and $\vee$I are applied outside of assertoric contexts. Second, the act-theorist internalist surely does not want to claim that asserting $A \wedge B$ is tantamount to asserting A and asserting B, for then the connective becomes logically irrelevant, and $\wedge E$ becomes homeless in act-theoretic propositional logic. The remaining interpretation is the claim that the assertion of a conjunction entails the assertion of its conjuncts. This distribution claim makes the assertion of A, B a dual *necessary* condition for the assertion of their conjunction. Of course, this condition should not be confused with *agglomeration* (which is evidently invalid): asserting A and asserting B together entails asserting $A \wedge B$. Instead the distribution claim requires that these two assertoric acts must conceptually precede the act of asserting $A \wedge B$.

The beginning and the end of the passage above makes it clear that a logical claim – that assertion distributes over conjunction – is argued for by means of a pragmatic maxim. For sure, if the assertability conditions for a conjunctive proposition are satisfied then it follows that the assertability conditions for either of the conjuncts are also satisfied. If you were allowed to perform the act of asserting a conjunction then that is a sufficient condition for being allowed to assert one or both of the conjuncts. Only doing so is something you may, or may not, decide to do. I can agree to the pragmatic point about assertability without having to agree to the logical point about an assertion entailing other assertions. I am tempted to speak of a confusion between entitlement and entailment.

The italicized passage also makes it clear that the internalist construes assertoric contexts as hyperintensional: a asserts A, A entails B, but a does not assert B. The thing, though, is that conjunction apparently plays by different rules: a asserts $A \wedge B$, and $A \wedge B$ entails A (B), so a asserts A (B). This suggests a tension between an extensional and a non-extensional conception of propositions. The extensional conception predicts closure of assertoric act types over conjunction, disjunction, etc. This yields the distribution of assertion over conjunction that Hanks is after; but also, the assertion of $A \vee B$, which he wishes to avoid.[12] The non-extensional conception rules out deductive closure over assertion. This gets in the way of the distribution claim for conjunction. On the upside, it blocks the transition from asserting A to asserting $A \vee B$, which is a restriction Hanks wants to incorporate into his propositional logic. It should also be fairly clear why nobody would want this: "I just completed asserting a conjunction, and I shall now proceed to factor out both conjuncts and assert them separately." This misconstrues what the assertion of a conjunctive proposition is all about (and turns you into a tedious speaker). This, however, could be interesting to pursue: "I just attributed the assertion of a conjunction to agent a, and I shall now proceed to factor out both conjuncts and attribute two further assertoric acts to a, one of asserting one and the other of asserting the other conjunct." But we would need a logical rule to make it valid to detach the conjuncts and attribute these two new assertoric acts. We cannot just detach A, B *con forza*, for the conjuncts occur within an asserted (hence non-extensional) context. Note that the rule in question would be one that *legitimates* a complex act, not *legislates* it.

Thus, I agree with Hanks that the assertion of A does not entail the assertion of $A \vee B$. I disagree that the assertion of $A \wedge B$ entails the assertion of A and of B. So, neither of these two entailments is valid, in my view:

$$A \wedge B \text{ is asserted} \Rightarrow A \ (B) \text{ is asserted}$$

---

[12]We should discard some easy, but irrelevant, counterexamples to detachment of conjuncts. For instance, "An empathetic midwife walks in the park, and he is whistling" is a *dynamic* case, while "The whistling midwife wakes up, and he puts on his trousers" is a *progressive* case, thus the conjuncts do not commute. One of the conjuncts is dependent on the other, so they cannot be detached as stand-alone propositions that are asserted in isolation from one another. See also [1], building on Grice.

$$A \text{ is asserted} \Rightarrow A \vee B \text{ is asserted}$$

And they are invalid for the same reason: the laws of extensional logic ($\vee E, \wedge I$) and modal/intensional logic (distribution, i.e., any variant of the K axiom) do not apply to hyperintensional contexts such as assertoric ones. By going hyperintensional, we are treading on partially uncharted territory, because new rules need to be laid down.

Note that, in arguing against both cases, I cannot invoke 'by parity of reasoning' as a strategy: the arguments for blocking/invalidating one do not carry over to the other one. There are two interlocking differences between the conjunctive and the disjunctive case one could point to. One could reasonably argue that the individual assertability of A, B is a necessary condition for the assertability of $A \wedge B$, whereas there would be no point in arguing that the assertability of $A \vee B$ is a necessary condition for the assertability of A. And relatedly, it matters (philosophically, if not formally) that one rule is an *introduction* rule and the other one an *elimination* rule. The (attempted) application of an elimination rule to assertoric contexts may have something going for it, because it trims the content of what it is applied to, whereas the (attempted) application of an introduction rule brings in fresh content. It is easy to argue intuitively against the transition from asserting that the sun is shining to asserting that the sun is shining or lead melts at 325°C, because the fresh disjunct introduces new concepts which have nothing to do with the sun shining and which may be beyond the ken of the agent making the original assertion. It is less intuitively obvious that the transition from asserting that the sun is shining and lead melts at 325°C to asserting that the sun is shining and asserting that lead melts at 325°C should be invalid. Still my main objection stands: conjunction elimination is being appealed to outside its jurisdiction.

Let us scrutinize the differences between internalism and externalism a bit further. For the internalist, the turnstile is about *assertion*; for the Fregean externalist (2a), about *knowledge*. Frege's turnstile is essential to his logic because of how he thinks of logic. Logic is primarily about inference, and inference serves to generate inferential knowledge, i.e., new knowledge extracted from old knowledge exclusively by means of valid deductive rules. The premises must be not only true but known to the agent drawing an inference before a new, known truth can be extracted from them. An inference is a transition from known truths to known truths. Hence, for instance:

$$\frac{\vdash (A \to B) \quad \vdash A}{\vdash B} \to \text{E}$$

and

$$\frac{\vdash (A \wedge B)}{\vdash A \; (\vdash B)} \; \wedge\mathrm{E}$$

Note that, in keeping with the Frege-Geach point, the antecedent and the consequent go unasserted while the conditional is asserted.[13] Of course, logic cannot enjoin anyone to apply $\rightarrow$E or $\wedge$E or any other rule. A rule issues the warrant that if the premises are known (hence true, by factivity) then the rule will also make the conclusion known (because you acknowledge the conclusion as true). This is the direction from premises via rule to conclusion: how to grow your *knowledge* by means of logic.[14] We can also go in the opposite direction: from conclusion via rule to premises. Then we use the rule and the premises to *justify* our assertion that we know the proposition that is the conclusion: "I know B because I know that the rule is valid and I know that each of the premises A_i is true. So, the proposition that figures as conclusion must be true, too."

For the modern-day externalist (2b), $\vdash A$ is a *pragmatic context* in which A is asserted. Eliminating $\vdash$ from $\vdash A$ leaves the proposition A. The proposition is a *logical* object that obeys a certain formal *semantics*. For both kinds of internalist, $\vdash B$ is a proposition that is already a *pragmatic* context. If B is $\langle \mathbf{a}, \mathbf{F} \rangle$ then $\vdash B$ is the proposition that is the act type of predicating F of a. More specifically, an agent's tokening this type is tantamount to the agent's going through the multi-step process of identifying property F, identifying individual a, and predicating F of a. For the 'inverted' internalist (3), this act of predication is imbued with assertoric force. Eliminating $\vdash$ from $\vdash \langle \mathbf{a}, \mathbf{F} \rangle$ leaves the sub-propositional rump $\langle \mathbf{a}, \mathbf{F} \rangle$. The structured hyperproposition $\vdash \langle \mathbf{a}, \mathbf{F} \rangle$ decomposes into the three atomic act types $\vdash$, $\mathbf{a}$, $\mathbf{F}$ which recompose into the molecular act type of assertorically predicating F of a.

Here is how the act-theoretic internalist scales up from atomic to molecular propositions. The conjunctive proposition $Fa \wedge Gb$ has this form:[15]

$$\vdash \langle (\vdash)\mathbf{a}, \mathbf{F} \rangle, \vdash \langle \mathbf{b}, \mathbf{G} \rangle), \mathbf{Conj} \rangle$$

The logical symbols involved are:

---

[13]See [9, §14].

[14]The *paradox of inference* rears its head here. See [3] for a solution within the hyperpropositional framework of TIL.

[15]See[9, §4.3], [8, §3].

- ⊢   *the act type of assertoric predication*[16]

- ↑   *the act type of target-shifted predication, which serves to make act types such as propositions available as subjects of predication*[17]

- **Conj**   *the act type of applying the conjunction relation to a pair of propositions*[18]

The relevant fragment of his signature is this:

- **a, b**   *act types that refer to the individuals a, b*

- **F, G**   *act types that express the properties F, G*

The notation "$\langle \ldots, \ldots \rangle$" represents an act-theoretic structure whose ultimate parts are one-step act types which when executed terminate in an individual, a property, etc. If we apply the notation below (which is mine and not Hanks's), this formula represents a conjunctive proposition:

$$\vdash_\uparrow \langle (\vdash \mathbf{A}, \vdash \mathbf{B}), \mathbf{Conj} \rangle$$

The key thing to note is that the constituent conjuncts occur with their assertoric force 'on'. A central feature of act-theoretic internalism is that assertoric force is suspended (or 'cancelled', as Hanks calls it) in all molecular propositions apart from the conjunctive ones. This is for exactly the same reason that Frege stipulated that, e.g., the antecedent and consequent of a conditional occur unasserted. Therefore, the corresponding formula for a disjunctive proposition would be this:

$$\vdash_\uparrow \langle (\sim\vdash \mathbf{A}, \sim\vdash \mathbf{B}), \mathbf{Disj} \rangle$$

where the tilde, '$\sim$', represents suspension.[19] The logical effect of $\sim$ is to contextualize a proposition to a cancellation context, in which the assertoric force of the

---

[16]See [9, §25*ff*].

[17]See [9, p.99].

[18]My [12] takes issue with the fact that **Conj** appears to be defined for pairs of *true* propositions only. This constraint makes perfect sense, as far as the *assertion* of a conjunction goes (if you are asserting a conjunction then you are committed to both conjuncts being true), but certainly not as regards the *formation* of a conjunctive proposition. This objection carries over, *mutatis mutandis*, to **Disj**.

[19]The tilde appears to be a logically superfluous symbol. It serves to indicate that the context in its scope is a cancellation context. But whether a context is a cancellation context is inherent to the context in question and not something the tilde can dictate. This objection is not unlike Peano's objection in [14] that Frege's turnstile is redundant, because the very position a proposition is embedded in already indicates whether the proposition occurs asserted (*cf.* "la varia posizione

proposition is suspended. The proposition is not judged to be true (or false), as it is not being judged at all. Put differently, the sort of act type that is a proposition cannot be tokened in a cancellation context. Using standard terminology usually applied to sentences, the proposition/act type is mentioned and not used.

This is how Hanks summarizes his position:

> On this view, to assert a conjunction is to predicate the relation _____ *and* _____ *are both true* of two propositions, and to assert a disjunction is to predicate the relation *either* _____ *or* _____ *is true* of two propositions.[9, p.106]

The rationale for the first claim must be that since asserting A and asserting B are a jointly necessary condition for asserting $A \wedge B$, then asserting $A \wedge B$ is a sufficient condition for asserting A and asserting B. Hence, in Hanks's framework distribution is not contentious, but trivial, in virtue of how rich an assertion-involving notion of conjunction he is advocating.

However, in my view there is nothing incoherent about someone asserting a conjunctive proposition *without* thereby either asserting or having already asserted (in point of conceptual priority) both conjuncts severally. Such an agent, in making the assertion, treats the conjunctive proposition as a unit, disregarding its truth-conditions and their impact on the truth-conditions of the individual conjuncts. I take this to be an option, as soon as we go hyperintensional and do not relate agents to truth-conditions. I just need this mere theoretical possibility to make my case that conjuncts do not automatically detach from an asserted conjunction.

Here is a way to make this point with regard specifically to internalism. The notation $\vdash_\uparrow \langle (\vdash \mathbf{A}, \vdash \mathbf{B}), \mathbf{Conj} \rangle$ obscures the fact that the act type **Conj** includes the act-theoretic propositions $\vdash \mathbf{A}$, $\vdash \mathbf{B}$ in its scope. **Conj** cannot be the extensional function $\wedge$ taking a pair of truth-values to a truth-value, but must be a non-extensional operator which turns its operand ($\vdash \mathbf{A}$, $\vdash \mathbf{B}$) into a non-extensional context. Therefore, we cannot apply $\wedge$E to get us from $\vdash_\uparrow \langle (\vdash \mathbf{A}, \vdash \mathbf{B}), \mathbf{Conj} \rangle$ to

---

che puó avere in una formula una proposizione indica completamente ció che di essa si afferma"). Peano explains that where Frege uses "$\vdash a$" to express that a is true, he himself uses the unadorned "a" ("é vera la a"). In Peano's own ideography (the *Formulario*), the proposition $(a \supset b) \supset c$ "non indica la verità di a, b, c né di $a \supset b$, ma solo la verità della relazione indicata fra queste proposizioni". That is, it is asserted that $(a \supset b)$ implies c. The reason is that the right-hand '$\supset$' is the main connective of the context $(a \supset b) \supset c$ as well as of the context in which $(a \supset b) \supset c$ is embedded. The fact that $(a \supset b) \supset c$ is a sub-formula of itself thus suffices for the implication $(a \supset b) \supset c$ to occur asserted. The absence of the turnstile does not detract from the proposition occurring asserted, and the presence of the turnstile would add nothing. Hence, $(a \supset b) \supset c$ and $\vdash (a \supset b) \supset c$ are indiscernible in point of assertoric force. I am indebted to an anonymous reviewer for pointing out the similarity between the objections and for directing me toward [14].

$\vdash A, \vdash B$. The sort of rule required would be a hyperintensional counterpart of $\wedge$E that would apply to conjuncts embedded within an assertoric context. However, I see no cogent reason to develop one such rule. In case propositions are created in the image of assertoric acts (i.e., propositions being, by and large, indiscernible from judgements) then any of the rules pertaining to propositions that belong to theories heeding the Frege-Geach point may not apply to act-theoretic propositions. I have been assuming throughout that acts cannot be logically coerced. Therefore, a logical investigation of acts would probably need to be a *presuppositional* one: performing act X logically presupposes having already performed acts X_i. This still does not help the internalist to a validation of their distribution claim, though. We can happily grant that the assertability of a conjunction presupposes the assertability of its conjuncts without granting that asserting a conjunction presupposes that its conjuncts have already been asserted.

3   Assertion and conjunctive/disjunctive propositions

Above I described the 'inverted' variant of act theory (3), which combines force-imbued propositions with force suspension. I want to contrast this position with one at the other end of the spectrum, namely TIL (2b). TIL is a fiercer form of semantic realism than Dummett ever attributed to Frege, because TIL exceeds truth-conditional semantics, and because TIL does not make the turnstile (i.e., assertion or judgement) part and parcel of its logic or semantics. This second Part demonstrates how TIL handles the assertion of a conjunctive/disjunctive proposition, and what follows, or rather does not follow, logically from such an assertion. Two questions will be addressed here. The first question breaks down into two halves:

- *What is a conjunctive proposition? What is a disjunctive proposition?*

So does the second question:

- *How can a conjunctive, or disjunctive, proposition occur asserted? What follows logically from an asserted conjunctive, or disjunctive, proposition?*

For the concepts and notation specific to TIL, please refer to the Appendix.

First-order propositional logic offers sixteen binary truth-functions (functions from a pair of truth-values to a truth-value). One of these truth-functions is *conjunction*, and it is standardly denoted by '$\wedge$'. Its truth-table is such that only if $v(A) = v(B) = 1$ is $v(A \wedge B) = 1$. The formula "$A \wedge B$" is a *conjunctive* one in virtue of '$\wedge$' being its main connective. But thanks to the functional completeness of this logic, there is a multitude of syntactically different formulas that share the same truth-table without being conjunctive formulas. The syntax of first-order propositional logic reveals nothing about the logical structure, if any, of its propositions. This logic trades exclusively in *truth-values*. Intensional logic trades in *truth-conditions*, entailment taking truth-conditions to truth-conditions. We can impose various constraints so as not to multiply truth-conditions, e.g., in terms of idempotence and commutativity. But when being told that A and $A \wedge A$, or $A \wedge B$ and $B \wedge A$, are idempotent, we remain at the level of syntactic transformations and are none the wiser about in what way truth-conditions might conceivably be said to be conjunctive, disjunctive, etc. Thus, if propositions are equal to truth-conditions, we are none the wiser about in what way *propositions* are conjunctive, disjunctive, etc. In modal logic, a contingent proposition is simply some subset of logical space. So, since symbols such as '$\wedge$' and '$\vee$' represent truth-functions, what is the upward logical path from truth-functions to propositions? Tichý's answer:

> The [conjunction or disjunction] sign that occurs in the corresponding formula indicates a constituent of the expressed propositional construction, not of the denoted proposition. A sentence [or formula] is not a picture of the proposition it denotes but of a particular construction of that proposition. [18, p.516]

A *construction* is a structured hyperintension, more specifically, a fine-grained logical procedure detailing how various logical objects interact so as to produce an output object, such as a truth-condition. A formula like "$A \wedge B$" or "$A \vee B$" counts as a rudimentary picture of a structured hyperproposition, which is a construction of (i.e., a logical path leading toward) a truth-condition. The following are a conjunctive and a disjunctive propositional construction:

$$\lambda w \lambda t \,[^0\wedge\, [\lambda w \lambda t \,[A_{wt}]]_{wt} \,[\lambda w \lambda t \,[B_{wt}]]_{wt}]$$

$$\lambda w \lambda t \,[^0\vee\, [\lambda w \lambda t \,[A_{wt}]]_{wt} \,[\lambda w \lambda t \,[B_{wt}]]_{wt}]$$

which can be reduced thus:

$$\lambda w \lambda t \,[^0\wedge\, A_{wt} \, B_{wt}]$$

$$\lambda w \lambda t \,[^0\vee\, A_{wt} \, B_{wt}]$$

Types: $\wedge, \vee/(ooo)$; $A, B/*_1 \to o_{wt}$; $w/*_1 \to \omega$; $t/*_1 \to \tau$.

The mapping which the turnstile represents can be typed as a logical object and, therefore, be processed logically. Though '$\vdash$' is not part of the ideography of TIL, we can nonetheless assign a type to what we stipulate to be its denotation.

$\vdash$ represents, in (2) and (3), the *act* of making an assertion, but in the interest of comparison we can treat it as the *result* of performing an assertoric act, which is the empirical property pertaining to hyperpropositions of having been asserted (by someone at some moment at some world). The general type is this: $\vdash /(o*_n)_{\tau\omega}$. Hence, the hyperproposition

$$\lambda w \lambda t \, [{}^0\vdash_{wt} \, {}^0A]$$

should be glossed as "At $\langle w, t \rangle$, A has been asserted". The sentence "The conjunction of A, B has been asserted" goes into:

$$\lambda w \lambda t \, [{}^0\vdash_{wt} \, {}^0[\lambda w \lambda t \, [{}^0\wedge \, A_{wt} \, B_{wt}]]]$$

The Trivialization of the conjunctive hyperproposition that occurs asserted is crucial. Thanks to Trivialization, the hyperproposition occurs embedded in a hyperintensional context. Therefore, it is controlled by a hyperintensional logic. TIL's hyperintensional logic centres around the notion of procedural isomorphism, which in its current version is defined in terms of α-conversion together with β-reduction by value.[20] This criterion of co-hyperintensionality lays down whether any given pair of hyperintensions are intersubstitutable. Moreover, different closure principles for knowledge, belief, assertion, etc., need to be selected for different species of epistemic, doxastic, assertoric, etc., contexts. For instance, it will matter whether we are modelling human agents or computer programs. To get a specific logic off the ground, it must be specified which logical rules the agents master and under what circumstances they apply them.

Let us begin with the easier case: why an asserted atomic hyperproposition A does not entail an asserted disjunctive hyperproposition. Since we will be drawing inferences, we want to operate on *arbitrary* empirical indices (i.e., world and time variables), so λE is our first move, which yields this *open* construction:[21]

$$[{}^0\vdash_{wt} \, {}^0A]$$

From this the following does *not* follow:

$$[{}^0\vdash_{wt} \, {}^0[\lambda w \lambda t \, [{}^0\vee \, A_{wt} \, B_{wt}]]]$$

The reason is because we have not specified the rule that each agent who asserts A must apply $\vee$I and go on to assert $A \vee B$. We could, of course, decide to add such a

---

[20] See [2].

[21] A construction containing at least one *free* occurrence of a variable is an *open* construction.

rule. But doing so would yield a logic of assertion that would be an unrealistic and irrelevant modelling of what finite agents (natural or artificial) could or should do.

What follows from an asserted conjunctive proposition? In particular, is this a valid inference, where the conclusion might equally well be $[^0\vdash_{wt} \, ^0B]$?

$$\frac{[^0\vdash_{wt} \, ^0[\lambda w \lambda t \, [^0\wedge \, A_{wt} \, B_{wt}]]]}{[^0\vdash_{wt} \, ^0A]}$$

No, it is not. I do not think it is the case that an agent asserts two propositions separately and then goes on to form their conjunction, which then counts as an asserted conjunctive proposition. Nor do I think that in the act of asserting a conjunctive proposition an agent thereby necessarily asserts the conjuncts. Instead I believe that the agent may assert a conjunctive proposition as a self-contained unit, as it were, which tracks the syntactic structure of the uttered or inscribed sentence pretty closely, but is unsusceptible to any rules from extensional or intensional logic. This phenomenon is captured formally by the fact that in the premise the asserted hyperproposition occurs *Trivialization-bound*: $^0[\lambda w \lambda t \, [^0\vee \, A_{wt} \, B_{wt}]]$ This has the effect that every constituent *subconstruction* also occurs Trivialization-bound, which in turn means that they occur in the *displayed* as opposed to *executed* mode ('mentioned' as opposed to 'used'). Of course, a Trivialization-bound construction can be made amenable to logical operations. We do this, for instance, when quantifying into hyperintensional attitude contexts. But doing so requires the deployment of some additional functions, none of which is mentioned in the invalid inference above. The thing about quantifying-in is that it does not generate a new attitude from an existing attitude, but instead spells out a logical feature of the attitude in question. If, on the other hand, I were to make the additions required to validate the above inference, I would be generating a new attitude or act from an existing attitude or act. And this is exactly what I have been arguing against, on the grounds that logic in and by itself cannot force an agent to perform an additional act or adopt an additional attitude. To see this, pretend you are designing a program that will make your (futuristic) AI detach the conjuncts of each conjunction it asserts and proceed to assert the conjuncts individually. Then the program should not only (obviously) contain the rule of ∧E, but also the *command* to invariably apply the rule to each conjunction it asserts and subsequently assert the conjuncts severally. Thus, as I have argued, distribution of assertion over conjunction is a highly conditional matter. It is not an absolute or trivial matter of fact that "any account of the necessary and sufficient conditions for [the assertion of a conjunction] will surely also hold for the utterances of each conjunct", as the act-theoretic internalist would have us believe.

Appendix

I follow the relevant definitions as formulated in, e.g., [2], [4], [11], [13], to which I refer for further details.

Definition 1 (*construction*).

(i) *Variables* x, y, ... are *constructions* that *construct* objects (elements of their respective ranges) dependently on a valuation v; they *v-construct*.

(ii) Where X is an object whatsoever (an extension, an intension or a *construction*), ^0X is the *construction Trivialization*. ^0X *constructs* (*displays*) X without any change of X.

(iii) Let X, $Y_1,...,Y_n$ be arbitrary *constructions*. Then *Composition* $[X\,Y_1...Y_n]$ is the following *construction*. For any valuation v, the *Composition* $[X\,Y_1...Y_n]$ is *v-improper* if at least one of the *constructions* X, $Y_1,...,Y_n$ is *v-improper* by failing to *v-construct* anything, or if X does not *v-construct* a function that is defined at the *n*-tuple of objects *v-constructed* by $Y_1,...,Y_n$. If X does *v-construct* such a function then $[X\,Y_1...Y_n]$ *v-constructs* the value of this function at the *n*-tuple.

(iv) The (λ-)*Closure* $[\lambda x_1...x_m\,Y]$ is the following *construction*. Let x_1, x_2, ..., x_m be pair-wise distinct variables and Y a *construction*. Then $[\lambda x_1\,...\,x_m\,Y]$ *v-constructs* the function f that takes any members $B_1,...,B_m$ of the respective ranges of the variables $x_1,...,x_m$ into the object that is $v(B_1/x_1,...,B_m/x_m)$-*constructed* by Y(if there is such an object), where $v(B_1/x_1,...,B_m/x_m)$ is like v except for assigning B_1 to x_1, ..., B_m to x_m.

(v) The *Double Execution* ^2X is the following *construction*. Where X is any entity, the *Double Execution* ^2X is *v-improper* if X is not itself a *construction*, or if X does not *v-construct* a *construction*, or if X *v-constructs* a *v-improper construction*. Otherwise, let X *v-construct* a *construction* Y and Y *v-construct* an entity Z; then ^2X *v-constructs* Z.

(vi) Nothing is a *construction*, unless it so follows from (i) through (v).

Definition 2 (*simple type*). Let B be a *base*, where a base is a collection of pair-wise disjoint, non-empty sets. Then:

(i) Every member of B is an elementary *type of order 1 over B*.

(ii) Let α, β_1, ..., β_m ($m > 0$) be types of order 1 over B. Then the collection (α $\beta_1\,...\,\beta_m$), of all *m*-ary partial mappings from $\beta_1 \times\,...\,\times \beta_m$ into α, is a functional *type of order 1 over B*.

(iii) Nothing is a *type of order 1 over B* unless it so follows from (i) and (ii).

Remark. For the purposes of natural-language analysis, we are currently assuming the following base of ground types, which form part of the ontological commitments of TIL:

o: the set of truth-values $\{\mathbf{T}, \mathbf{F}\}$;
ι: the set of individuals (the universe of discourse);
τ: the set of real numbers (doubling as times);
ω: the set of logically possible worlds (the logical space).

Definition 3 (*ramified hierarchy of types*).

$\mathbf{T_1}$ (*types of order 1*). See Definition 2.
$\mathbf{C_n}$ (*constructions of order n*)

(i) Let x be a variable ranging over a type of order n. Then x is a *construction of order n over B*.
(ii) Let X be a member of a type of order n. Then ^0X, ^2X are *constructions of order n over B*.
(iii) Let $X, X_1,..., X_m$ $(m > 0)$ be *constructions* of order n over B. Then $[X\, X_1...X_m]$ is a *construction of order n over B*.
(iv) Let $x_1, ..., x_m, X$ $(m > 0)$ be *constructions* of order n over B. Then $[\lambda x_1...x_m\, X]$ is a *construction of order n over B*.
(v) Nothing is a *construction of order n over B* unless it so follows from $\mathbf{C_n}$ (i)-(iv).

$\mathbf{T_{n+1}}$ (*types of order $n + 1$*) Let $*_n$ be the collection of all constructions of order n over B. Then:

(i) $*_n$ and every type of order n are *types of order $n + 1$*.
(ii) If $0 < m$ and $\alpha, \beta_1,...,\beta_m$ are types of order $n + 1$ over B, then $(\alpha\, \beta_1\, ...\, \beta_m)$ (see $\mathbf{T_1}$ (ii)) is a *type of order $n + 1$ over B*.
(iii) Nothing is a *type of order $n + 1$* over B unless it so follows from $\mathbf{T_{n+1}}$ (i) and (ii).

Notational conventions. '$y \to \alpha$' means that variable y ranges over the type α. If C is a construction, then '$C \to \alpha$' means that C is typed to construct an entity of type α. That an object a is of a type α is denoted 'a/α'. Thus, for instance, '$C/*_n \to \iota$' means that the construction C is of order n (i.e., belongs to type $*_n$) and is typed to construct an individual. For the variables w, t this holds: $w \to \omega$, $t \to \tau$. If $C \to \alpha_{wt}$ then the frequently used Composition $[[C\, w]\, t] \to \alpha$ will be written as 'C_{wt}' for short.

Definition 4 (*subconstruction*). Let C be a construction. Then:

(i) C is a *subconstruction of C*.

(ii) If C is ^0X or ^2X and X is a construction, then X is a *subconstruction of C*.

(iii) If C is $[X\,X_1\ldots X_n]$ then $X, X_1, \ldots, X_n$ are *subconstructions of* C.

(iv) If C is $[\lambda x_1 \ldots x_n\,Y]$ then Y is a *subconstruction of C*.

(v) If A is a *subconstruction of B* and B is a *subconstruction of C* then A is a *subconstruction of C*.

(vi) A construction is a *subconstruction of C* only if it so follows from (i) – (v).

Definition 5 (*displayed vs executed mode of occurrence of a construction*). Let C be a construction and D a *subconstruction* of C. Then:

(i) If C is identical to ^0X and D is identical to X, then the *occurrence of D and of all the subconstructions of D are displayed in C*.

(ii) If D is *displayed* in C and C is a *subconstruction* of a construction E such that E is not identical to ^2F for any construction F, then the *occurrence of D and of all the subconstructions of D are displayed in E*.

(iii) If D is identical to C, then the *occurrence of D is executed in C*.

(iv) If C is identical to $[X_1\,X_2 \ldots X_m]$ and D is identical to one of the constructions X_1, X_2, ..., X_m, then the *occurrence of D is executed in C*.

(v) If C is identical to $[\lambda x_1 \ldots x_m\,X]$ and D is identical to X, then the *occurrence of D is executed in C*.

(vi) If C is identical to ^2X and D is identical to X, then the *occurrence of D is executed in C*.

(vii) If C is identical to ^{20}X such that X is typed to *v-construct* an object of a type of order 1, and D is identical to X, then the *occurrence of D is executed in C*.

(viii) If an occurrence of D is executed in a construction E such that this occurrence of E is executed in C, then the *occurrence of D is executed in C*.

Remark. This otherwise inductive definition does not contain a closure clause and is insofar open-ended. The reason is that the further cases not defined in the preceding clauses are indeterminate/undecidable. The final definition of displayed and executed occurrences is still work in progress.

References

[1] Carston, R. (1993), 'Conjunction, explanation and relevance', *Lingua* 90, pp. 27-48.

[2] Duží, M. (2019), 'If structured propositions are logical procedures then how are procedures individuated?', *Synthese* 196, pp. 1249-1283.

[3] Duží, M. (2010), 'The paradox of inference and the non-triviality of analytic information', *Journal of Philosophical Logic* 39, pp. 473-510.

[4] Duží, M., Jespersen, B., Materna, P. (2010), *Procedural Semantics for Hyperintensional Logic*, Heidelberg: Springer-Verlag.

[5] Frege, G. (1893), *Grundgesetze der Arithmetik*, Jena: Hermann-Pohle-Verlag.

[6] Frege, G. (1891/1986), 'Funktion und Begriff', in: *Funktion, Begriff, Bedeutung*, G. Patzig (ed.), Göttingen: Vandenhoeck & Ruprecht.

[7] Geach, P. (1965), 'Assertion', *Philosophical Review* 74, pp. 449-65.

[8] Hanks, P. (2019), 'On cancellation', *Synthese* 196, pp. 1385-1402.

[9] Hanks, P. (2015), *Propositional Content*, Oxford: Oxford University Press.

[10] Hanks, P. (2011), 'Structured propositions as types," *Mind* 120, pp. 11-52.

[11] Jespersen, B. (2015), 'Qualifying quantifying-in', in: *Quantifiers, Quantifiers, Quantifiers*, A. Torza (ed.), *Synthese Library* 373, pp. 241-69.

[12] Jespersen, B. (ms.), 'Conjunction and the inverted Frege-Geach point', *under review*.

[13] Kosterec, M. (2020), 'Substitution contradiction, its resolution and the Church-Rosser Theorem in TIL', *Journal of Philosophical Logic* 49, pp. 121-133.

[14] Peano, G. (1895), 'Recensione: G. Frege, *Grundgesetze der Arithmetik, begriffsschriftlich abgeleitet*', *Rivista di Matematica* 5, pp. 122-124. Reprinted in *Opere scelte* 1, Edizione cremonese (1957).

[15] Pedriali, W.B. (2017), 'The logical significance of assertion', *Journal of the History of Analytic Philosophy* 5, pp. 1-22.

[16] Schaar, M. van der (2018), 'Frege on judgement and the judging agent', *Mind* 127, pp. 25-50.

[17] Tichý, P. (1988), *The Foundations of Frege's Logic*, deGruyter.

[18] Tichý, P. (1986), 'Constructions', *Philosophy of Science* 53, pp. 514-34.

[19] Tichý, P. (1979), 'Questions, answers, and logic', *American Philosophical Quarterly* 15, pp. 275-84.

Received 24 March 2020

Normalisation for Bilateral Classical Logic with some Philosophical Remarks

Nils Kürbis

University of Łódź

nils.kurbis@filhist.uni.lodz.pl

Abstract

Bilateralists hold that the meanings of the connectives are determined by rules of inference for their use in deductive reasoning with asserted and denied formulas. This paper presents two bilateral connectives comparable to Prior's *tonk*, for which, unlike for *tonk*, there are reduction steps for the removal of maximal formulas arising from introducing and eliminating formulas with those connectives as main operators. Adding either of them to bilateral classical logic results in an incoherent system. One way around this problem is to count formulas as maximal that are the conclusion of reductio and major premise of an elimination rule and to require their removability from deductions. The main part of the paper consists in a proof of a normalisation theorem for bilateral logic. The closing sections address philosophical concerns whether the proof provides a satisfactory solution to the problem at hand and confronts bilateralists with the dilemma that a bilateral notion of stability sits uneasily with the core bilateral thesis.

1 Introduction

It is a commonly held view that the meanings of the expressions of a language are determined by the use its speakers make of them. One way of giving substance to this view is to propose that that use can be systematised for the hypothetical project of constructing a theory of meaning for a language in terms of the conditions of the correct assertibility of sentences containing the expressions. This is the course taken by Dummett [2, 3]. *Bilateralism*, by contrast, is the view that the meanings of expressions are determined not only in terms of the conditions for the correct assertibility of sentences containing them, but by these in tandem with the conditions for

I thank Julien Dutant, Dorothy Edgington, Keith Hossack, Guy Longworth and Mark Textor for discussions about assertion and denial and audiences in Lecce, Łódź and Stirling for their comments.

their correct deniability. The view was proposed in response to Dummett by Price, who 'takes the fundamental notion for a recursive theory of sense to be not assertion conditions alone, but these in conjunction with rejection, or *denial* conditions' [20, 162]. We may distinguish the two views by calling the former *unilateralism*.

Unilateralism and bilateralism provide alternative forms for a theory of meaning for an entire language, but I will here only consider their restrictions to the logical constants of propositional logic. In that region of language, Dummett's insights coupled with important contributions by Prawitz have lead to the development of *proof-theoretic semantics*, an alternative to truth-theoretic semantics. Whereas in the latter the meanings of the logical constants are given in terms of their contributions to the truth conditions of sentences containing them, the principal tenet of proof-theoretic semantics is that their meanings are determined by the use of such sentences in deductive arguments.

In this paper I will present a problem for bilateral proof-theoretic semantics in the form of bilateral connectives that are comparable to Prior's *tonk*. But whereas *tonk* can be excluded from unilateral logic on principled grounds that form part of the philosophical background of proof-theoretic semantics, the issue is more involved in the case of bilateralism. The main part of the paper contains a proof of a normalisation theorem for a system of bilateral classical logic. This provides a solution of sorts to the problem, but it also has certain philosophical drawbacks. In particular, the proof appeals to an unrestricted version of a bilateral principle of non-contradiction, while Rumfitt requires this principle to be restricted to atomic premises. Secondly, the solution is based on a redefinition of the notion of a maximal formula, and it may be objected that the solution therefore merely constitutes a change of subject. I conclude that it would appear that the best solution appeals to bilateral analogues of Prawitz's inversion principle. These are desirable in any case and for independent reasons. Appeal to such principles, however, endangers the core thesis of bilateralism and threatens collapse it into unilateralism.

2   A System of Bilateral Classical Logic

Proof-theoretic semantics along Dummett's and Prawitz's lines arguably does not go any further than intuitionist logic. From their perspective, the rules governing classical negation are defective. Advocates of bilateralism claim that this situation is rectified in their framework. They recommend the use of systems of natural deduction with two kinds of rules: For each connective **c**, there are assertive rules specifying the grounds for and consequences of asserting a formula with **c** as main operator, and rejective rules specifying the grounds for and consequences of denying such a formula. The most prominent such system has been proposed by Rumfitt [22], building on work by Smiley [23].[1] Rumfitt's system is intended to satisfy Dummett's requirements for when the rules of inference governing a connective specify its meaning: they do so if they are *in harmony* or, more precisely, *stable* [1, Chapters 11-13]. The aim is to provide 'a direct specification of the senses of the connectives in terms of their deductive use' [22, 805], where the premises and conclusions of rules of inference are assertions and denials.

Formulas in the system $\mathfrak{B}$ of bilateral classical logic are *signed* by $+$, indicating asserted formulas, or $-$, indicating denied ones. $\bot$ indicates the incoherence that arises from asserting and denying the same formula. Deductions do not begin with $\bot$. Lower case Greek letters α, β range over signed formulas, ϕ may also be $\bot$. α^* designates the result of reversing α's sign from $+$ to $-$ or conversely. The terminology follows Rumfitt, and the rules of $\mathfrak{B}$ are his [22, 800ff].

Deductions in $\mathfrak{B}$ have the familiar tree shape, with the (discharged or undischarged) assumptions at the top-most nodes or leaves and the conclusion at the bottom-most node or root. Every assumption in a deduction belongs to an *assumption class*, marked by a natural number, different numbers for different assumption classes. Formula occurrences of different types must belong to different assumption classes. Formula occurrences of the same type may, but do not have to, belong to the same assumption class. Discharge of assumptions is marked by a square bracket around the formula: $[\alpha]^i$, where i is a label for the assumption class to which α belongs. If the assumption is discharged, the label is repeated at the application of the rule. The formulas in an assumption class are discharged all together or not at all. Empty assumption classes are permitted for vacuous discharge, when a rule that allows for the discharge of assumptions is applied with no assumptions being discharged. The conclusion of a deduction is said to depend on the undischarged assumptions of the deduction. Similar terminology is applied to subdeductions of deductions.

Upper case Greek letters Σ, Π, Ξ, possibly with subscripts or superscripts, denote deductions. Often some of the assumptions and the conclusion of the deduction are mentioned explicitly at the top and bottom of Σ, Π, Ξ. Using the same designation more than once to denote subdeductions of a deduction means that these subdeductions are exact duplicates of each other except that assumption classes may be different: the deductions have the same structure, and at every node formulas of the same type are premises and conclusions of applications of the same rules.[2]

Definition 1 (Deduction in $\mathfrak{B}$).

---

[1]Humberstone proposed a similar system at the same time as Rumfitt [10].

[2]The layout of natural deduction used here follows [25].

(i) The formula occurrence $+ A\,^n$ is a deduction in $\mathfrak{B}$ of $+ A$ from the undischarged assumption $+ A$, and $- A\,^n$ is one of $- A$ from the undischarged assumption $- A$, where n marks the assumption class to which $+ A, - A$ belong.

(ii) If Σ, Π, Ξ are deductions in $\mathfrak{B}$, then so are the following, where the conclusion depends on the undischarged assumptions of Σ, Π, Ξ except those in assumption classes i and j:

$$+ \wedge I: \quad \frac{\overset{\Pi}{+ A} \quad \overset{\Sigma}{+ B}}{+ A \wedge B} \qquad\qquad + \wedge E: \quad \frac{\overset{\Pi}{+ A \wedge B}}{+ A} \quad \frac{\overset{\Sigma}{+ A \wedge B}}{+ B}$$

$$- \wedge I: \quad \frac{\overset{\Pi}{- A}}{- A \wedge B} \quad \frac{\overset{\Sigma}{- B}}{- A \wedge B} \qquad - \wedge E: \quad \frac{\overset{\Xi}{- A \wedge B} \quad \overset{[- A]^i}{\overset{\Pi}{\phi}} \quad \overset{[- B]^j}{\overset{\Sigma}{\phi}}}{\phi} \; i,j$$

$$+ \vee I: \quad \frac{\overset{\Pi}{+ A}}{+ A \vee B} \quad \frac{\overset{\Sigma}{+ B}}{+ A \vee B} \qquad + \vee E: \quad \frac{\overset{\Xi}{+ A \vee B} \quad \overset{[+ A]^i}{\overset{\Pi}{\phi}} \quad \overset{[+ B]^j}{\overset{\Sigma}{\phi}}}{\phi} \; i,j$$

$$- \vee I: \quad \frac{\overset{\Pi}{- A} \quad \overset{\Sigma}{- B}}{- A \vee B} \qquad\qquad - \vee E: \quad \frac{\overset{\Pi}{- A \vee B}}{- A} \quad \frac{\overset{\Sigma}{- A \vee B}}{- B}$$

$$+ \supset I: \quad \frac{\overset{[+ A]^i}{\overset{\Pi}{+ B}}}{+ A \supset B} \; i \qquad\qquad + \supset E: \quad \frac{\overset{\Pi}{+ A \supset B} \quad \overset{\Sigma}{+ A}}{+ B}$$

$$- \supset I: \quad \frac{\overset{\Pi}{+ A} \quad \overset{\Sigma}{- B}}{- A \supset B} \qquad\qquad - \supset E: \quad \frac{\overset{\Pi}{- A \supset B}}{+ A} \quad \frac{\overset{\Sigma}{- A \supset B}}{- B}$$

$$+ \neg I: \quad \frac{\overset{\Pi}{- A}}{+ \neg A} \qquad\qquad + \neg E: \quad \frac{\overset{\Pi}{+ \neg A}}{- A}$$

$$-\neg I: \quad \frac{\begin{array}{c}\Pi\\+A\end{array}}{-\neg A} \qquad\qquad -\neg E: \quad \frac{\begin{array}{c}\Pi\\-\neg A\end{array}}{+A}$$

$$\text{Reductio:} \quad \frac{\begin{array}{c}[\alpha^*]^i\\\Pi\\\bot\end{array}}{\alpha}\, i \qquad\qquad \text{Non-Contradiction:} \quad \frac{\begin{array}{cc}\Pi & \Sigma\\\alpha & \alpha*\end{array}}{\bot}$$

(iii) Nothing else is a deduction in $\mathfrak{B}$.

Rumfitt calls reductio and non-contradiction *co-ordination principles*. They have the character of structural rules required by the formal framework of bilateral logic to regulate the interaction between $+$, $-$ and $\bot$.[3]

According to Rumfitt, non-contradiction must be restricted to atomic premises [22, 815f]. His reason is that on a bilateral account of meaning, only the atomic sentences are co-ordinated primitively by non-contradiction: it is a consequence of how their use is specified in terms of the conditions of their correct assertibility and deniability. That the complex sentences are also so co-ordinated is a consequence of co-ordination at the atomic level and how the meanings of the connectives are specified by their assertive and rejective rules. By contrast, I will not impose this restriction on non-contradiction.

It is generally considered to be a necessary requirement for a system of natural deduction to be satisfactory from the perspective of proof-theoretic semantics that deductions in it *normalise*. In unilateral logic, a deduction is in *normal form* if it contains no *maximal formulas*, where a maximal formula is one that is the conclusion of an introduction rule and major premise of an elimination rule for its main connective. This definition carries over to $\mathfrak{B}$, only that maximal formulas are signed by $+$ or $-$. In section 4 it will be shown that such formulas, and further undesirable ones, can indeed be removed from deductions in $\mathfrak{B}$. The proof of this more general result requires the unrestricted version of non-contradiction.

---

[3]An alternative version of reductio avoids the use of $\bot$: from $\Gamma, \alpha^* \vdash \beta$ and $\Delta, \alpha^* \vdash \beta^*$, infer $\Gamma, \Delta \vdash \alpha$. An intuitionist bilateral logic has been formalised in [12]. It fulfils the requirements Rumfitt imposes on a satisfactory bilateral logic, and hence the claim that only classical bilateral logic can do so is false. For an informal argument against the view that bilateralism inevitably leads to classicism, see [13]. The stipulation that nothing can be both asserted and denied addresses the problem with negation in intuitionist logic noted in [11].

3 Some Principles of Proof-Theoretic Semantics

Proof-theoretic semantics has its roots in a comment of Gentzen's, who formulated the rudiments of a theory of meaning for the connectives:

> The introductions constitute, so to speak, the "definitions" of the symbols concerned, and the eliminations are in the end only consequences thereof, which could be expressed thus: In the elimination of a symbol, the formula in question, whose outer symbol it concerns, may only "be used as that which it means on the basis of the introduction of this symbol". [8, 189]

Gentzen's comment is the foundation of Prawitz's *inversion principle*: 'an elimination rule is, in a sense, the inverse of the corresponding introduction rule: by an application of an elimination rule one essentially only restores what had already been established if the major premise of the application was inferred by an application of an introduction rule [...;] nothing is "gained" by inferring a formula through introduction for use as a major premiss in an elimination.' [19, 33f] Prawitz proposes the normalisability of deductions as a formal criterion for when the inversion principle is met.

According to Dummett, the meanings of expressions are determined by two aspects of their use, their contributions to the grounds for asserting sentences in which they occur and to the consequences of asserting such sentences [1, 211ff]. The connectives are a particularly clear cases of how this insight may be applied. The introduction rules for a connective specify the *canonical grounds* for deriving a formula with that connective as main operator, and its elimination rules specify the *canonical consequences* that follow from such a formula. For the rules governing a connective to determine its meaning completely, the two aspects of their use must be stable.

Prawitz's inversion principle captures the thought that the elimination rules for a connective **c** should not licence the deduction of more formulas from a formula with **c** as main operator than are justified by the grounds of its assertion as specified by its introduction rule. This is what's wrong with Prior's *tonk* [21,]:

$$tonk \ I: \quad \frac{A}{A \, tonk \, B} \qquad\qquad tonk \ E: \quad \frac{A \, tonk \, B}{B}$$

The elimination rule of *tonk* licences the derivation of too many consequences from *AtonkB* relative to its introduction rule. Maximal formulas of the form *AtonkB* cannot be removed:

$$\frac{A}{AtonkB}$$
$$\frac{AtonkB}{B}$$

The rules for *tonk* do not satisfy Prawitz's inversion principle.

Prawitz's inversion principle is not enough for meaning-theoretical purposes. Consider a connective with the introduction rule of conjunction but only one of its elimination rules. Something is missing: its elimination rule does not permit the use of the connective in all the ways one should be able to use it relative to its introduction rule.

Prawitz's inversion principles spells out the notion of harmony. Dummett's notion of stability consists in harmony together with a suitable convers. The latter, as Moriconi and Tesconi note [17, 111], is provided by an inversion principle of Negri's and von Plato's: 'Whatever follows from the direct grounds for deriving a proposition must follow from the proposition' [18, 6]. The elimination rules for a connective should licence the deduction of all the consequences from a formula with that connective as main operator that are justified relative to its introduction rules.[4]

Notice that *tonk* satisfies Negri's and von Plato's inversion principle: whatever follows from the direct grounds for deriving $AtonkB$ follows from $AtonkB$. Consequently, as Prawitz's inversion principle is tied to normalisation, it is a notion interesting enough to be considered by itself, whereas a suitable converse of Prawitz's inversion principle, such as Negri's and von Plato's, is usually considered only in combination with Prawitz's.

If both inversion principles are satisfied, stability obtains and the elimination rules for a connective license the deduction of all and only the consequences from a sentence with the connective as main operator that are justified by the grounds for deriving it as specified by its introduction rules.

4   Bilateral Dissonance

Consider the connective *conk*:

$+conkI$: $\quad \dfrac{+\,A \qquad +\,B}{+\,AconkB}$ $\qquad +conkE$: $\quad \dfrac{+\,AconkB}{+\,A} \qquad \dfrac{+\,AconkB}{+\,B}$

$-conkI$: $\quad \dfrac{-\,A \qquad -\,B}{-\,AconkB}$ $\qquad -conkE$: $\quad \dfrac{-\,AconkB}{-\,A} \qquad \dfrac{-\,AconkB}{-\,B}$

conk means trouble. Given reductio, the assertion of any formula follows from the assertion of any formula:

---

[4]For more on inversion principles, see [16].

$$\frac{\overline{-\ AconkB}\ ^1}{-\ A \qquad\qquad +\ A}$$
$$\frac{\bot}{\dfrac{+\ AconkB}{+\ B}\ _1}$$

The denial of any formula also follows from the denial of any formula:

$$\frac{\overline{+\ AconkB}\ ^1}{+\ A \qquad\qquad -\ A}$$
$$\frac{\bot}{\dfrac{-\ AconkB}{-\ B}\ _1}$$

Notice that *conk* permits the restriction of non-contradiction to atomic premises.

Next consider the connective *honk*:

$$+honkI: \quad \frac{-\ A \qquad +\ B}{+\ AhonkB} \qquad +honkE: \quad \frac{+\ AhonkB}{-\ A} \qquad \frac{+\ AhonkB}{+\ B}$$

$$-honkI: \quad \frac{+\ A \qquad -\ B}{-\ AhonkB} \qquad -honkE: \quad \frac{-\ AhonkB}{+\ A} \qquad \frac{-\ AhonkB}{-\ B}$$

honk, too, means trouble. Given reductio, the assertion of any formula follows from the denial of any formula:

$$\frac{\overline{-\ AhonkB}\ ^1}{+\ A \qquad\qquad -\ A}$$
$$\frac{\bot}{\dfrac{+\ AhonkB}{+\ B}\ _1}$$

The denial of any formula follows from the assertion of any formula:

$$\frac{\overline{+\ AhonkB}\ ^1}{-\ A \qquad\qquad +\ A}$$
$$\frac{\bot}{\dfrac{-\ AhonkB}{-\ B}\ _1}$$

honk also permits the restriction of non-contradiction to atomic premises.

The rules for *conk* and *honk* appear to be just as good as the rules for the connectives of 𝔅. They combine old rules in novel ways. Like *tonk*, *conk* combines rules for conjunction and disjunction, only this time they are bilateral rules and all assertive rules for conjunction and all rejective rules for disjunction are used. *honk* combines the rejective rules for implication with assertive rules that would be correct for another connective. But unlike *tonk*, *conk* and *honk* have the rather unusual feature that although adding them to $\mathfrak{B}$ gives an incoherent system, maximal formulas that arise from concluding $A conk B$ or $A honk B$ by an introduction rule and using them as major premises of an elimination rule may be removed from deductions by the same reduction procedures that remove such maximal formulas with conjunction, disjunction or implication as main connectives.

The rules for the connectives of $\mathfrak{B}$ satisfy bilateral versions of the inversion principles. The assertive elimination rules for a connective of $\mathfrak{B}$ license the deduction of all and only the consequences from an asserted sentence with the connective as main operator that are justified by the grounds for deriving it as specified by its assertive introduction rules. The rejective elimination rules for a connective of $\mathfrak{B}$ licence the deduction of all and only the consequences from a denied sentence with the connective as main operator that are justified by the grounds for deriving it as specified by its rejective introduction rules. Unlike *tonk*, the rules for *conk* and *honk* also satisfy these bilateral inversion principles.[5]

It is evident where the problem lies. In the bilateral framework, it is not enough that inversion principles balance the grounds and consequences of asserting a formula and others balance the grounds and consequences of denying a formula. There also needs to be a sort of stability between the assertive and the rejective rules for a connective, a kind of inversion that balances the grounds and consequences of the assertion and the denial of a formula.

The issue can also be put in terms of the question why the rejective and the assertive rules for a connective of $\mathfrak{B}$ are rules for the same connective. What is it, for instance, that makes the assertive rules for the symbol $\wedge$ and the rejective rules for the symbol $\wedge$ rules for *conjunction*? What justifies the use of the same symbol in both cases? We are, of course, able to recognise that the two sets of rules are intended to be rules for the same connective. But this depends on our previous understanding of the connectives, while the aim was to specify their meanings completely in terms of the rules governing them. It should not be down to our grasp of their meanings that we can recognise which rules belong to which connective, but solely down to

---

[5]Gabbay has also proposed a connective that satisfies the bilateral inversion principles but leads to incoherence [7]. *conk* and *honk*, however, are worse than Gabbay's connective, as they satisfy an additional requirement concerning the proper subformulas of premises and conclusions of rules of inference. An exposition of the precise nature of this requirement and why it may reasonably be imposed on rules that are to determine the meanings of the connectives they govern completely is the subject of a piece currently in preparation.

the meaning-theoretical framework. Without addressing this question, bilateralists cannot claim that the meanings of the connectives of $\mathfrak{B}$ are determined completely by the rules of inference governing them, and should this be their objective, they have no right to use the same symbol in the two sets of rules governing a connective.

Inversion principles that link the assertive and the rejective rules for a connective would answer the question raised in the previous paragraphs. There is, however, also another possible diagnosis of what has gone wrong in the four deductions, and this leads to a result of independent interest. In each of them, a complex formula is the conclusion of reductio and major premise of an elimination rule. Reductio provides grounds for the assertion and denial of formulas. These should be in harmony with the consequences of asserting and denying them as specified by the respective elimination rules for their main connective. This motivates the demand that formulas that are conclusion of reductio and major premise of an elimination rule should be removable from deductions.

Furthermore, inferences by non-contradiction draw consequences from formulas which should be in harmony with the grounds for deriving them. Finally, reductio and non-contradiction should presumably be in harmony with each other, too, although this has nothing to do with the connectives, but rather with the formal framework of bilateral logic.

These formulas also count as maximal in the normalisation theorem for deductions in $\mathfrak{B}$ that is proved in the next section.

5   Normalisation for $\mathfrak{B}$

This section contains a proof of a normalisation theorem for deductions in $\mathfrak{B}$.[6]

Definition 2. The *degree* of a signed formula $+ A$ or $- A$ is the number of connectives occurring in A.

$\perp$ is not a signed formula and gets degree 0.

Definition 3. A *maximal signed formula* is an occurrence of a formula in a deduction that is one of the following:
(a) conclusion of an introduction rule and major premise of an elimination rule;

---

[6]The reader is invited to compare it with Stålmarck's proof of normalisation for unilateral classical logic [24]. There is some resemblance, if − is read as negation. However, as $\mathfrak{B}$ has a larger number of operational rules than Stålmarck's system, certain complications that arise in Stålmarck's proof do not arise here. In particular, there is no need to consider assumption contractions separately from reduction steps for maximal formulas. The larger number of rules also requires reduction steps for which there are no equivalents in Stålmarck's proof.

(b) conclusion of reductio and major premise of an elimination rule;

(c) conclusion of reductio and premise of non-contradiction;

(d) conclusion of an introduction rule and premise of non-contradiction the other premise of which is also the conclusion of an introduction rule.

For brevity, I will mostly use 'maximal formula' instead of 'maximal signed formula'.

To distinguish the four kinds of maximal formulas, I will call those of kind (a) *maximal formulas with introduction and elimination rules* or *i/e maximal formulas*; those of kind (b) *maximal formulas with reductio and elimination rules* or *r/e maximal formulas*; those of kind (c) *maximal formulas with reductio and non-contradiction* or *r/nc maximal formulas*; and those of kind (d) *maximal formulas with introduction rules and non-contradiction* or *i/nc maximal formulas*.

Formulas of the third and fourth kind are clearly 'maximal' in some sense, even though the philosophical reasons for requiring the removability of maximal formulas of the first (and perhaps the second) kind may not apply to them. They have been included here to ensure that deductions in normal form have the subformula property. For Rumfitt, i/nc maximal formulas do not arise, as he restricts non-contradiction to atomic premises. The reduction steps to remove r/e maximal formulas where the elimination rule is $+ \vee E$ or $- \wedge E$ require the general version of non-contradiction.

Definition 4 (Segment, Length and Degree of a Segment, Maximal Segment).

(a) A *segment* is a sequence of two or more formula occurrences $C_1 \ldots C_n$ in a deduction such that C_1 is not the conclusion of $+ \vee E$ or $- \wedge E$, C_n is not the minor premise of $+ \vee E$ or $- \wedge E$, and for every $i < n$, C_i is minor premise of $+ \vee E$ or $- \wedge E$ and C_{i+1} its conclusion.

(b) The *length* of a segment is the number of formula occurrences of which it consists, its *degree* is their degree.

(c) A segment is *maximal* if and only if its last formula is major premise of an elimination rule or premise of non-contradiction.

I will say that the formula occurrence C_i is *on* segment $C_1 \ldots C_n$. A segment is above another one in a deduction if its last formula is above the other's first formula. I will speak of segments being the premises or conclusions of the rules of which their last or first formulas are premises or conclusions.

Prawitz only counts a segment as maximal if it begins with the conclusion of an introduction rule [19, 49]. The more general notion used here is also used by Troelstra and Schwichtenberg in the proof of normalisation for intuitionist logic [25, 179]. For philosophical reasons, the more general notion is called for, as it must be ensured that $+ \vee E$ and $- \wedge E$ do not introduce grounds for the derivation of formulas that are not in balanced by the elimination rules for their main operators. This is irrespective of how the first formula of the segment is derived.[7]

Definition 5. A deduction is in *normal form* if it contains neither maximal formulas nor maximal segments.

The reduction steps to be given next remove maximal formulas and maximal segments from deductions. Applying then in the systematic fashion specified in the proof of the normalisation theorem transforms any deduction into a deduction in normal form. $\rightsquigarrow$ indicates that the deduction to its left or above it is transformed into the deduction on its right or below it. I will call the deduction to which a reduction step is applied the *original deduction* and the result of the application the *reduced deduction*. A formula in square brackets between two deductions

$$\Pi$$
$$[A]$$
$$\Sigma$$

means that the deduction on top is used to conclude all formulas in the assumption class $[A]$.

(A) *Permutative Reduction Steps for Maximal Segments*
The lower application of the elimination rule or of non-contradiction is permuted upwards to conclude with a minor premise of $+\vee E$ or $-\wedge I$. Here are two examples, the others being similar.

(1) The maximal segment consists of formula occurrences of the form $+\,C\vee D$, the last of which is concluded by $-\wedge E$:

$$
\begin{array}{ccccccc}
 & & [-A]^i & [-B]^j & & & \\
 & \Xi_1 & \Pi_1 & \Pi_2 & [+C]^k & [+D]^l & \\
 & -A\wedge B & +C\vee D & +C\vee D & \Sigma_1 & \Sigma_2 & \\
 \cline{3-4}
 & & +C\vee D & {}_{i,j} & \phi & \phi & {}_{k,l}\\
 \cline{3-6}
 & & & \phi & & & \\
 & & & \Xi_2 & & &
\end{array}
$$

$$\rightsquigarrow$$

$$
\begin{array}{cccccc}
[-A]^i & [+C]^{k_1} & [+D]^{l_1} & [-B]^j & [+C]^{k_2} & [+D]^{l_2}\\
\Pi_1 & \Sigma_1 & \Sigma_2 & \Pi_2 & \Sigma_1 & \Sigma_2\\
+C\vee D & \phi & \phi & +C\vee D & \phi & \phi
\end{array}
$$

If some occurrence of ϕ forms part of a maximal segment in the original deduction, the permutative reduction step increases its length in the reduced deduction. In the proof of the normalisation theorem a strategy will be given to avoid increasing the length of a maximal segment of the same or higher degree than the one shortened or removed: in a nutshell, apply the reduction step to the rightmost segment of highest degree first. Furthermore, it needs to be ensured that the reduction step does not duplicate maximal formulas and segments of highest degree in Σ_1 and Σ_2: to do so it is applied to a topmost maximal segment of highest degree, one above which there is none other of highest degree.

(2) The right premise of non-contradiction is conclusion of $- \wedge E$:

$$
\begin{array}{ccc}
& & [- A]^i \quad [- B]^j \\
& \Xi_1 & \Pi_1 \quad \Pi_2 \\
\Sigma & - A \wedge B & \alpha^* \quad \alpha^* \\
\hline
\alpha & & \alpha^* \quad {}_{i,j} \\
& \bot & \\
& \Xi_2 &
\end{array}
$$

$$\rightsquigarrow$$

$$
\begin{array}{cccc}
& & [- A]^i & [- B]^j \\
& \Sigma \quad \Pi_1 & \Sigma \quad \Pi_2 \\
\Xi_1 & \alpha \quad \alpha^* & \alpha \quad \alpha^* \\
- A \wedge B & \bot & \bot \quad {}_{i,j} \\
\hline
& \bot & \\
& \Xi_2 &
\end{array}
$$

The reduction step shortens the right segment, but if the left premise of non-contradiction is a maximal formula or the last formula of a segment, it duplicates it. As α and α^* have the same degree, it needs to be ensured that the step actually reduces the complexity of the deduction. So for the purpose of the proof of normalisation, the right premise of reductio will be counted as having a degree of one higher than the left premise, if both premises are maximal. This decides the question to which premise of reductio a reduction step is applied first in this and other cases.

(B) *Reduction Steps for Maximal Formulas*
(a) *Reduction Steps for Maximal Formulas with Introduction and Elimination Rules*
These are not essentially different from those for intuitionist logic given by Prawitz, except that now a $+$ or $-$ is carried along in front of formulas, and there are additional reduction steps for the signed negations of formulas. The reduction steps for

maximal formulas of the forms $+ A \vee B$ and $- A \wedge B$ are similar to those Prawitz gives for disjunctions, those for maximal formulas of the forms $- A \vee B$, $+ A \wedge B$ and $- A \supset B$ are similar to those Prawitz gives for conjunctions, those for maximal formulas of the form $+ A \supset B$ are similar to those Prawitz gives for implications, and the reduction steps for maximal formulas of the forms $+ \neg A$ and $- \neg A$ are evident enough. Applying such a reduction step may introduce new maximal formulas and segments into the reduced deduction, but they are of lower degree than the maximal formula removed from the original deduction. In cases of maximal formulas of the form $+ A \supset B$, $+ A \vee B$ and $- A \wedge B$, the reduced deduction may contain multiple copies of subdeductions of the original deduction: to avoid multiplying maximal formulas or segments of the same or higher degree than the one removed, in the proof of the normalisation theorem the reduction steps are applied to maximal formulas of highest degree such that no maximal formulas or segments of highest degree stand above them or above the minor premises of the elimination rule of which they are the major premises.

(b) *Reduction Steps for Maximal Formulas with Reductio and Elimination Rules*
In the first three reduction steps below, if $+ A$ or $- A$ is major premise of an elimination rule or premise of non-contradiction in Σ, the reduction step introduces a new r/e or r/nc maximal formula of lower degree than the one removed, which presents no problem for the proof of normalisation. In the fourth case, a more difficult issue arises.

(1) The r/e maximal formula has the form $+ A \wedge B$:

$$
\begin{array}{c}
[- A \wedge B]^i \\
\Pi \\
\dfrac{\bot}{+ A \wedge B}\ i \\
\dfrac{}{+ A} \\
\Sigma
\end{array}
\qquad \rightsquigarrow \qquad
\begin{array}{c}
\dfrac{[- A]^i}{[- A \wedge B]} \\
\Pi \\
\dfrac{\bot}{+ A}\ i \\
\Sigma
\end{array}
$$

If any occurrences of $- A \wedge B$ in the assumption class $[- A \wedge B]^i$ of the original deduction are major premises of $- \wedge E$, then the reduction step introduces new i/e maximal formulas into the reduced deduction that have the same degree as the r/e maximal formula removed from the original deduction. Remove them as part of the present reduction step by applying the reduction step for i/e formulas of the form $- A \wedge B$ to each of them immediately after the transformation above: this creates at worst new maximal formulas of lower degree than the ones removed. Similarly if any occurrences of $- A \wedge B$ in the assumption class $[- A \wedge B]^i$ of the original deduction are premises of non-contradiction the other premise of which is also derived by an introduction rule: then new i/nc maximal formulas are introduced into the deduction, which are removed immediately after the transformation above as part of the step, and then, as the reduction procedures for such formulas to be given below show, at worst maximal formulas of lower degree arise.

The case where $+ B$ has been derived by $+ \wedge E$ is similar, and so are the cases for r/e maximal formulas of the form $- A \vee B$.

(2) The r/e maximal formula has the form $- \neg A$:

$$
\begin{array}{c}
[+ \neg A]^i \\
\Pi \\
\dfrac{\bot}{- \neg A}\ i \\
\dfrac{}{+ A} \\
\Sigma
\end{array}
\qquad \rightsquigarrow \qquad
\begin{array}{c}
\dfrac{[- A]^i}{[+ \neg A]} \\
\Pi \\
\dfrac{\bot}{+ A}\ i \\
\Sigma
\end{array}
$$

As in case (1), the reduction step may introduce new i/e or i/nc maximal formulas of the same degree as the r/e maximal formula removed, and this is dealt with in the same way: apply the relevant reduction steps immediately after the transformation above as part of the reduction step for r/e maximal formulas of the form $- \neg A$.

The case for r/e maximal formulas of the form $+ \neg A$ is similar.

(3) There are three options for maximal formulas arising from reductio and elimination rules for implication:

(i) The r/e maximal formula has the form $+ B \supset A$:

$$
\begin{array}{c}
[- B \supset A]^i \\
\Pi \\
\dfrac{\bot}{+ B \supset A}\ i \qquad \dfrac{\Xi}{+ B} \\
\dfrac{\qquad\qquad\qquad}{+ A} \\
\Sigma
\end{array}
\qquad \rightsquigarrow \qquad
\begin{array}{c}
\dfrac{\overset{\Xi}{+ B} \qquad [- A]^i}{[- B \supset A]} \\
\Pi \\
\dfrac{\bot}{+ A}\ i \\
\Sigma
\end{array}
$$

The reduction step may introduce new i/e or i/nc maximal formulas of the same degree as the r/e maximal formula removed, and this is dealt with as in previous cases.

(ii) The r/e maximal formula has the form $- B \supset A$ and $- A$ is concluded:

$$
\begin{array}{c}
[+ B \supset A]^i \\
\Pi \\
\dfrac{\bot}{- B \supset A}\ i \\
\dfrac{}{- A} \\
\Sigma
\end{array}
\qquad \rightsquigarrow \qquad
\begin{array}{c}
\dfrac{[+ A]^i}{[+ B \supset A]} \\
\Pi \\
\dfrac{\bot}{- A}\ i \\
\Sigma
\end{array}
$$

The reduction step may introduce new i/e or i/nc maximal formulas of the same degree as the r/e maximal formula removed, and this is dealt with as in previous cases.

(iii) The r/e maximal formula has the form $- A \supset B$ and $+ A$ is concluded:

$$
\begin{array}{c}
[+ A \supset B]^i \\
\Pi \\
\dfrac{\bot}{\dfrac{- A \supset B}{+ A}}\ i \\
\Sigma
\end{array}
\quad \rightsquigarrow \quad
\begin{array}{c}
\dfrac{[+ A]^i \qquad [- A]^{ii}}{\dfrac{\bot}{\dfrac{+ B}{[+ A \supset B]}}\ i} \\
\Pi \\
\dfrac{\bot}{+ A}\ ii \\
\Sigma
\end{array}
$$

The reduction step may introduce new i/e or i/nc maximal formulas of the same degree as the r/e maximal formula removed, and this is dealt with as in previous cases.[8]

(4) The r/e maximal formula has the form $+ A \vee B$:

$$
\begin{array}{c}
[- A \vee B]^i \\
\Xi \\
\dfrac{\bot}{+ A \vee B}\ i \qquad
\begin{array}{c}[+ A]^{ii}\\ \Pi_1 \\ \alpha\end{array} \qquad
\begin{array}{c}[+ B]^{iii}\\ \Pi_2 \\ \alpha\end{array} \\
\hline
\alpha \quad ii,iii \\
\Sigma
\end{array}
$$

$$\rightsquigarrow$$

$$
\begin{array}{c}
\begin{array}{cc}
\begin{array}{c}[+ A]^i \\ \Pi_1 \\ \alpha \end{array} \quad [\alpha^*]^{iii} \\
\dfrac{\bot}{- A}\ i
\end{array}
\qquad
\begin{array}{cc}
\begin{array}{c}[+ B]^{ii} \\ \Pi_2 \\ \alpha \end{array} \quad [\alpha^*]^{iii} \\
\dfrac{\bot}{- B}\ ii
\end{array} \\
\hline
[- A \vee B] \\
\Xi \\
\dfrac{\bot}{\alpha}\ iii \\
\Sigma
\end{array}
$$

---

[8] If non-contradiction is restricted to atomic premises, then the reduction step is incomplete: if A is not atomic, the application of non-contradiction must be replaced by applications of non-contradiction to atomic subformulas of A. This, however, poses no difficulty, as A is of lower degree than the r/e maximal formula removed.

If α in the original deduction is $\perp$, non-contradiction is not applicable, but also not necessary: conclude $- A$ and $- B$ directly by reductio in the reduced deduction.

The reduction step for r/e maximal formulas of the form $- A \vee B$ is similar.

Π_1 and Π_2 get multiplied as many times as there are assumptions in assumption class $[- A \vee B]^i$, so it must be ensured that when choosing a maximal formula to which to apply the reduction step, Π_1 and Π_2 contain no maximal formulas or segments of highest degree. The same strategy indicated for i/e maximal formulas works here: choose a maximal formula of highest degree such that no maximal formula or segment of highest degree stands above it or above the minor premises of the elimination rule of which it is the major premise.

The reduction step may introduce new i/e or i/nc maximal formulas of the same degree as the r/e maximal formula removed, and this is dealt with as in previous cases. There are also three further cases to be considered.

First, if α is major premise of an elimination rule in Σ, the reduction step may introduce an r/e maximal formula of unknown degree into the reduced deduction. In that case, however, α is the last formula of a maximal segment in the original deduction. To show that any deduction can be brought into one in normal form, the proof of the normalisation theorem describes a method that systematically removes all maximal formulas and segments from a deduction, beginning with those of highest degree: thus if the reduction step is applied as part of this process, α cannot be of higher degree than $+ A \vee B$. In the reduction steps for maximal segments and i/e maximal formulas it was noted that they are applied to maximal formulas of highest degree such that no maximal formulas of highest degree stand above them or the minor premises of the rule of which they are major premises. We need to ensure that in case the occurrence of α in Σ is the last formula of a maximal segment of the same degree as $+ A \vee B$ or forms part of such a maximal segment that continues in Σ, then the relevant permutative reduction step is applied to the segment first. The procedure indicated in the permutative reduction steps works here, too. If both have no maximal formulas or segments of highest degree above them, we apply the relevant reduction step to the rightmost one first, that is to one of which α forms part in this case. It'll be made more precise what 'rightmost' means in the proof of the normalisation theorem.

Second, if α is the conclusion of reductio in Π_1 or Π_2, the reduction step may introduce an r/nc maximal formula of unknown degree into the reduced deduction. In that case, the application of non-contradiction in the reduced deduction is redundant and dropped from the reduction step. For example, suppose the last application of a rule in Π_1 is reductio. Then Π_1 has a subdeduction that derives $\perp$ from assumption classes $[+ A]^i$ and $[\alpha^*]$, so conclude $- A$ directly by reductio, discharging formulas in the assumption class $[+ A]^i$, and assign the formulas in the assumption class $[\alpha^*]$ in

Π_1 to the new assumption class iii and discharge them at the application of reductio that concludes with the α on top of Σ. Similarly if the last application of a rule in Π_2 is reductio, and if that is the last rule in both.

Third, if α is the last formula of a segment in Π_1 or Π_2, then the reduction step introduces a new maximal segment into the deduction: remove it by permuting the application of non-contradiction upwards as described in the permutative reduction steps above as part of the reduction step. There remains one troublesome case to be taken care of: if the first formula of the segment is concluded by reductio in the original deduction, permuting non-contradiction upwards introduces an r/nc maximal formula of unknown degree into the reduced deduction. A version of the strategy of the previous paragraph works in this case, too. If the first formula of the segment is derived by reductio, we already have a subdeduction Π_1' of Π_1 of $\perp$ from $[+ A]^i$ and $[\alpha^*]$ or a subdeduction Π_2' of Π_2 of $\perp$ from $[+ B]^{ii}$ and $[\alpha^*]$. So conclude $- A$ or $- B$ directly by reductio without the redundant step of non-contradiction and assign the formulas in the assumption class $[\alpha^*]$ of Π_1' or Π_2' to assumption class iii, discharging them at the lower application of reductio marked in the reduction step. This leaves those assumptions in Π_1' or Π_2' undischarged that were discharged by applications of $+ \vee E$ or $- \wedge E$ that gave rise to the segments in Π_1 or Π_2: so insert these applications before continuing with Σ, using the conclusion α of the lower application of reductio as the required minor premise. If α is on a segment in Σ, this increases its length. But notice that such a segment is either not maximal or of lower degree than the r/e maximal formula removed, by the choice of the strategy of choosing maximal segments or formulas in the proof of normalisation.

(c) *Reduction Steps for Maximal Formulas with Reductio and Non-Contradiction*
There are two options to be considered.

(1) The assumption discharged by the application of reductio is not premise of non-contradiction. I give as an example the case where the left premise of non-contradiction is a denial derived by reductio:

$$
\begin{array}{cc}
& [+ A]^i \\
& \Pi \\
\Sigma & \dfrac{\perp}{- A}\ i \\
\dfrac{+ A \qquad\qquad}{\quad} \\
\dfrac{\perp}{\Xi}
\end{array}
\quad\leadsto\quad
\begin{array}{c}
\Sigma \\
[+ A] \\
\Pi \\
\perp \\
\Xi
\end{array}
$$

If any of the formulas in the assumption class $[+ A]^i$ is the major premise of an elimination rule in Π and $+ A$ is the conclusion of an introduction rule or of reductio in Σ, then the reduction step introduces i/e or r/e maximal formulas into the reduced deduction. However, in this case both premises of non-contradiction are maximal, and the right one will be counted as one degree higher as the left one, and so the maximal formula created by the reduction step is of lower degree than the one removed. Similarly if $+A$ is conclusion of $+\vee E$ or $-\wedge E$ in Σ and any of the formulas in the assumption class $[+A]^i$ is major premise of an elimination rule in Π: the new maximal segment is of degree one lower than the formula removed.

If the situation is the mirror image of the one displayed and reductio concludes the left premise of non-contradiction, then the right premise is not conclusion of an elimination rule, as that one would be removed first.

(2) The assumption discharged by reductio is premise of non-contradiction, say it is the left one:

$$
\cfrac{\Sigma\quad\cfrac{\cfrac{[+A]^i\quad\cfrac{\Pi'}{-A}}{\bot}}{\cfrac{\Pi''}{\cfrac{\bot}{-A}}\,i}}{\cfrac{+A\qquad}{\bot}}\;\;\rightsquigarrow\;\;\cfrac{\cfrac{+A\quad\cfrac{\Sigma\quad\Pi'}{-A}}{\bot}}{\cfrac{\Pi''}{\cfrac{\bot}{\Xi}}}
$$

This reduction step does only what it is supposed to do: it removes one maximal formula and introduces no complications.

(d) *Reduction Steps for Maximal Formulas with Introduction Rules and Non-Contradiction*

Two examples should suffice, the other cases being similar or obvious.

(1) One premise is derived by $+\wedge I$, the other by $-\wedge I$:

$$
\cfrac{\cfrac{\Pi_1\quad\Pi_2}{+A\quad+B}}{\cfrac{+A\wedge B}{}}\quad\cfrac{\Sigma}{\cfrac{-B}{-A\wedge B}}\;\;\Bigg/\;\;\cfrac{\bot}{\Xi}\;\;\rightsquigarrow\;\;\cfrac{\cfrac{\Pi_2}{+B}\quad\cfrac{\Sigma}{-B}}{\cfrac{\bot}{\Xi}}
$$

(2) One premise is derived by $+\supset I$, the other by $-\supset I$:

$$
\cfrac{\cfrac{[+A]^i\\ \Pi\\ +B}{+A\supset B}\,i\quad\cfrac{\Sigma_1\quad\Sigma_2}{\cfrac{+A\quad-B}{-A\supset B}}}{\cfrac{\bot}{\Xi}}\;\;\rightsquigarrow\;\;\cfrac{\cfrac{\Sigma_1\\ [+A]\\ \Pi\\ +B}{}\quad\cfrac{\Sigma_2}{-B}}{\cfrac{\bot}{\Xi}}
$$

Applying the reduction steps may introduce new maximal formulas into the reduced deduction, but they are of lower degree than the maximal formulas removed from the original deduction. Choice of maximal formula to which to apply the step avoids duplicating maximal formulas of highest degree in Σ.

This completes the reduction steps for maximal formulas.

(C) *Simplification Conversions*
Applications of $+ \vee E$ and $- \wedge E$ with empty assumption classes are redundant and may be removed from deductions.

This completes the description of the transformations of deductions applied in normalisation.

The degree of a maximal formula or segment that is the right premise of reductio the left premise of which is also a maximal formula or segment is the degree of the formula (on the segment) plus 1. For all others, it is the degree of the formula (on the segment). This also settles the question to which premise reduction steps for i/nc maximal formulas are applied, although this is of comparatively minor significance.

Definition 6 (Rank of a Deduction). The *rank* of a deduction Π is the pair $\langle d, l \rangle$ where d is the highest degree of any maximal formula or segment in Π, and l is the sum of the number of maximal formulas and the sum of the lengths of all maximal segments in Π. If there are no maximal formulas or segments in Π, its rank is 0.

Ranks are ordered lexicographically: $\langle d, l \rangle < \langle d', l' \rangle$ iff either $d < d'$ or $d = d'$ and $l < l'$.

As we have been rather explicit about the considerations necessary to ensure that the complexity of a deduction is decreased in applying the reduction steps, the proof of normalisation itself can thankfully be brief. All that remains is to explicate the notion of a 'rightmost' maximal formula or segment. Here we follow Prawitz [19, 50].

Theorem 1. *Any deduction Π of α from Γ in $\mathfrak{B}$ can be brought into a deduction in normal form of α from some of Γ.*

Proof. By induction over the rank of deductions and applying the reduction steps. Take a maximal formula or maximal segment of highest degree such that (i) no maximal formula or segment of highest degree stands above it in the deduction, (ii) no maximal formula or segment of highest degree stands above a minor premise of the elimination rule of which the maximal formula or segment is the major premises, and (iii) no maximal segment of highest degree contains a formula that is minor premise of the elimination rule of which the maximal formula or maximal segment is the major premise. This reduces the rank of the deduction. *Q.e.d.*

6 Philosophical Assessment

There are at least two reasons why not everyone will be satisfied that the proof of section 5 solves the philosophical problems of section 4:

(1) It appeals to non-contradiction in its general form.
(2) The definition of 'maximal signed formula' merely changes the topic.

Let's look at each charge in turn

In reduction step (B.b.4), the one for r/e maximal formulas of the form $+ A \vee B$ and $- A \wedge B$, non-contradiction is applied to arbitrary formulas α. According to Rumfitt, if α is not atomic, the inference from α and α^* to $\perp$ needs to be replaced by applications of non-contradiction to atomic subformulas of α. The difficulty is that this may introduce new maximal formulas of unknown degree into the deduction. Consider the construction that shows how to replace premises of the form $C \vee D$ by C and D:

$$\frac{+ C \vee D \qquad - C \vee D}{\perp}$$

$$\rightsquigarrow$$

$$\frac{+ C \vee D \qquad \dfrac{+ C \ ^{i} \quad \dfrac{- C \vee D}{- C}}{\perp} \qquad \dfrac{+ D \ ^{i} \quad \dfrac{- C \vee D}{- D}}{\perp} \ ^{i}}{\perp}$$

Suppose in the original deduction displayed in the reduction step (B.b.4), α is a disjunction on a segment that is the conclusion of $+ \vee I$ or $- \wedge I$. If this segment is major premise of $+ \vee E$ or $- \wedge E$ or non-contradiction, all is fine: either α has lower degree than $- A \vee B$ or its segment is removed first. If it is not, however, then the procedure for removing complex premises of non-contradiction introduces maximal formulas of unknown degree into the reduced deduction.

The bilateralist who insists on restricting non-contradiction to atomic premises requires a different proof of normalisation from the one given here. Alternatively, the bilateralist could treat $\wedge$ and $\vee$ as defined in terms of $\supset$ and $\neg$. One might also wonder whether the restriction of non-contradiction to atomic premises is an essential element of bilateralism. It is according to Rumfitt, but the current considerations may constitute a recommendation to drop it.

Another option that solves the problems of section 4 would be to restrict reductio to atomic conclusions. Ferreira observes that once non-contradiction is restricted to atomic premises, there may be no good reason not to restrict reductio correspondingly [4,]. Rumfitt's reasons for restricting non-contradiction seem to apply just as well to reductio. Reductio is a rule of the same kind as non-contradiction, a structural rule concerning the co-ordination of assertion and denial.

Ferreira shows, however, that the resulting logic is not classical and contains neither $+ A \lor \neg A$ nor $- A \land \neg A$ as theorems. This may not be so much a defect of bilateralism, as rather the surprising or interesting result that the correct logic of bilateralism is not classical logic, but a constructive logic with strong negation. This is the position for which Wansing argues [26,]. The current considerations may add support to this line of thought. It certainly has something to be said for it. It was noted by Gibbard that dropping reductio and non-contradiction altogether from $\mathfrak{B}$ gives a constructive logic with strong negation [9]. Reading $-$ as $\neg$ and ignoring $+$, it is Nelson's logic of constructible falsity, also discussed by Prawitz [19, 96f]. While Wansing's logic adds further connectives, which require additional reduction steps, the proof of section 5 also gives normalisation theorems for logics arising from $\mathfrak{B}$ by dropping non-contradiction and reductio or restricting both to atomic formulas.

In as much as bilateralism was supposed to justify classical logic, however, this line of argument is problematic. Much of the motivation for bilateralism is to overcome Dummettian objections to classical logic, in particular that the rules for classical negation are not stable. Many bilateralists will therefore prefer a different route to excluding *honk* and *conk*.

Now for changing the subject. The requirement that r/e maximal formulas be removable from deductions is rather different from the similar requirement on i/e maximal formulas. The latter provides a formal criterion for fulfilment of Prawitz's inversion principle. Stability is a relation between the operational rules for a connective, its introduction and elimination rules. The unilateral approach locates any defects in rules for connectives in the operational rules governing them. The notion of a maximal signed formula incorporates a relation between one rule and all elimination rules. That one rule is a structural rule, concerning the formal framework of bilateralism, and so the notion of a maximal signed formula incorporates aspects of a rather different kind than those on which proof-theoretic semantics was originally built.

This objection does, I think, show something, but not that something is wrong with the present notion of a maximal signed formula. It rather exhibits a shortcoming of bilateralism. There must obtain some balance in the inferential powers of reductio and the other rules. If the rather obvious way of capturing that balance employed here is objectionable, so much the worse for bilateralism.

Where I would agree is that the solution proposed here does not really go to the heart of the matter of what is wrong with *conk* and *honk*. The problem with *tonk* lies in the mismatch of its introduction and elimination rules. One would expect a comparable diagnosis of the problem with *conk* and *honk* from bilateralism: it lies in a mismatch of their assertive and their rejective rules. Locating the problem with *conk* and *honk* in reductio is not to the point. One should expect bilateral inversion principles that provide a general basis on which to diagnose mismatch of operational rules, just as the inversion principles in the unilateral context do, where these cut across the divide of assertive and rejective rules.

7 Conclusion

The most promising solution to the problem of section 4 would be to formulate a bilateral notion of stability that incorporates bilateral inversion principles and a notion harmony between the assertive and the rejective rules of the connectives.

One proposal of how to do this has been formulated by Francez [5]. His notion of *vertical harmony* holds between assertive introduction and elimination rules and rejective introduction and elimination rules, while *horizontal harmony* holds between assertive and rejective introduction rules. Francez modifies horizontal harmony slightly in a later paper, where it is also noted that it provides a notion of harmony between rejective and assertive elimination rules [6]. Another proposal is by the present author [14].[9]

There are, however, reasons to believe that adopting a bilateral notion of stability would be counterproductive for the bilateralist.

In the unilateral framework, there are two aspects of the use, and thus meaning, of the connectives in deductive arguments: one is captured by the introduction rules and the other by the elimination rules for a connective. These aspects must be in harmony, or more precisely stable, and satisfy the inversion principles. Following Gentzen, the introduction rules for the connectives define their meanings, and the elimination rules are consequences thereof. Following Dummett and Prawitz, they are consequences in the sense that they are determined from the introduction rules by the inversion principles. As stability is a requirement on rules that are to define the meanings of the connectives completely, the process could be reversed and the elimination rules taken as prior and the introduction rules determined from them.

Transpose this to the bilateral case. The motivating thesis of bilateralism is that the meanings of the expressions of a language are determined by the conditions of the correct assertibility and the correct deniability of sentences of which they form part. The bilateralist agrees that stability must obtain between the introduction

---

[9]Both proposals allow the bilateralist to rule out the bilateral intuitionist logic of [12], the rules of which, it must be admitted, are not as nicely symmetrical as those of $\mathfrak{B}$.

and elimination rules for the connectives. Let's follow Gentzen again and pick the introduction rules as those that define the meaning of a connective, while its elimination rules are consequences of them by the bilateral notion of stability. *honk* and *conk* show that we cannot simply lay down assertive and rejective introduction rules for a connective. They, too, must be balanced by the bilateral notion of stability. But this means that only one kind of introduction rules defines the meaning of the connective, and the other is a consequence by bilateral stability.

In the absence of a principled way of deciding between the two kinds of introduction rules, we might as well pick the assertive introduction rules as defining the meanings of the connectives, all others being determined from them by bilateral stability. And now the situation looks awkward for the bilateralist. The bilateralist claims that the meanings of the connectives are defined by the assertive and rejective rules governing them. A closer look into the matter reveals that they are defined by the assertive introduction rules. That is exactly the thesis of the unilateralist. All introduction rules of unilateral logic are assertive.

Nothing hangs on the choice of assertive introduction rules as defining meaning. To rule out *conk* and *honk*, and to emulate the notion of stability of the unilateral approach, the bilateralist needs inversion principles that determine the three other sets of rules for a connective from any given one. Still, it is only one aspect of the use of the connective that defines its meaning, the others being consequences by stability, not two of them, as claimed by the bilateralist. It is not the assertive rules in tandem with rejective rules that determine the meaning of a connective, but only one half of one of those two aspects – either the assertive introduction rules, or the assertive elimination rules, or the rejective introduction rules, or the rejective elimination rules – the rest being determined by bilateral stability. Thus it looks as if adopting a bilateral notion of stability means that the characteristically bilateral thesis on how meanings are determined is effectively abandoned, and bilateralism collapses into a form of unilateralism.

This looks like a dilemma for bilateralists. Formulate a bilateral notion of stability, or else face *conk* and *honk*. But if you do the former, face giving up bilateralism.

References

[1] Michael Dummett. *The Logical Basis of Metaphysics*. Cambridge, Mass.: Harvard University Press, 1993.

[2] Michael Dummett. What is a Theory of Meaning? (I). In *The Seas of Language*, pages 1–33. Oxford: Clarendon, 1993.

[3] Michael Dummett. What is a Theory of Meaning? (II). In *The Seas of Language*, pages 34–93. Oxford: Clarendon, 1993.

[4] Fernando Ferreira. The co-ordination principles: A problem for bilateralism. *Mind*, 117(468):1051–1057, 2008.

[5] Nissim Francez. Bilateralism in proof-theoretic semantics. *Journal of Philosophical Logic*, 43(2/3):239–259, 2014.

[6] Nissim Francez. Bilateralism does provide a proof theoretic treatment of classical logic (for non-technical reasons). *Journal of Applied Logics*, 5(8):1653–1662, 2018.

[7] Michael Gabbay. Bilateralism does not provide a proof theoretic treatment of classical logic (for technical reasons). *Journal of Applied Logic*, 25(S):108–122, 2017.

[8] Gerhard Gentzen. Untersuchungen über das logische Schließen. I. *Mathematische Zeitschrift*, 39(2):176–210, 1934.

[9] Peter Gibbard. Price and Rumfitt on rejective negation and classical logic. *Mind*, 111(442):297–303, 2002.

[10] Lloyd Humberstone. The revival of rejective negation. *Journal of Philosophical Logic*, 29(4):331–381, 2000.

[11] Nils Kürbis. What is wrong with classical negation? *Grazer Philosophische Studien*, 92(1):51–85, 2015.

[12] Nils Kürbis. Some comments on Ian Rumfitt's bilateralism. *Journal of Philosophical Logic*, 45(6):623–644, 2016.

[13] Nils Kürbis. Bilateralist detours: From intuitionist to classical logic and back. *Logique et Analyse*, 60(239):301–316, 2017.

[14] Nils Kürbis. Bilateral inversion principles. *To be published*, 2019.

[15] Nils Kürbis. *Proof and Falsity. A Logical Investigation*. Cambridge University Press, 2019.

[16] Peter Milne. Inversion principles and introduction rules. In Heinrich Wansing, editor, *Dag Prawitz on Proofs and Meaning*, pages 189–224. Cham, Heidelberg, New York, Dordrecht, London: Springer, 2015.

[17] Enrico Moriconi and Laura Tesconi. On inversion principles. *History and Philosophy of Logic*, 29(2):103–113, 2008.

[18] Sara Negri and Jan von Plato. *Structural Proof-Theory*. Cambridge University Press, 2001.

[19] Dag Prawitz. *Natural Deduction*. Stockholm, Göteborg, Uppsala: Almqvist and Wiksell, 1965.

[20] Huw Price. Sense, assertion, Dummett and denial. *Mind*, 92(366):161–173, 1983.

[21] Arthur Prior. The runabout inference ticket. *Analysis*, 21(2):38–39, 1961.

[22] Ian Rumfitt. "Yes" and "No". *Mind*, 109(436):781–823, 2000.

[23] Timothy Smiley. Rejection. *Analysis*, 56(1):1–9, 1996.

[24] Gunnar Stålmarck. Normalisation theorems for full first order classical natural deduction. *Journal of Symbolic Logic*, 52(2):129–149, 1991.

[25] A.S. Troestra and H. Schwichtenberg. *Basic Proof Theory.* Cambridge University Press, 2 edition, 2000.

[26] Heinrich Wansing. A more general general proof theory. *Journal of Applied Logic*, 25(S):23–46, 2017.

Received 22 December 2019

Extended Syllogistics in Calculus CL

Jens Lemanski

FernUniversität in Hagen, Germany, Institute of Philosophy
jens.lemanski@fernuni-hagen.de

Abstract

Extensions of traditional syllogistics have been increasingly researched in philosophy, linguistics, and areas such as artificial intelligence and computer science in recent decades. This is mainly due to the fact that syllogistics is seen as a logic that comes very close to natural language abilities. Various forms of extended syllogistics have become established. This paper deals with the question to what extent a syllogistic representation in CL diagrams can be seen as a form of extended syllogistics. It will be shown that the ontology of CL enables numerically exact assertions and inferences.

Keywords: Diagrammatic Reasoning, Extended Syllogistics, Ontology, Natural Logic, Generalized Quantifiers.

1 Introduction

'Calculus CL' is a specific kind of logic diagram which is named in honor of the so-called 'Cubus Logicus', designed by the German philosopher Johann Christian Lange. At the beginning of the 18th century, Lange published his plans, based on diagrammatic reasoning, to build a logic machine. His idea was based on the combination of all features of diagrams which were well known during that time, esp. tree diagrams, spatial logic diagrams (esp. Eulerian diagrams), and squares of opposition.

One difference between Lange's original cubus logicus and its modern version CL already affects the technical use of Lange's method: Lange's idea was to have a calculus for extended syllogistics, propositional logic, modal syllogistics, etc. This should be achieved by only one mechanical device. Unlike Lange's machine, the modern diagrammatic interpretation CL requires various diagrams to depict a specific ontology and to prove inferential reasoning. This is one of the reasons why the name 'Calculus *CL*' was chosen instead of 'cubus logicus': the new name draws attention to Lange's original idea of a logic machine, on the one hand, but it indicates that the presented type of diagram is used in various other forms, on the other hand.

Furthermore, *CL* is not limited to techniques of 18th century-logic, although Lange extended traditional Aristotelian and Stoic logic in a unique way. In the last years, Lange diagrams and especially *CL* were used in various context of modern logic such as ontological reasoning [9], analogical reasoning [17], oppositional geometry [10], bitstring semantics [22] and formal systems of propositional logic [6].

In the following, I will ask the question of what kind of extended syllogistics can be implemented in *CL*. I will argue that an exact description of the ontology of *CL* leads to the use of numerically exact assertions and inferences. Numerically exact syllogistics means extending the traditional Aristotelian logic by adding expressions such as **There is exactly this and that** *p* **that is this and that** *q*. Assertions such as these can be represented and calculated by *CL* diagrams. Section 2 outlines some differences between traditional and modern extended syllogistics. In contrast to most approaches in extended syllogistics, *CL* is based on a specific ontology. This is shown in Section 3. Concerning this specific ontology, *CL* comes closer to traditional Aristotelian syllogistics. Taking a specific ontology as a basis, this allows *CL* to apply exact assertions (Section 4) that are in turn the basis of numerically exact syllogisms (Section 5). The aim of the following paper is not to set up a formal system of *CL*, but to explain how one kind of extended syllogistics can be applied in a *CL* diagram.

2 Syllogistics and Extendend syllogistics

In this section, I will give a short overview of what syllogistics is and then explain what modern extensions of syllogistics have been researched in recent years. In presenting traditional syllogistics, I will not mainly refer to Aristotle, as there are numerous themes and topics that are controversial in research on Aristotle. I will, therefore, outline a modern form of syllogistics as it was developed in the (early) modern age, e.g. in [8].

Due to the scope of this section, I will also not be able to present all the extensions of syllogistics that have been made, esp. in the last 50 years. My goal is to draw attention to the topics and theses of syllogistics and extended syllogistics that are taken over in *CL* or are treated differently in *CL*.

2.1 Doctrines of Traditional Syllogistics

Traditional syllogistics consists of three consecutive doctrines based on various paradigmatic books. Usually, all three doctrines (D1–D3) are understood compositionally, so that D1 is a prerequisite for D2 and D3 is a composition of D1+D2.

(D1) The first doctrine describes the function and structure of concepts. This doctrine goes back to Porphyrius' *Isagoge*, Seneca's *Ad Lucilium 58*, Aristotle's *Categoriae*, and *Metaphysica*. It describes mainly high-order concepts, so-called 'categories', such as substance, quantity, quality, etc, and certain functions applied on categories, so-called 'predicables', such as genus, species, difference, etc. With the help of categories and predicables, ontologies can be formed in such a way that they describe the conceptual structure and content of reality. All further doctrines of syllogistics are composed of these ontologies.

(D2) Next, assertions are formed from concepts. This doctrine goes back mainly to Aristotle's *De Interpretatione* and various commentaries. An assertion is an affirmative or negative connection of a subject and a predicate in a proposition, which is true if the referents of the subject and predicate terms are in the same positive or negative relationship as the proposition indicates. Due to the quality, an assertion can either be affirmative or negative; and due to the quantity it can either be universal or particular. Altogether there are four types of assertions, which are classified by letters according to Petrus Hispanus' mnemonics: *A* (universal affirmative), *I* (particular affirmative), *E* (universal negative), *O* (particular negative). Furthermore, assertions can form oppositional relations and can be transformed by rules of conversion.

(D3) Assertions are used (D3.1) to form inferences and (D3.2) to formulate proofs, which in turn can be applied to other sciences. (D3.1) Assertoric and modal inferences are described in Aristotele's *Analytica Priora*. Inferences called 'syllogisms' usually consist of three concepts that form three assertions, two of which are the premises and one is the conclusion. (D3.2) Proofs that specific combinations of premises and conclusions constitute a valid assertoric syllogism are formulated in Aristotle's *Analytica Posteriora*: If so-called 'imperfect syllogisms' can be reduced to perfect ones with the help of the theory of conversion, the validity of the imperfect syllogisms is proven. According to Theophrastus and Galenus, these syllogisms are usually classified into four figures, and according to Petrus Hispanus these syllogisms are assigned mnemonic names such as Barbara, Celarent, etc.

2.2 Topics of Traditional Syllogistics

In the following, I describe some topics from D1–D3 that are important in both extended syllogistics and *CL*.

According to D2, the four types of assertions are composed of a subject (S) and a predicate (P) as follows:

(A-assertion)	SaP	$\forall x\ S(x) \rightarrow P(x)$	All S is P.
(E-assertion)	SeP	$\forall x\ S(x) \rightarrow \neg P(x)$	No S is P.
(I-assertion)	SiP	$\exists x\ S(x) \wedge P(x)$	Some S is P.
(O-assertion)	SoP	$\exists x\ S(x) \wedge \neg P()x$	Some S is not P.

These four assertions are in several oppositional relationships, namely contrariety (CT), contradiction (CD), subalternation (SB), and subcontrariety (SCT):

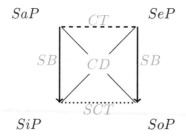

- For all CD-relations it holds that both assertions cannot be true together and cannot be false together.

- For all CT-relations it holds that both assertions cannot be true together but both can be false together.

- For all SCT-relations it holds that both assertions can be true together but both cannot be false together.

- For all SB-relations it holds that the universal assertion entails the particular one.

According to D3, a syllogism consists of three assertions. Each concept occurs exactly twice, in distinct assertions of a syllogism. The concept that is found only in the premises and thus connects S and P in the conclusion is called 'medius' M. The four perfect syllogisms are characterized above all by the fact that they express all four types of assertions in the conclusion:

(A-conclusion)	Barbara	$MaP, SaM \vdash SaP$
(E-conclusion)	Celarent	$MeP, SaM \vdash SeP$
(I-conclusion)	Darii	$MaP, SiM \vdash SiP$
(O-conclusion)	Ferio	$MeP, SiM \vdash SoP$

2.3   Topics of Extended Syllogistics

Traditional syllogistics is usually regarded as a fragment of predicate logic and is thus limited in its expressiveness. In the following, we will refer to any form of syllogistics as extended that increases the expressivity of traditional syllogistics and thus goes beyond the designated fragment of predicate logic. In contrast to artificial systems of formal logic, syllogistics is considered to be very close to natural language (see [1], [12]). This is of particular interest in the field of artificial intelligence, language processing, and ontology engineering. After all, the goal here is to understand or simulate natural human abilities. Furthermore, extended syllogistics are often considered as independent logical systems, which are not regarded as fragments of e.g. first-order logic, even if both have similarities.

Similar to an economic cycle, there are 'periods of expansion and recession' in the history of extended syllogistics. Already in Late Antiquity and the Middle Ages, there were various extensions of syllogistics such as relational syllogisms in Galen, Buridan et al. (cf. [16]). In the 17th and early 18th century there was another period of expansion, as can be seen for example in the logics of Leibniz or Lange (cf. [8]). After a long period of recession, several authors of the 19th century simultaneously criticized that *A*-assertions, in particular, were formulated in an unclear manner, since it is not obvious whether **every S** is identical either to only **some P** or to **all P** (cf. [13]). This has aroused the suspicion that the four assertions of traditional syllogistics fall too short.

Since the middle of the 19th century two approaches to extended syllogistics developed from this criticism (cf. (cf. [4]): Hamilton added quantifiers at the predicate position, such as $(\forall x \, (Sx) \rightarrow \forall y \, (Py)) \rightarrow x = y$. De Morgan, Boole and Jevons introduced numerical syllogisms including counting quantifiers at the subject and predicate position. This makes it possible to form definite assertions (using **at least, at most**) or exact assertions (using **exactly**) such as **45 *S* (or more) are each of them one of exactly 70** *P* $(\exists_{\geqslant 45} x \, Sx \wedge \exists_{70} x \, Px)$.

With the beginning of Frege-Russell-Whitehead logic, syllogistics were often criticized in the late 19th or early 20th century of not being able to express relations or to make assertions that went beyond the categorical judgements. Syntax and semantics did not appear to be clearly separated from each other, thus preventing axiomatization. Furthermore, syllogistics were accused of the fact that concepts in assertions are only 1-place predicates, and previously developed extensions of the syllogistics that include 2-place predicates were ignored for a long time.

Since the middle of the 20th century, these misinterpretations were gradually relativized (cf. [21]). A turning point came with Mostowski's seminal paper *On a Generalization of Quantifiers* in which a natural generalization of the logical quantifiers were studied. This work was continued by numerous logicians and linguists in fields such as natural semantics, generalized quantifier theory, Montague grammar, term logic, or natural logic ([7]).

Since the 1980s one can speak of another peak period of (extended) syllogistics, which has not only lasted until today but is increasing (cf. e.g. [23]). One reason for this is that traditional syllogistics can not only be extended but can also be interpreted as a kind of 'natural logic' (cf. [14], [21]). Syllogistics today is seen as the prototype of a logic that comes closest to the natural linguistic abilities of concept users. From this perspective, syllogistics has considerable potential for technical areas in which human faculties are transferred to artificial systems (cf. [16]).

I refer here to three possible extensions of assertoric syllogistics, which are relevant for the following considerations. All three types of extensions can be combined with each other:

1) *Relational Syllogistics* is a kind of extended syllogistics using transitive verbs between concepts, which can be reflexive, symmetric, transitive, etc.

2) *Nummerical Syllogistics* is a kind of extended syllogistics in which numerical quantifiers such as `at least`, `at most`, `exactly` for cardinals k are involved.

3) *Syllogistic Systems* are a form of extended syllogistics in which axiomatizations and rules are applied in order to obtain a sound and complete system or to increase the expressiveness compared to other logic systems.

In recent years, for example, Pratt-Hartmann has researched on *numerically definite syllogistic, numerically definite relational syllogistics* and has shown that there is a *relational syllogistic* that is sound and complete [18], [20]. Furthermore, Van Rooij has argued that there is an extended syllogistic that can complete the fragment of propositional logic [21]. From this, it can be concluded that a very large fragment of predicate logic can already be described with extended syllogistics today. Moss goes even further and interprets various extensions of syllogistics as independent fragments of natural language [15].

In the following, an extended syllogistics for CL is given, which differs in several points from other current approaches. While most of today's syllogistics start directly with assertions (D2), inferences, or proofs (D3.1, D3.2), CL is based on a previously defined ontology (D1), which is described in Section 3. Based on this ontology CL concretizes the numerically definite syllogistic (using `at least`, `at most`) with exact quantifiers (using `exactly`). This will be shown in Sections 4 and 5.

3 Ontologies in *CL*

Section 3.1 first defines the basic terms needed to construct and describe the ontology of a *CL* diagram. In Section 3.2, two *CL* diagrams are constructed and explained as examples.

3.1 Construction of a *CL* Diagram

A *CL* diagram has the shape of a square (2-d) or a cube (3-d) in order to represent a well-structured ontology (cf. [5]). Various geometric forms can be applied to this square or cube in order to analyse, depict, or prove logical relations given in the ontology. In what follows, I will speak only of 2-d, not of 3-d *CL* diagrams. But one can imagine a 3-d *CL* as a regular hexahedron including at least two squares that work together in a similar way as described below for one square.

In general, a 2-d *CL* diagram consists of several rows of boxes that can have different sizes and represent the extension of concepts or sets (called 'classes') and objects or members (called 'basics'). The latter is displayed in small solid boxes on the bottom row of the square diagram, the former in larger solid boxes in the upper rows. Since the classes consist of basics, the basics contained in classes are represented in all higher rows by dotted lines inside of solid boxes. A solid box is one in which all sides of the box are represented by solid lines. Dotted boxes are rectangles with four dotted lines; however, several or even all dotted lines can be covered by solid lines of a solid box. Since basics can be contained in classes but contain nothing themselves, classes and basics can be easily distinguished: Classes are solid boxes that contain dotted boxes. Basics are solid boxes that do not contain dotted boxes. The following definitions repeat in brief what has been said:

Definition 1 (Solid and dotted boxes). *A solid box in a CL diagram is a rectangle bounded by four solid lines. A dotted box in a CL diagram is a rectangle bounded by four dotted lines. If the line of a dotted box is at the same position as the line of a solid box, then only the line of the solid box is visible in the CL diagram.*

Definition 2 (Basics and classes). *Basics are solid boxes at the bottom row of the CL diagram which do not contain dotted boxes. Basics are represented in all higher rows by dotted boxes. Classes are solid boxes (at all other rows above the row of basics) which contain two or more dotted boxes.*

Example. Let ☐ and ☐ be two adjacent basics at the bottom row of the *CL* diagram, then they are represented by two adjacent dotted boxes, ⬚ and ⬚ , at all higher rows in the diagram directly above the two given solid boxes. Since in all higher rows at least two adjacent and represented basics form a class, ⬚ is a class including ⬚ and ⬚ , which represent itselves □ and □.

Two valid types of *CL* diagrams can be built, namely regular and irregular diagram types. Both diagram types differ in the number of basics in the bottom row of the square. The more basics a *CL* diagram has, the more classes can be constructed in higher rows.

A regular *CL* diagram is based on 2^n basics. Any *CL* diagram that is not based on 2^n basics can be called irregular. According to definition (2), basics are illustrated by small boxes of solid lines on the lowest row of the diagram. In a regular *CL* diagram each 2-tuple of basics form one higher class which is depicted by a solid box on the row above both basics. This box has the same horizontal length as the pair of basics given below. Every class in a higher row is a formation of two classes which are given below. In sum, each row has half as many boxes as the number of boxes in the row below. However, all boxes in a row are distributed on the same length as all boxes of basics together on the bottom row.

Since the number of basics determines the size of the *CL* diagram, the following principle can be stated: the more basics there are, the more classes the *CL* diagram contains and the larger it is. However, there is no limit to the size of the *CL* diagram, which means that *CL* diagrams can be enlarged at will.

Numerous mathematical methods and models can be used to describe the ontology of a *CL* diagram. In the following, I will first use the descriptions already introduced and add some definitions:

Definition 3 (Matrix). *Each basic forms a column with a solid box at the bottom row and dotted boxes directly above. In each higher row, the dotted boxes form a row of classes, which is supplemented by further rows above, so that a square is formed. Each solid and all dotted boxes can thus be described by the row index m and a column index n of a m × n-matrix:*

$$a_{mn} = \begin{bmatrix} a_{11} & a_{12} & \cdots & a_{1n} \\ a_{21} & a_{22} & \cdots & a_{2n} \\ \vdots & \vdots & \ddots & \vdots \\ a_{m1} & a_{m2} & \cdots & a_{mn} \end{bmatrix}.$$

Since each class in a row comprises several columns, classes are described by a row vector.

Definition 4 (Unique Identifier). *Unique identifiers are symbols to represent boxes or the set of boxes given in a CL diagram.*

For example, for each solid box, a unique identifier, such as the Roman alphabet with capital letters, $\Sigma = \{A, B, C...\}$, is given. For each dotted box, the row and column index, as given in Def. 3, can be used as a unique identifier.

3.2 Examples of CL diagrams

In what follows, two examples, for regular square CL diagrams are shown. **Ex1** will illustrate a larger diagram with 8 basics and **Ex2** a smaller one with 4 basics, which will be used as an example in all following sections.

Ex1 shows a diagram including eight basics for $m = 4$, i.e. $H = [a_{41}], I = [a_{42}], J = [a_{43}], \ldots, O = [a_{48}]$, which is thus called a CL^8 diagram. The row vector for $m = 3$, including four solid boxes as classes, comprises $D = [a_{31}\ a_{32}], \ldots, G = [a_{37}\ a_{38}]$; for $m = 2$, two solid boxes as classes comprises $B = [a_{21}\ a_{22}\ a_{23}\ a_{24}], C = [a_{25}\ a_{26}\ a_{27}\ a_{28}]$; one solid box as the highest class comprises all columns or basics, i.e $A = [a_{11} \ldots a_{18}]$. The dotted boxes represent the basics contained in classes: For example, the basic H is given in a_{41} and furthermore comprised in D as a_{31}, in B as a_{21} and in A as a_{11}.

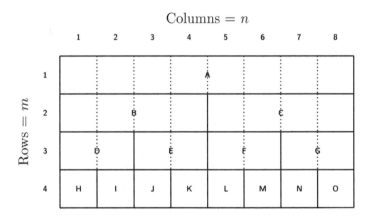

Ex1. A regular CL^8 diagram

Ex2 shows a CL^4 diagram including four basics for $m = 3$, i.e. $D = [a_{31}], E = [a_{32}], F = [a_{33}], G = [a_{34}]$. Since CL^4 is half as large as CL^8, described above, the number of classes as well as the row vector of each class are also smaller: for $m = 2$, two classes are given, i.e. $B = [a_{21}\ a_{22}], C = [a_{23}\ a_{24}]$; for $m = 1$, the highest class comprises four dotted boxes, $A = [a_{11}\ a_{12}\ a_{13}\ a_{14}]$, which correspond with all four basics $D = a_{31}, E = a_{32}, F = a_{33}, G = a_{34}$ or two classes $B = a_{21\ 22}, C = a_{23\ 24}$.

Compared to all other regular CL diagrams, CL^4 has a very low expressivity in terms of basics and classes. Nevertheless, it has sufficient scope to illustrate assertions and inferences of extended syllogistics in regular CL diagrams without loss of generality (see [6]). We will therefore only use CL^4 as an example in the following since all results can be easily transferred to other regular CL diagrams.

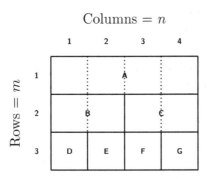

Ex2. A regular CL^4 diagram

4 Assertions in CL

Assertions can be expressed in CL by using arrows of straight lines between solid boxes of the ontology. The reading direction of an assertion follows the arrow from shaft to head.

In the following, I will first define how arrows can be used correctly in CL (syntax). In a second step, I will show that the arrows can be read as assertions (semantics).

4.1 Syntax of Arrows

First, it is important to show how to draw an arrow in a CL diagram.

Definition 5 (Arrows). *In a CL diagram, arrows should be drawn by using a straight line in such a way that the arrow shaft is located in one solid box and the arrowhead in another solid box.*

Next, the direction of the arrows in the ontology must be defined. In doing so, an even more precisely defined notation is introduced.

Definition 6 (Arrow direction). *In the ontology of CL, four directions of arrows are possible to draw: bottom-up ($\uparrow\uparrow$), top-down ($\uparrow\downarrow$), horizontal ($\rightleftarrows$), and transversal arrows ($\nearrow$).*

Now it is necessary to differentiate more precisely of what is a well-formed diagram by using arrows.

Definition 7 (Well-formed diagram). *For each arrow, the longest straight line between two solid boxes that corresponds to its orientation has to been drawn. In detail:*

(C1) For all vertical ($\uparrow\uparrow$, $\uparrow\downarrow$) and horizontal arrows ($\rightleftarrows$), the straight line of the arrowhead and shaft must form an angle of $90°$ with the most distant sides of the two solid boxes involved.

(C2) For all transversal arrows ($\nearrow$), the straight line of the arrowhead and shaft must form an internal angle of less than $90°$ with the most distant vertices of the two solid boxes involved such a way that the main diagonal or anti-diagonal of the matrix are given.

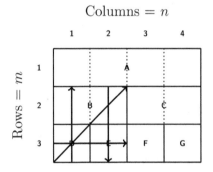

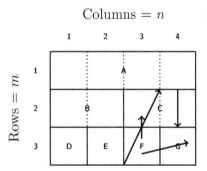

Ex3. Well-formed diagram   **Ex4.** Not well-formed diagram

Examples. In the CL^4 diagram of **Ex3** the arrows between the solid boxes of B $[a_{21}\ a_{22}]$, D $[a_{31}]$, $and E$ $[a_{32}]$ show well-formed assertions or a well-formed diagram.

 $\uparrow\uparrow$ from a_{31} to a_{21} fullfils C1 with a $90°$ angle at the arrow shaft of a_{31} and the arrow head in a_{21}.

$\uparrow\downarrow$ from a_{22} to a_{32} fullfils C1 with a $90°$ angle at the arrow shaft of a_{22} and the arrow head in a_{32}.

$\rightleftarrows$ from a_{31} to a_{32} fullfils C1 with a $90°$ angle at the arrow shaft of a_{31} and the arrow head in a_{32}.

$\diagup$ from a_{31} to a_{22} fullfils C2 with a $45°$ internal angle at the arrow shaft of a_{31} and the arrow head in a_{22} such that $\diagup$ forms the antidiagonal $[a_{31}\ a_{22}]$.

The arrows of the boxes $C\ [a_{23}\ a_{24}]$, $F\ [a_{33}]$, and $G\ [a_{34}]$ in **Ex4** are not well-formed for several reasons:

- The bottom-up arrow ($\uparrow\uparrow$) in a_{33} and a_{23} is neither does not form a $90°$ angle with the most distant sides of F and C and thus does not fullfil C1.

- The arrow shaft and the head of $\uparrow\downarrow$ in a_{24} are in one and the same box and thus do not fulfill C1.

- The arrow going from a_{33} to a_{34} does not meet the most distant sides of the two solid boxes. Furthermore, it is not obvious whether the arrow is horizontal ($\rightleftarrows$) or transversal ($\diagup$).

- The line of the transversal arrow ($\diagup$) creates at head and shaft an internal angle less than $90°$ with the most distant vertices of a_{33} and a_{23}. However, since a_{23} is not a solid box, but rather a dotted, C2 is not fulfilled. Furthermore, $[a_{33}\ a_{23}]$ is not an antidiagonal, thus the arrow is not a well-formed assertion.

4.2 Semantics of Arrows

Having now learned how to draw arrows in the ontology of CL, we must now define what the arrows mean.

Definition 8 (Meaning of Arrows). *The solid box including a fletching of a straight line arrow indicates the 1st concept (subject) and the solid box with the arrowhead the 2nd concept (predicate) of an assertion.*

Since the arrows depict the 1st and 2nd term of an assertion, it is possible to translate the arrows into A-,E-,I-,O-assertions. Instead of the mnemonic between the two terms to be connected (SaP, SeP, SiP, SoP) it is also possible to place the arrow symbols ($\uparrow\uparrow$, $\uparrow\downarrow$, $\rightleftarrows$, $\diagup$), which correspond to the diagrammatic representation.

The orientation of an arrow in the CL diagram corresponds to the orientation of arrows or lines in the square of opposition, as given in Section 2.2: Vertical arrows are positive ($\uparrow\uparrow$, $\uparrow\downarrow$), all other arrows are negative ($\rightleftarrows$, $\diagup$).

Definition 9 (Translatability). *A straight line arrow fulfilling C1 or C2 can be translated as follows:*

$S \uparrow\uparrow P = SaP$

`All S is P.`

$S \rightleftarrows P = SeP$

`No S is P.`

$S \uparrow\downarrow P = SiP \lor SaP$

`Some S is P. or All S is P.`

$S \diagdown P = SoP \lor SeP$

`Some S is not P. or No S is P.`

In total there are six different interpretations of the four types of arrows or assertions. This is due to the fact, that the particular assertions are ambiguous. However, in contrast to Euler diagrams for example (cf.[2]), this ambiguity of particular assertions is minimized to two possibilities. The individual arrows are explained in more detail below:

$S \uparrow\uparrow P$ is a universal affirmative assertion (SaP) if arrows are drawn (in accordance to C1) between all dotted boxes of one solid box bottom-up to some dotted boxes corresponding to another solid box.

$S \rightleftarrows P$ is an universal negative assertion (SeP) if a horizontal arrow is drawn between two solid boxes according to C1.

$S \uparrow\downarrow P$ is a particular affirmative assertion (SiP) if the arrow shaft is in some but not all dotted boxes of one solid box and leads to some other dotted boxes of another solid box. However, if only one bottom-up arrow can be drawn between a basic and a class, no distinction can be made between particular and universal affirmative assertions. In this case $SiP = SaP$.

$S \diagdown P$ is a particular negative assertion (SoP) if a transversal arrow is drawn between two solid boxes according to C2 and vertical arrows can be drawn between both boxes. If no vertical arrows can be drawn between two solid boxes connected by a transversal arrow, $SoP = SeP$.

Note: In our logical notation, double bottom-up arrows ($\uparrow\uparrow$) are used when all possible arrows from all dotted boxes of a lower solid box are drawn in the diagram. If the arrow shafts are only in some dotted boxes of a solid box, this is represented in the notation by bottom-up/top-down arrows ($\uparrow\downarrow$). For example, if only one bottom-up arrow of is drawn from one dotted box of B to one dotted box of A, although B has two dotted boxes, $\uparrow\downarrow$ is used. If the second possible bottom-up arrow is drawn from B to A, $\uparrow\uparrow$ is used.

Example. With the help of the definitions given so far, we are now able to identify the well-formed diagrams of **Ex3**, describe them with the *CL* notation, and translate them into the well-known notation of traditional syllogistics.

$$D \uparrow\uparrow B = DaB \qquad\qquad D \rightleftarrows E = DeE$$
All D is B.   No D is E.

$$B \uparrow\downarrow E = BiE \qquad\qquad D \diagup B = DoB$$
Some B is E.   Some D is not B.

If we would end up with this result, however, we would only have designed a diagrammatic description of a traditional syllogism using a concrete ontology. But the goal is to show that the diagrammatic description of *CL* provides much more exact information that goes beyond traditional syllogistics.

And a more precise description is certainly necessary, since according to traditional syllogistics (see Section 2.2) the assertions $D \uparrow\uparrow B$ and $D \diagup B$ are contradictions. To put it in other words, how can it be possible that both assertions, *All D is B* and *Some D is not B*, should be true at the same time? This is not possible for classic syllogistics. But already in diagram **Ex3**, we see that both assertions refer to two different dotted boxes of *B*. Therefore we need an exact description of *CL* in order to solve the problem of alleged contradictions.

5   Exact Assertions and Inferences in *CL*

Since a *CL* ontology can be described exactly with the $m \times n$-matrix, exact information about which sets of basics and classes are related to each other in assertions can be given. In order to read numerically exact assertions from *CL*, the unique identifiers of the solid boxes $\{A, B, \ldots\}$ and the dotted boxes corresponding to the respective basics in brackets are described. Section 5.1 first explains how to describe assertions in *CL* exactly. Since syllogisms consist of assertions, this technique can also be applied to syllogisms in Section 5.2.

5.1   Exact Assertions

So far we have only been able to make logical assertions about solid boxes. Quantifiers were applied to solid boxes, which were themselves identified by letters of the Roman alphabet. But now we would like to make these assertions more precise by identifying which components of these solid boxes we mean exactly. So we have to add unique identifiers also for the dotted boxes in logical assertions.

Definition 10 (Exact Assertion). *An exact assertion in CL adds to the unique identifiers of the solid boxes also the unique identifiers for dotted boxes.*

In the method used here, the additional information are the components of the $m \times n$-matrix.

Example. We take again the above-given diagram **Ex3** and add the unique identifiers of the $m \times n$-matrix:

$$D[a_{31}] \uparrow\uparrow B[a_{21}] \qquad D[a_{31}] \rightleftarrows E[a_{32}]$$

$$B[a_{22}] \uparrow\downarrow E[a_{32}] \qquad D[a_{31}] \diagup B[a_{22}]$$

There are several possible ways to translate this notation into common language. A simple paraphrase would be to specify the general assertions of syllogistics with the help of exact terms:

- $D[a_{31}] \uparrow\uparrow B[a_{21}] = $ All D is B means here exactly that $[a_{31}]$ of D is $[a_{21}]$ of B.

- $B[a_{22}] \uparrow\downarrow E[a_{32}] = $ Some B is E means here exactly that $[a_{22}]$ of B is $[a_{32}]$ of E.

- $D[a_{31}] \rightleftarrows E[a_{32}] = $ No D is E means here exactly that $[a_{31}]$ of D is not $[a_{32}]$ of E.

- $D[a_{31}] \diagup B[a_{22}] = $ Some D is not B means here exactly that $[a_{31}]$ of D is not $[a_{22}]$ of B.

We can now see not only from the diagram but also from the notation that the CD-problem mentioned at the end of Section 4.2 does not apply at all. In the CL diagram used here, $D \uparrow\uparrow B$ and $D \diagup B$ are not contradictions, since (some) B means something different in both assertion, once $B[a_{21}]$ and once $B[a_{22}]$. In Hamiltonian Syllogistics, for example, we could say something like All D is some B, but these D are not some other B, in order to avoid the CD-problem. This would have already extended traditional syllogistics (see [4], [19]).

In CL, however, we can say diagrammatically and with the help of the $m \times n$-matrix even more precisely what some (other) B's are: All D are these B, namely $B[a_{21}]$, and some D are other B, namely $B[a_{22}]$. But since we can describe not only the predicate term but also the subject term exactly in CL, All D is B and Some D is not B actually mean as much as $D[a_{31}]$ is $B[a_{21}]$ and $D[a_{31}]$ is not $B[a_{22}]$.

5.2 Exact Inferences

With the help of the technique given in Section 5.1, syllogisms can now be described exactly. For this we have to consider the following definition.

Definition 11 (Syllogistic Patterns). *A syllogism in CL is a pattern of three arrows, such that all six ends of the arrows must meet in three different solid boxes.*

Example. The perfect syllogisms mentioned in Section 2.2 are now taken as examples. We substitute basics and classes for the three terms so that the definitions of syllogistic patterns and translatability are fulfilled.

The definition of syllogistic patterns gives us several possibilities to apply patterns such as Barabara, Celarent, Darii, Ferio, etc. to one ontology of a *CL* diagram. In the following, we give all the possibilities for the four modes mentioned in a CL^4 diagram, but only the first of them (marked with *) is shown diagrammatically:

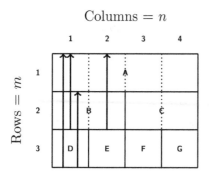
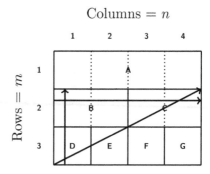

Modus Barbara*   Modus Celarent*

Barbara $(MaP, SaM \vdash SaP)$
$B[a_{21}\ a_{22}] \uparrow\uparrow A[a_{11}\ a_{12}], D[a_{31}] \uparrow\uparrow B[a_{21}] \vdash D[a_{31}] \uparrow\uparrow A[a_{11}]^*$
$B[a_{21}\ a_{22}] \uparrow\uparrow A[a_{11}\ a_{12}], E[a_{32}] \uparrow\uparrow B[a_{22}] \vdash E[a_{32}] \uparrow\uparrow A[a_{12}]$
$C[a_{23}\ a_{24}] \uparrow\uparrow A[a_{13}\ a_{14}], F[a_{33}] \uparrow\uparrow C[a_{23}] \vdash F[a_{33}] \uparrow\uparrow A[a_{13}]$
$C[a_{23}\ a_{24}] \uparrow\uparrow A[a_{13}\ a_{14}], G[a_{34}] \uparrow\uparrow C[a_{24}] \vdash G[a_{34}] \uparrow\uparrow A[a_{14}]$

Celarent $(MeP, SaM \vdash SeP)$
$B[a_{21}\ a_{22}] \rightleftarrows C[a_{23}\ a_{24}], D[a_{31}] \uparrow\uparrow B[a_{21}] \vdash D[a_{31}] \nearrow C[a_{23}\ a_{24}]^*$
$B[a_{21}\ a_{22}] \rightleftarrows C[a_{23}\ a_{24}], E[a_{32}] \uparrow\uparrow B[a_{22}] \vdash E[a_{32}] \nearrow C[a_{23}\ a_{24}]$
$C[a_{23}\ a_{24}] \rightleftarrows B[a_{21}\ a_{22}], F[a_{33}] \uparrow\uparrow C[a_{23}] \vdash F[a_{33}] \nearrow B[a_{21}\ a_{22}]$
$C[a_{23}\ a_{24}] \rightleftarrows B[a_{21}\ a_{22}], G[a_{34}] \uparrow\uparrow C[a_{23}] \vdash G[a_{34}] \nearrow B[a_{21}\ a_{22}]$

Darii $(MaP, SiM \vdash SiP)$

$D[a_{31}] \uparrow\uparrow B[a_{21}], A[a_{11}] \uparrow\downarrow D[a_{31}] \vdash A[a_{11}] \uparrow\downarrow B[a_{21}]^*$

$E[a_{32}] \uparrow\uparrow B[a_{22}], A[a_{12}] \uparrow\downarrow E[a_{32}] \vdash A[a_{12}] \uparrow\downarrow B[a_{22}$

$F[a_{33}] \uparrow\uparrow C[a_{23}], A[a_{13}] \uparrow\downarrow F[a_{33}] \vdash A[a_{13}] \uparrow\downarrow C[a_{23}]$

$G[a_{34}] \uparrow\uparrow C[a_{24}], A[a_{14}] \uparrow\downarrow G[a_{34}] \vdash A[a_{14}] \uparrow\downarrow C[a_{24}]$

Ferio $(MeP, SiM \vdash SoP)$

$B[a_{21}\ a_{22}] \rightleftarrows C[a_{23}\ a_{24}], A[a_{11}] \uparrow\downarrow B[a_{21}] \vdash A[a_{11}] \diagup C[a_{23}\ a_{24}]^*$

$B[a_{21}\ a_{22}] \rightleftarrows C[a_{23}\ a_{24}], A[a_{12}] \uparrow\downarrow B[a_{22}] \vdash A[a_{12}] \diagup C[a_{23}\ a_{24}]$

$B[a_{21}\ a_{22}] \rightleftarrows C[a_{23}\ a_{24}], A[a_{11}\ a_{12}] \uparrow\downarrow B[a_{21}\ a_{22}] \vdash A[a_{11}\ a_{12}] \diagup C[a_{23}\ a_{24}]$

$D[a_{31}] \rightleftarrows E[a_{32}], B[a_{21}] \uparrow\downarrow D[a_{31}] \vdash B[a_{21}] \diagup E[a_{32}]$

$F[a_{33}] \rightleftarrows G[a_{34}], C[a_{23}] \uparrow\downarrow F[a_{33}] \vdash C[a_{23}] \diagup G[a_{34}]$

Modus Darii*   Modus Ferio*

6   Conclusion and Discussion

In the previous sections, we have shown how CL diagrams can be used to develop numerically exact syllogisms. Similar to traditional syllogistics, we first focused on the ontology. With the help of this conceptual structure, we could define what assertions are and how they can be represented in a CL diagram using straight arrows. With the two types of unique identifiers (the Roman alphabet and the $m \times n$-matrix), we were able to make assertions about entire basics and classes on the one hand and to name certain elements of the classes precisely on the other. This technique can be described as a kind numerical syllogistics (see Section 2.3) using a quantifier similar to `exactly`. Thus, I propose to call the extended syllogistics that can be read from CL diagrams a 'numerically exact syllogistics'.

The approach presented here is not yet a formal system for which soundness and completeness have been proved. This is one of the next steps, once all definitions and rules have been formulated. A further step would be to highlight whether and how the inferential technique given in Section 5.2 could be useful in computational understanding of natural language. However, the task in this paper was only to answer the question of which kind of extended syllogistics can be derived from CL diagrams.

For example, by looking at the first three syllogistic patterns of Ferio given in Section 5.2, it is striking that in traditional syllogistics they would indicate only one inference about the same three classes, i.e. $BeC, AiB \vdash AoC$. In the CL^4 diagram, however, one can see that the same inference in traditional syllogistics can be represented by three exact patterns. (In larger regular CL diagrams there would be correspondingly more applications of the Ferio-pattern). The difference between traditional syllogistics and the extension in CL is mainly the interpretation of AiB: The first two specified Ferio-pattern contain only partial expressions of B, whereas the third pattern is applied to the whole of B. Since B is composed of two dotted boxes, $[a_{21}\ a_{22}]$ can again be used to determine which part of the class is involved in diagrammatic reasoning. This can be the part of B that corresponds to D or that corresponds to E:

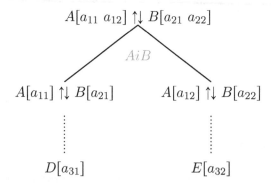

In our everyday language, we are often confronted with such assertions as AiB, where we have to or want to distinguish exactly which relation between A and B we are talking about. Some A are B can mean all B or even some B. If it is some B, the question may arise as to which constituents of B it is exactly, and if it is all B, the question of what there is in B may arise. At least since Quine, we know that these questions posed from a logical point of view can become unpleasant if we have not thought about our ontology before. And in CL, ontology, first of all, means the commitment to a certain arrangement of basics and classes.

7 Acknowledgement

This paper was presented at the conference *Assertion and Proof (WAP 2019)*, Lecce (Italy), 12–14 September 2019. I would like to thank the organizers and participants of the conference for their comments and discussions on the topic presented here. I would also like to thank Reetu Bhattacharjee as well as two anonymous reviewers for their suggestions that have helped to improve the paper.

References

[1] van Benthem J. (1986) Natural Logic. In: Essays in Logical Semantics. Studies in Linguistics and Philosophy, vol 29. Springer, Dordrecht, pp. 109–119.

[2] Bernhard, P. (2008) Visualizations of the Square of Opposition, Logica universalis 2 , 31–41.

[3] Hacker, E., Parry, W. (1967) Pure numerical Boolean syllogisms, Notre Dame Journal of Formal Logic 8:4, pp. 321–324.

[4] Heinemann, A.S. (2014) Quantifikation des Prädikats und numerisch definiter Syllogismus. Die Kontroverse zwischen Augustus De Morgan und William Hamilton: Formale Logik zwischen Algebra und Syllogistik. mentis, Münster.

[5] Jansen, L. (2008) Classifications. In: Munn, K., Smith, B. (Eds.) Applied Ontology: An Introduction. Ontos, Heusenstamm, pp. 159–173.

[6] Jansen, L., Lemanski, J. (2020) Calculus *CL* as a Formal System. In: A.-V.Pietarinen, P. Chapman, L. Bosveld-de Smet, V. Giardino V., J. Corter, S. Linker (Eds.) Diagrammatic Representation and Inference. Diagrams 2020 (LNCS 12169). Springer, Cham, pp. 445–460.

[7] Keenan, E.L., Westerståhl, D. (2011) Generalized Quantifiers in Linguistics and Logic. In: J. van Benthem, A. ter Meulen (Eds.): Handbook of Logic and Language, 2nd ed., Elsevier, Amsterdam, pp. 859–910.

[8] Lange, J.C. (1714) Inventvm novvm quadrati logici vniversalis, Müller, Giessen (Gissae-Hassorum).

[9] Lemanski, J. (2018) Automated Reasoning and Ontology Editing with Calculus *CL*. In: P. Chapman, G. Stapleton, A. Moktefi, S. Perez-Kriz, F. Bellucci (Eds.): Diagrammatic Representation and Inference. 10th International Conference, Diagrams 2018, Edinburgh, UK, June 18-22, 2018, Proceedings (LNAI 10871). Springer, Cham, pp. 752–756.

[10] Lemanski, J. (2017) Oppositional Geometry in the Diagrammatic Calculus *CL*. In: South American Journal of Logic 3:2, pp. 517–531.

[11] Lemanski, J. (2020) Euler-type Diagrams and the Quantification of the Predicate. In: Journal of Philosophical Logic 49 (2020), pp. 401–416.

[12] Angeli, G., Manning, C.D. (2014) NaturalLI: Natural Logic Inference for Common Sense Reasoning. Proceedings of the 2014 Conference on Empirical Methods in Natural Language Processing (EMNLP). Stroudsburg PA: ACL, pp. 534–545.

[13] Moktefi, A. (2020) Schopenhauer's Eulerian diagrams. In: J. Lemanski (Ed.), Langage, Logic, and Mathematics in Schopenhauer. Cham, Birkhäuser, pp. 121–146.

[14] Pratt-Hartmann, I., Moss, L. S. (2009) Logics for the Relational Syllogistic. Review of Symbolic Logic 2:4, pp. 647–683.

[15] Moss, L. S. (2008) Completeness Theorems for Syllogistic Fragments. In: F. Hamm, S. Kepser (Eds.) Logics for Linguistic Structures Berlin: de Gruyter, pp. 143–175.

[16] Nilsson, J. F. (2015) In pursuit of natural logics for ontology-structured knowledge bases. In The Seventh International Conference on Advanced Cognitive Technologies and Applications.

[17] Barbot, N., Miclet, L., Prade, H., Gilles, R. (2019) A New Perspective on Analogical Proportions. In: Kern-Isberner G., Ognjanović Z. (Ed.) Symbolic and Quantitative Approaches to Reasoning with Uncertainty. ECSQARU 2019. LNCS, vol. 11726. Springer, Cham, pp. 163–174.

[18] Pratt-Hartmann, I.(2008) On the Computational Complexity of the Numerically Definite Syllogistic and Related Logics. The Bulletin of Symbolic Logic 14:1, pp. 1–28.

[19] Pratt-Hartmann, I. (2011) The Hamiltonian Syllogistic. Journal of Logic, Language and Information 20, pp. 445–474.

[20] Pratt-Hartmann, I. (2013) The Relational Syllogistic Revisited. Linguistic Issues in Language Technology, 9:10, pp. 1–35.

[21] van Rooij, R. (2010). Extending Syllogistic Reasoning. In Logic, Language and Meaning, pp. 124–132.

[22] Schang, F., Lemanski, J. (2021) A Bitstring Semantics for Calculus CL. In Vandoulakis, I., Beziau, J.-Y. (Eds.): Studies in Universal Logic. Birkhäuser, Basel.

[23] Sommers, F., Englebretsen, G. (2000) An Invitation to Formal Reasoning: The Logic of Terms. Ashgate, Aldershot et al.

Received 25 February 2020

Assertions of Counterfactuals and Epistemic Irresponsibility

Vittorio Morato*

*Department of Philosophy, Sociology, Education and Applied Psychology
University of Padua, Italy*
vittorio.morato@unipd.it

Abstract

So-called reverse Sobel sequences seem to pose a problem for the variably strict semantics for counterfactuals. The existence of such sequences is taken by some scholars to be the main evidence in favour of an alternative, dynamic approach to the semantics of counterfactuals. According to Moss, however, a pragmatic approach to RSSs would be able to save the standard account. Central to her proposal there is a principle called 'principle of epistemic irresponsibility', according to which it is irresponsible to assert a counterfactual which contradicts a possibility that has become salient in the discourse. While agreeing on a pragmatic take on RSSs, in this paper I discuss the EI principle and highlight some problems of it.

Keywords: Counterfactual conditionals, assertion, dynamic semantics, salience

1 Introduction

This paper is about a problem concerning the assertion of counterfactual conditionals. In particular, it is about a problem posed to the standard variably strict semantics developed in [7] and [3] by so-called *reverse Sobel sequences* (RSSs). The existence of such sequences is taken by some scholars to be the main evidence in favour of an alternative *semantic* approach to counterfactuals, such as the dynamic strict conditional approach developed in [8] and [1].

*I would like to thank Massimiliano Carrara, Daniele Chiffi and Ciro De Florio for their patience, an anonymous referee of the *Journal of Applied Logics* for valuable comments and audiences at the Lecce conference on assertion and proof (2019) and at the Padua conference on speech acts (2019) for their feedbacks and suggestions.

According to [5], however, a pragmatic approach to RSSs would be able to save the standard account. Central to her proposal there is a principle called 'principle of epistemic irresponsibility', according to which it is irresponsible to assert a counterfactual which contradicts a possibility that has become salient in the discourse. By means of this principle, the infelicity of RSS can be explained without any need to abandon the standard semantics. A pragmatic approach to RSSs is preferable to a semantic approach and Moss is right when she claims that some linguistic evidences go against a semantic approach to RSS. However, the principle of epistemic irresponsibility on which Moss's account is based is problematic and in this paper, I am going to explain why.

2   Sobel and sequences: Strict and variably strict analyses

The variable strict approach to counterfactuals is reputed for being able to explain the felicity of so-called *Sobel sequences* (SSs). Consider the following two sequences of counterfactuals:

(1)   (a)   If the USA had thrown their nuclear weapons into the sea, there would have been war.

     (b)   But if the USA and all other nuclear powers had thrown their weapons into the sea, there would have been peace.

(2)   (a)   If Sophie had gone to the parade, she would have seen Pedro.

     (b)   But if Sophie had gone to the parade and been stuck behind a tall person, she would not have seen Pedro.

Intuitively, given some plausible background assumptions (for example, that Sophie does not push away people in front of her or that aliens make war to humans after the dismantling of nuclear weapons), asserting these two pairs of counterfactuals seems felicitous.

The way in which the variably strict account explains the felicity of asserting such sequences is *semantical*. 2 and 1 are felicitous, because they are consistent, both sets – 1a and 1b and 2a and 2b – are true.

Consider the logical form of an SS (i.e., the logical form of each constituent of an SS):

(3)   (a) $\phi > \psi$

(b) $\phi \wedge \chi > \neg\psi$.

To claim that 3a and 3b are consistent is to claim that $>$ is a conditional for which 'strenghtening the antecedent' does *not* hold: from $\phi > \psi$ it does not necessarily follow $\phi \wedge \chi > \psi$ for arbitrary χ.

Stalnaker-Lewis semantics for counterfactuals is able to explain how strengthening the antecedent could fail for a conditional. A variably strict conditional, $\phi \boxrightarrow \psi$, quantifies over all ϕ-worlds *most similar* to the world of evaluation and claims that, in those worlds, the consequent is true. But the $\phi \wedge \chi$-worlds most similar to the actual world are not necessarily all part of the most similar ϕ-worlds. Some of them may be less similar to the actual world than $\phi \wedge \neg\chi$ worlds. In such worlds, ψ could be false. So if $\phi > \psi$ is $\phi \boxrightarrow \psi$, the constituents of 3 can be true together. The felicity of 1 and 2 is thus semantically explained.

The felicity of SSs cannot be explained, at least through consistency, by those analyses of conditionals for which strengthening the antecedent holds. A strict conditional analysis of counterfactuals is a case in point. A (non-variably) strict conditional quantifies over all worlds where the antecedent ϕ is true and claims that, in those worlds, the consequent is true. But, if there are some worlds where $\phi \wedge \chi$ is true, such worlds form a subset of the set formed by the worlds where ϕ is true, so if ψ is true in the latter, it cannot be false in the former. So, if $\phi > \psi$ is $\Box(\phi \rightarrow \psi)$, the constituents of 3 cannot be true together; they are *inconsistent*. If SSs forms inconsistent pairs, then the felicity of asserting 1 and 2 cannot be explained semantically in this account.

Both the success and the failure to explain the felicity of both 1 and 2 rest on the following principles that relate the semantic dimension of consistency/inconsistency to the pragmatic dimension of felicity/infelicity of an assertion:

(Cons → Fel) If ϕ and ψ are consistent, ψ can felicitously be asserted after ϕ.

(Incons → Infel) If ϕ and ψ are inconsistent, then ψ cannot felicitously be asserted after ψ.

The idea behind the first principle is that, given an assertion of ϕ, consistency with ϕ is a condition for felicitous assertability after ϕ in a discourse. Of course, not everything consistent with ϕ should be asserted, but consistency seems to be a sufficient condition for the *possibility* of assertion.[1]

---

[1]The principle should not be confused with a principle that claims that consistency is sufficient for felicitous assertion. **Cons → Fel** claims that, if ϕ and ψ are consistent, then somone is allowed to – in the sense that one can – assert them together felicitously. In effect, the principle states that consistency is a sufficient condition for asserta-*bility*, not for assertion.

The second principle is stronger: it claims that after an assertion of ϕ, nothing that is inconsistent with it should be asserted.

These principles commit the two views under discussion to a general position with respect to SSs.

(Cons → Fel) commits the variably strict analysis to the view that SS can be felicitous. From this it follows that the account does not exclude the existence of *infelicitous* SSs. Consider, for example, the following:

(4) (a) If Sophie had gone to the parade, she would have seen Pedro.

    (b) # But if Sophie had gone to the parade and she had worn a red T-shirt, she would not have seen Pedro.

Given some obvious background assumptions (i.e., that wearing red t-shirts does not make it difficult to see people at parades), the variably strict analysis is perfectly able to explain the infelicity of such a sequence.

(Incons → Infel) commits instead the (non-variably) strict conditional analysis to the view that SS can never be felicitous.

3   Reverse Sobel sequences

The problem for the variably strict analysis comes in the case we reverse the order of assertion in 1 and 2 to obtain 5 and 6:

(5) (a) If the USA and all the nuclear powers had thrown their weapons into the sea, there would have been peace.

    (b) # But if the USA had thrown their nuclear weapons into the sea, there would be war.

(6) (a) If Sophie had gone to the parade and been stuck behind a tall person, she would not have seen Pedro.

    (b) # But if Sophie had gone to the parade, she would have seen Pedro.

These are *reverse Sobel sequences*, and in both cases they seem infelicitous to assert.

The explanation of the felicity of 1 and 2 through consistency is not able to explain the infelicity of 5 and 6. Consistency is in fact a symmetric relation, so if $\phi > \psi$ is consistent with $\phi \wedge \chi > \neg\psi$, also the opposite is true. In the absence of any other semantic or pragmatic mechanism, the variably strict account predicts thus that 5 and 6 should be felicitous, contrary to linguistic intuitions.

Notice that, in view of **(Cons → Fel)**, the variably strict account is committed to the general view that RSS can be felicitous. This does not exclude the capacity of the variably strict account to explain cases of infelicitous RSSs. The problem, however, is local: the same pair of counterfactuals that are felicitous in the guise of an SS are infelicitous in the guise of an RSS.

The opposite is true for the variably strict account. Also inconsistency is a symmetric relation. So, in the absence of any other semantic or pragmatic mechanism, the strict account rightly predicts that 5 and 6 are infelicitous.

Notice further that, given **(Incons → Infel)**, the strict conditional account is committed to the general view that RSS cannot be felicitous.

With respect to 1, 5, 2, and 6, the situation is thus the following:

- The variably strict account accounts for the felicity of 2, and 1 and it does not account for the infelicity of 5, and 6

- The strict conditional account accounts for the infelicity of 5, and 6 and it does not account for the felicity of 1, and 2

As we have seen, however, according to the variably strict account, there can be felicitous and infelicitous SSs and RSSs, while, according to the strict conditional account, there cannot be felicitous SSs and RSSs.

With respect to our examples, the variably strict account needs some story to explain the infelicity of 5 and 6 (given that it treats 1 and 2 as felicitous), while the strict conditional account needs some story to explain the felicity of 1 and 2 (given that it treats 5 and 6 as infelicitous).

This seems a symmetrical situation, but if we look at the general commitments of the theories, some differences emerge: the variably strict account is compatible with the existence of felicitous and infelicitous RSSs and SSs, while the strict conditional account excludes the possibility of felicitous SSs *and* RSSs.

The real problem for the strict conditional analysis is that of explaining the felicity of SSs, not the infelicity of RSSs (even though we will discover that the way in which the strict conditional analysis explains the infelicity of RSSs is probably too strong).

4 The dynamic approach

von Fintel [8] and Gillies [1] have independently proposed a *dynamic* variant of the strict conditional analysis that apparently is able to explain both the felicity of SSs like 1 and 2 and the infelicity of their reversed versions, 5 and 6.

In their analysis, the assertion of a counterfactual in a discourse is able to update the context in which subsequent counterfactuals are evaluated. Updating a context means that, when a counterfactual is asserted in a context, at least some antecedent worlds (i.e., worlds where the antecedent is true) become accessible. After the context is updated, counterfactuals behave semantically like strict conditionals within this set of contextually determined worlds.

According to Gillies, the context is taken to be a *hyperdomain*, which is a class of domains nested by the subset relation and ordered by a similarity relation with the world of evaluation. Similarity thus plays a role even in these new strict conditional analyses. When a counterfactual $\phi > \psi$ is asserted, the context gets updated in the sense that the hyperdomain includes now the closest ϕ-worlds. Given that the hyperdomain is a set of nested subdomains, now every domain in the hyperdomain contains at least a ϕ-world. The counterfactual behaves like a strict conditional with respect to this hyperdomain. This mechanism is able to explain the infelicity of 6: the prior assertion of 6a makes contextually available parade-worlds where Sophie is stuck behind someone tall and thus worlds where she does not see Pedro.[2] These worlds are more distant to the world of evaluation than worlds where Sophie goes to the parade and has a clear view. Given that domains are nested, now every domain contains at least a world where Sophie goes to the parade and is stuck behind someone tall. In these conditions, 6b is thus false. With the hyperdomain already enlarged to contain 'parade-and-stuck-behind-someone-tall' worlds, the assertion of 6, requiring that there are parade-worlds, has no further effect on the hyperdomain. But within such domains, 6b cannot be true, because not all worlds in the hyperdomain where Sophie goes to the parade are worlds where she sees Pedro. The reverse Sobel sequence is thus infelicitous, because 6b is false. The dynamic approach is thus able to explain the infelicity of the RSS and such infelicity of assertion is explained semantically by the falsity of 6b. We have seen however, that the infelicity of RSSs through the falsity of the second component of the sequence is something that the simple strict conditional approach was already capable to explain.

The real advantage of the new dynamic strict conditional approach is that it is now able to explain the felicity of SSs. When 2a is asserted, the worlds where Sophie goes to the parade and has a clear view become contextually accessible. By assumption these are the parade-worlds most similar to the world of evaluation. With respect to this hyperdomain 2a is thus true. When 2b is asserted, the hyperdomain is enlarged to include the less similar parade-worlds where Sophie is stuck behind someone tall. Given that the domains are nested this assures that in every

---

[2]The fact that Sophie does not see Pedro in worlds where she is stuck behind someone tall is an assumption: we are assuming that, in normal conditions, Sophie, when stuck behind someone tall, does not push this person away or always change position, etc.

domain of the hyperdomain there is now a parade-world where Sophie is stuck behind someone tall. In all these worlds, Sophie does not see Pedro, so in every nested domain of the hyperdomain, Sophie does not see Pedro. Thus 2b comes out true. Of course, there will be parade-and-stuck worlds where Sophie is able to see Pedro, but such worlds are less similar to the actual world than parade-and-stuck worlds where Sophie does not see Pedro, so they do not belong to any subdomain of the hyperdomain.

5 Felicitous RSSs

According to the dynamic strict conditional analysis, RSSs are always infelicitous: once the context is updated with the antecedent worlds of the first counterfactual, it cannot be 'shrunk' to accomodate the second counterfactual. A semantic approach to the infelicity of RSSs makes their infelicity a systematic feature of their assertion.

It seems, however, that there are felicitous RSSs, and furthermore it seems that infelicitous RSSs can be made felicitous. The existence of felicitous RSS is the main evidence [5] uses against the dynamic approach and in favour of her pragmatic approach.

Consider the following case. Assume that Anne is a friend of Sophie's and a very good friend of Pedro's and that every time Pedro sees Anna, he invites her, and whoever is with her, to drink something. Assume further that Pedro is an otherwise shy guy who usually does not invite people to drink, unless they are very good friends.

Given such a scenario, consider thus the following RSS:

(7) (a) If Sophie had gone to the parade with Anne, she would have gone to drink something with Pedro.

  (b) But if Sophie had gone to the parade, she would not have gone to drink something with Pedro.

This sequence is of the following form:

- $\phi \wedge \chi > \psi$

- $\phi > \neg\psi$

and it is a felicitous RSS. For the dynamic approach it should be counted as infelicitous. The assertion of the first counterfactual expands the context to worlds where Sophie goes to the parade with Anne and in such worlds, given the assumptions, Pedro would invite her, and Anne to drink. At this point, there is no way to assert 7b in a felicitous way, because it has no chance of being true. So the dynamic approach counts 7 as infelicitous because 7b is false (if 7a is true).

One of the cases mentioned in [5, p. 575] is the following:

(8) (a) If Sophie had gone to the parade and been shorter than she actually is, she would not have seen Pedro.

   (b) But if Sophie had gone to the parade, she would have seen Pedro.

Even in such a case the RSS seems felicitous, and this is perfectly compatible with the variably strict conditional analysis (in this case, the corresponding SS is also felicitous).

Not only are there felicitous RSSs, but at least some RSSs can be made felicitous. Consider for example this modified version of 5:[3]

(9) (a) If all the nuclear powers had thrown their weapons into the sea, there would have been peace.

   (b) But if *only* the USA had thrown their weapons into the sea, there would have been war.

In such a case, the application of an 'exhaustive' operator *only* is able to rescue the sequence and make it felicitous. According to [5, p. 575] the felicity of a SS could be rescued even by adding up some material aimed to show that the possibility raised by the first component of the sequence is not really salient:

(10) (a) If Sophie had gone to the parade and been stuck behind a tall person, she would not have seen Pedro.

   (b) *But hey, listen upâ I am telling you:* if she had gone, she would have seen him.

All these linguistic data point to a pragmatic solution to the problem of RSSs, and as we have seen, the variably strict account seems better suited than the dynamic account to deal with it.

6 The principle of epistemic irresponsibility

According to [5, p. 158], the variably strict account can explain the infelicity of RSSs and, in particular, the infelicity of RSSs whose corresponding SSs are felicitous, by means of a pragmatic principle which she calls 'principle of epistemic irresponsibility.'

---

[3]The example and the phenomenon, in the context of a much larger and systematic approach to sequences of conditional and disjunctive sentences, is taken from [2].

Principle of epistemic irresponsibility, EI: It is epistemically irresponsible to utter sentence S in context C if there is some proposition ϕ and possibility μ such that when the speaker utters S:

- (i) S expresses ϕ in C
- (ii) ϕ is incompatible with μ
- (iii) μ is a salient possibility
- (iv) the speaker of S cannot rule out μ.

The EI principle is a general principle regulating assertability, and it is not intended to work only for counterfactual conditionals. In its generality, the principle claims that it is epistemically irresponsible to assert something if it is incompatible with a salient possibility that cannot be ruled out. The assumption is that, if an epistemically irresponsible assertion is effectively asserted, the result is an infelicitous assertion.

By means of this principle, it seems that the infelicity of RSSs can be explained. Take the case of Sophie, namely 6. According to Moss, after the assertion of the first counterfactual, the possibility that Sophie might be stuck behind someone tall, if she had gone to the parade, becomes salient. This possibility is incompatible with the proposition expressed by the second counterfactual, namely that if Sophie had gone to the parade, she would have seen Pedro. Furthermore, the speaker seems to be not in a condition, given the background assumptions, to rule out the possibility raised to salience expressed by the first counterfactual. In this situation, the principle implies that it is epistemically irresponsible to utter the second counterfactual, even though it is true (under the variably strict semantics). Thus the sequence is infelicitous.

Let us see how it works in the case of 5. As in the previous case, the first counterfactual raises to salience the possibility that the USA might have thrown their weapons into the sea, if all the other nuclear powers had thrown theirs. This possibility seems incompatible with what is expressed by the first and the second counterfactuals, given the background assumptions. In a situation where all powers throw their weapons, there would be peace, not war. Given that the possibility cannot be ruled out, it is epistemically irresponsible to assert 5b, and thus the sequence is infelicitous.

The principle is based on the assumption that the assertion of a counterfactual in a discourse makes a possibility salient. But what kind of possibility becomes salient after the assertion of a counterfactual? Given a counterfactual such as $\phi \wedge \chi > \psi$ (the first component of an RSS) one would expect that the possibility that becomes salient is simply $\phi \wedge \chi$ (so what becomes salient is really $\Diamond(\phi \wedge \chi)$). Indeed, it seems natural to assume that the assertion of a counterfactual raises to salience the possibility of its antecedent.

However, in Moss's approach the possibility that becomes salient is $\chi \diamondsuit\!\!\!\rightarrow \phi$. Choosing $\chi \diamondsuit\!\!\!\rightarrow \phi$ is not intuitive, but is very useful for the point Moss wants to make. Once $\chi \diamondsuit\!\!\!\rightarrow \phi$ is the salient possibility we can actually *prove*, within Lewis's logic of counterfactual VC, the incompatibility of $\chi \diamondsuit\!\!\!\rightarrow \phi$, $\phi \wedge \chi \,\square\!\!\!\rightarrow \psi$ and $\phi \,\square\!\!\!\rightarrow \psi$ (the second counterfactual of an RSS).[4] What is needed is just that $\diamondsuit\!\!\!\rightarrow$ be treated as the dual of $\square\!\!\!\rightarrow$ (as it is normally assumed in Lewis's approach) and that $\phi \wedge \chi \,\square\!\!\!\rightarrow \psi$ is a background assumption.

7 Problems for EI

In this section, I want to highlight some problems for the pragmatic approach developed by Moss. I will present three problems and comment on another one presented in [4].

The first problem is that, as anticipated in the previous section, it is not at all clear why it is $\phi \diamondsuit\!\!\!\rightarrow \chi$ that gets raised to salience. Moss suggests (p. 569) that often 'merely mentioning' the possibility of χ is sufficient to raise to salience $\phi \diamondsuit\!\!\!\rightarrow \chi$. But how is this supposed to work? Not many indications are given in the article, so one might ask: Is $\diamondsuit\chi$ the salient possibility that in turn makes $\phi \diamondsuit\!\!\!\rightarrow \chi$ salient? How exactly is $\diamondsuit\chi$ able to make salient $\phi \diamondsuit\!\!\!\rightarrow \chi$?

For a start, it is not clear why mentioning $\diamondsuit\chi$ in a discourse should make salient $\phi \diamondsuit\!\!\!\rightarrow \chi$. Take our examples. It is not at all clear why the mere mentioning that Sophie might be stuck behind someone tall should raise to salience also that she might be stuck behind someone tall *if she goes to the parade*, as it is not clear why the mere mentioning that all nuclear powers distinct from the USA might throw their weapons into the sea raises to salience the possibility that all other nuclear weapons might throw their weapons into the sea *if the USA throw their weapons*.

Maybe, $\diamondsuit\chi$ makes $\phi \diamondsuit\!\!\!\rightarrow \chi$ salient only if something of the form $\phi \wedge \chi \,\square\!\!\!\rightarrow \psi$ is previously asserted, but then it is not 'merely' mentioning the possibility of χ that makes $\phi \diamondsuit\!\!\!\rightarrow \chi$ salient. It is rather the combination of $\phi \wedge \chi \,\square\!\!\!\rightarrow \psi$ and $\diamondsuit\chi$ makes $\phi \diamondsuit\!\!\!\rightarrow \chi$.

But why? What are the 'laws' of salience that are relevant here? Is some proposition salient, if it is implied by other two salient propositions? Is salience closed under logical consequence? It seems difficult to think that salience works this way and in general Moss does not give any indications on how possibilities get salient in a context or what notion of salience is relevant.

---

[4]For the complete proof, see [5, p. 570].

The notion of 'salience' developed in cognitive-linguistic contexts ([6], for example) seems not applicable in such a case: one striking feature of salience, at least in these contexts, is that what becomes salient has to be *easy* to activate and with a minimal cognitive effort. When one asserts $\phi \wedge \chi \mathbin{\square\!\!\rightarrow} \psi$ is instead difficult to see how $\phi \mathbin{\diamondsuit\!\!\rightarrow} \chi$ could become, *easily and immediately*, the center of attention.

A further issue related to this problem is that, even granting that some possibility becomes salient after an assertion of $\phi \wedge \chi \mathbin{\square\!\!\rightarrow} \psi$, $\phi \mathbin{\diamondsuit\!\!\rightarrow} \chi$ does not seem just the right salient possibility. Note that, as far as we are try to understand what possibilities get raised to salience, felicitous and infelicitous RSS should be on a par. The mechanism of 'salience raising' should be independent on the felicity of the sequence. Consider thus a felicitous RSSs such as 8. If we follow Moss, the possibility that gets raised to salience when one asserts the first counterfactual ('If Sophie had gone to the parade and been shorter than she actually is, she would not have seen Pedro') should be 'Sophie might have been shorter, if she had gone to the parade,' but this seems wrong. It is definitely more plausible to think that the salient possibility, in this case, should be that Sophie might be shorter than she actually is, and in such a situation, she might have gone to the parade, something of the form $\diamondsuit(\phi \wedge \chi)$. But as we know, from $\diamondsuit(\phi \wedge \chi)$ it is not possible to derive the desired incompatibility between $\phi \wedge \chi \mathbin{\square\!\!\rightarrow} \psi$, $\phi \mathbin{\diamondsuit\!\!\rightarrow} \chi$ and $\phi \mathbin{\square\!\!\rightarrow} \psi$.

The second problem comes from an observation in [4]. Even conceding that the possibility raised to salience by $\phi \wedge \chi \mathbin{\square\!\!\rightarrow} \psi$ is of the form $\phi \mathbin{\diamondsuit\!\!\rightarrow} \chi$, the duality of $\mathbin{\diamondsuit\!\!\rightarrow}$ and $\mathbin{\square\!\!\rightarrow}$ is valid in Lewis's logic of counterfactuals, but not in Stalnaker's. So one can use the principle EI to prove the incompatibility among the three items *only if Lewis's semantics is adopted*. To avoid this, Moss claims that one could alternatively choose $\phi \mathbin{\square\!\!\rightarrow} \chi$ as the salient possibility; from this possibility, and the two counterfactuals of the RSS, a contradiction is now derivable in Stalnaker's system, but not in Lewis's one. According to K. Lewis, however, this is somewhat 'surprising', because it makes what is psychologically salient dependent on semantic theory. If we choose Lewis's logic, $\phi \mathbin{\diamondsuit\!\!\rightarrow} \chi$ gets raised to salience, while, if we choose Stalnaker's logic, $\phi \mathbin{\square\!\!\rightarrow} \chi$ gets raised to salience. Moss's approach is thus not sufficiently *theory-neutral*.

After this critical point, K. Lewis actually offers a way out to Moss. One could try to react by claiming that what gets raised to salience is the 'neutral' possibility that there is a $\phi \wedge \chi$-world *among the closest possible worlds*; this possibility can then be 'interpreted' either as $\phi \mathbin{\diamondsuit\!\!\rightarrow} \chi$ within Lewis's semantics and as $\phi \mathbin{\square\!\!\rightarrow} \chi$ in Stalnaker's semantics. The truth of $\diamondsuit\phi \wedge \chi$ in one of the closest possible worlds implies the truth of $\phi \mathbin{\diamondsuit\!\!\rightarrow} \chi$ in Lewis's logic and $\neg(\phi \mathbin{\square\!\!\rightarrow} \neg\chi)$ in Stalnaker's logic, according to Lewis.

But, again, why this salient possibility and not simply $\diamondsuit(\phi \wedge \chi)$? The constraint on similarity is probably intended to capture the idea that the possibility of $\phi \wedge \chi$ is salient. The thought is that the more a possibility is 'near' the actual world, the more is salient. The idea is that what becomes salient is the antecedent of the asserted counterfactual in the closest possible worlds.

But '$\phi \wedge \chi$ in the closest possible worlds' is a very strange thing to become salient. Why also an information about the similarity relation should be part of what becomes salient to a speaker? Should we attribute to speakers some explicit representation of the similarity relation? What becomes salient should be a plain proposition, not also a piece of information about the mechanics of the semantics we are using. In the end, claiming that '$\phi \wedge \chi$ in the closest possible worlds' becomes salient amount to the simpler claim that '$\phi \wedge \chi$' becomes salient.

There is then the further problem that the actual formulation of the EI principle has it that the incompatibility is among the salient possibility and the proposition expressed by an utterance, *not*, as it would be in this case, between *a consequence of the salient possibility* and the proposition expressed. If we first need to determine the consequences of what is salient, the application of the principle would become very difficult. In the case of RSS, for example, we would need first to calculate the consequence of '$\phi \wedge \chi$ in the closest possible worlds' (depending on one's favourite logic of counterfactuals) and then to calculate the incompatibility between such a consequence and the two counterfactuals of the RSS. And notice: all of this should happen before asserting the second counterfactual, that, in case the incompatibility obtained, should then be deemed as epistemically irresponsible to assert. I do not think that the capacity of performing such process is plausibly attributable to a competent speaker using counterfactuals. So I think that, contrary to what Lewis claims against her own point against EI, the problem of theory-neutrality is, in effect, a serious problem for EI.

The third problem is that, according to EI, the infelicity of an RSS is now basically grounded on the validity of a *proof*, because it is based on the capacity to derive a contradiction. This makes the infelicity 'systematic' in the sense that it is based on the logical forms of the counterfactuals involved and that of the salient possibility. In every case we have $\phi \wedge \chi \boxright \psi$ and $\phi \diamondright \chi$ and $\phi \boxright \psi$ a contradiction follows and thus infelicity of assertion.

Surely, the conditions under which these counterfactuals 'enter' the proof are pragmatic: salience in the case of $\phi \diamondright \chi$ and being part of the common ground in the case of $\phi \wedge \chi \boxright \psi$. But these conditions seem to be quite easily satisfied or not specified enough. On the one hand, Moss does not specify how $\phi \wedge \chi \boxright \psi$ comes to be part of the common ground; the mere assertion of it seems to be enough. On the other hand, as we have seen, the salience of $\phi \diamondright \chi$ could be obtained, for Moss, simply by making χ salient.

But if salience and common-groundedness are so easily obtained, then the incompatibility, being based on a logical proof, is also immediately obtained. Any RSS with the first counterfactual in the common ground is a sequence for which we can prove an incompatibility. There is only an 'event' that could block, at this point, the derivation of a contradiction: the capacity of the speaker of ruling out the possibility raised to salience.

But assume that the speaker is indeed able to rule out the possibility made salient by the first counterfactual. This would be a very strange situation: a situation where we have a counterfactual that is part of the common ground, but it is such that the possibility made salient by it is capable of being ruled out. One should rule out a possibility made salient by something one accepts as a common ground of the assertion.

But how something could enter the common ground in case the possibility that it makes salient is capable to being ruled out? It seems natural to expect that either a counterfactual becomes part of the common ground because the possibility that it makes salient is not capable of being ruled out or the counterfactual does not become part of the common ground because the possibility that it makes salient is capable of being ruled out.

So if the counterfactual enters the common ground, the incompatibility could be proved and from this the infelicity will follow. But then, just the acceptance of the first counterfactual is enough to make the RSS infelicitous. But then we have the very strange conclusion that the only way to make an RSS felicitous is not accepting the first counterfactual as part of the common ground. So, in effect, it does not seem possible to make RSS felicitous.

If this is true, there seems to be a tension in EI principle between condition (i) and condition (iv).

Another (related) point is the following. When we prove that an RSS is infelicitous we show that from $\phi \wedge \chi \boxright \psi$, $\phi \diamondsuitright \chi$, and $\phi \boxright \psi$ a contradiction follows. But then at least one of them has to be false. But consider the case of Sophie. In Lewis's approach, both 6a and 6b are true. To be false, then, is the salient possibility $\phi \diamondsuitright \chi$, namely it is false that there is some parade-world among the closest worlds to the actual world where Sophie gets stuck behind someone tall. But then, we are in a situation where a true counterfactual raises to salience a false possibility. This might not be *per se* wrong. Analogous phenomena exist: for example, a true proposition could have a false implicature so, one might argue, a true proposition might raise to salience a false proposition. The relation between of an expressed proposition with, respectively, an implicature and a salient proposition is different. The relation of the salient proposition with what is semantically expressed seems to be much stricter; as it is known, an implicature could have nothing to do with the proposition expressed and, in some cases, might be the only thing communicated in a discourse. What is salient seems to be intimately related to what is said, it depends on it and so it seems more reasonable to expect that it should also have the same truth-value or at least that the falsity of the salient proposition should have some effect on the epistemic status of the proposition expressed. In such a case, the functioning of the EI principle seems to be compromised.

8   Conclusions

In this paper, after reviewing some problems for a semantic approach, I have presented some problems for a pragmatic solution to the problem of RSSs.

As pointed out in [5] and [4], the dynamic semantics developed by Gillies and von Fintel does not work because it builds in the semantics of counterfactuals a mechanism of expansion of the domains whose existence is falsified by plain felicitous RSSs and infelicitous RSSs made felicitous.

The pragmatic solution developed by Moss is problematic because it is based on a principle, the EI principle, which is problematic. The principle is based on the idea that the assertion of a counterfactual makes salient some possibility, but it is not clear what possibility gets raised to salience and for what reasons. Furthermore, there seems to be some tensions between the main conceptual components of the principle, in particular between the idea that the explanation of the infelicity of an RSS is explained by proving an incompatibility and the notion of ruling out the salient possibility.

References

[1] Anthony S. Gillies. Counterfactual scorekeeping. *Linguistics and Philosophy*, 30:329–360, 2007.

[2] Michela Ippolito. Varieties of sobel sequences. *Linguistics and Philosophy*, OnlineFirst:1–39, 2019.

[3] David K. Lewis. *Counterfactuals*. Blackwell, Oxford, 1973.

[4] Karen S. Lewis. Counterfactual discourse in context. *Noûs*, 52:481–507, 2018.

[5] Sarah Moss. On the pragmatics of counterfactuals. *Noûs*, 46(3):561–586, 2012.

[6] Hans-Jörg Schmid. Entrenchment, salience, and basic levels. In D. Geeraerts and H. Cuyckens, editors, *The Oxford Handbook of Cognitive Linguistics*. Oxford University Press, 2010.

[7] R. Stalnaker. A theory of conditionals. In N. Rescher, editor, *Studies in Logical Theory*, pages 98–112. Blackwell, Oxford, 1968. Reprinted in [?, pp. 41–56].

[8] K. von Fintel. Counterfactuals in a dynamic context. In M. Kenstowicz, editor, *Ken Hale: a Life in Language*, chapter 3, pages 123–152. MIT Press, Cambridge MA, 2001.

Received 10 July 2020

A General Semantics for Logics of Affirmation and Negation

Fabien Schang*
Federal University of Goiás, Brazil
schangfabien@gmail.com

Abstract

A general framework for translating various logical systems is presented, including a set of partial unary operators of affirmation and negation. Despite its usual reading, affirmation is not redundant in any domain of values and whenever it does not behave like a full mapping. After depicting the process of partial functions, a number of logics are translated through a variety of affirmations and a unique pair of negations. This relies upon two preconditions: a deconstruction of truth-values as ordered and structured objects, unlike its mainstream presentation as a simple object; a redefinition of the Principle of Bivalence as a set of four independent properties, such that its definition does not equate with normality.

1 Introduction: Identifying Logics

The issue of how a logical system is to be characterized may receive a simple answer: by identifying the set of its theorems. Correspondingly, any two logics differ from each other whenever they do not include the same set of theorems. The point is, however, that any such set is infinite, and it is difficult to characterize anything in such a case. A way out is a comparative analysis of logics with respect to a reference or 'pattern': classical logic is supposed to be such one. Then non-classical logics are those logics that lack either of the characteristic theorems of classical logic.

Another trickier issue is the philosophical matter of meaning: Why do such logical systems differ with respect to the meaning of the 'same' logical constants, assuming that those are? Understanding such a difference is not only being able to explain where the disagreement comes from, but also being able to express this disagreement without favoring either of the logical points of view at stake. The disagreement may

---

*The author wants to thank the referee for their valuable comments.

be dubbed 'fundamental' or 'deep' otherwise, if the origin of the problem cannot even be expressed in common terms between the opposed speakers. A central assumption of the paper is that such a main disagreement between logical systems is about what 'truth' and 'falsity' mean according to these. One ensuing challenge is to find a way to illustrate such a discrepancy without favoring either of the particular readings of both truth-values, all the more that the relation between the latter need not be the same throughout the various logical systems. This semantic issue also leads to the underlying matter of translation between logical systems, in order to avoid the dead end of incommensurability.

How to translate some logic into another one, so that any 'source logic' may turn into a 'target logic' whilst making use of one common language to account for it? A number of works have been devoted to this topic of translation between logics.[1] There are cases in which one given logical system is translated into another one, so that the formal language of the latter becomes the reference. For instance, Béziau [4] claimed that "S5 is a paraconsistent logic, and so is first-order classical logic". In this example, the modal logic S5 becomes the formal system in terms of which the characteristic theorems of both paraconsistent and first-order classical logic may be translated. His account relied upon a *structural identity* between two kinds of formulas, namely: modal formulas of the form $\Box\varphi$, and quantified formulas of the form $\forall x\varphi$. Then Béziau showed that paraconsistent negation '$\neg_P$' is to be translated modally as $\neg\Box\varphi \leftrightarrow \Diamond\neg\varphi$ (in S5), just as Gödel [11] translated intuitionistic negation '$\neg_I$' modally as $\Box\neg\varphi \leftrightarrow \neg\Diamond\varphi$ (in S4).[2]

Now there can also be cases into one logic is translated into another one within a third common formal language that embraces both. Such a common pattern will be presented in the following to translate non-modal and modal logics; furthermore, logics will not be distinguished from each other according to the meaning of their respective negations but, rather, their respective *affirmations*.

The content of the paper is the following. In Section 2, the use of many-valued frames will be motivated. In Section 3, we argue that truth-values are structured objects and streamline their features into a general constructive process. In Section 4, the notion of bivalence is reassessed in the light of such constructive truth-values. In Section 5, a general logic of affirmation and negation is presented as a set of logical systems including distinctive affirmative operators. In Section 6, various logical systems are translated and some properties of them are proved into this general framework.

---

[1] See e.g. [5], [17].

[2] Note that the first author to show the translatability of paraconsistent logic into S5 was Jerzy Perzanowski, in his translation of [9]. I thank the referee for reminding this historical point.

2 Affirmation and negation

A deep discrepancy obtains about the meaning of truth-values, and this point will be of importance throughout the paper since the final semantic background crucially relies on a many-valued algebra. This discrepancy can be located around the status of a formula in logic, depending upon whether it should be treated broadly as a sentence or, much more narrowly, as a proposition in the Fregean sense of a thought that speakers may have in common. Therefore, clarifying this point is a precondition to the introduction of our semantic framework.

Take the sentence p, 'It is daylight'. What is the range of truth-values that can be assigned to p? From a classical point of view:

(T) p is *true* iff it is the case that it is daylight.
(F) p is *false* iff it is the case that it is not daylight.

Now two additional non-classical views may be supported in specific contexts of discourse, especially when the sentence is uttered at dawn or a dusk:

(B) p is *both true and false* iff it is the case that it is daylight and it is the case that it is not daylight.
(N) p is *neither true nor false* iff it is not the case that it is daylight and it is not the case that it is not daylight.

Two objections can be raised against the last two valuations. On the one hand, the sentence p has no truth-value because it is an incomplete *proposition*; rather, 'It is daylight at the time when this sentence is written out' is entitled to receive a truth-value. On the other hand, if both (B) and (N) may be said right when uttered at dawn or at dusk then p is true-and-false and neither-true-nor-false at once; such a valuation appears to be plainly absurd, however. Two replies can be given, as well. About the first point, the distinction between sentences and propositions relies upon the view that a truth-value is an *object* or, at the least, a standing property of propositions the speaker aims at. About the second point, (B) and (N) do make sense together if truth-value is taken to be a property relative to different *viewpoints*. Also, a side-effect of the putative objective truth is *deflationism*, according to which the predicate 'true' adds nothing substantial to the information afforded by a proposition. That is: the affirmative proposition 'φ' means the same as 'φ is *true*', and the negative proposition '$\neg\varphi$' means the same as 'φ is *false*'. If so, then how to account for formulas whose truth-value stands beyond the basic values of truth and falsehood?

One helpful reference to combine the two views of truth-value as an object and a predicate is von Wright's logic of truth [23]. There the author introduces truth-values either as unary operators X prefixed to sentences or as the usual operands X of these sentences, such that $v(\mathrm{X}\varphi) = X$. Moreover, von Wright argues that some imprecise or vague contexts of utterance (like the aforementioned one about daylight) may variably lead to the interpretations (B) and (N) of non-classical truth-values. Thus, the end of rainfall can be considered as a 'glutty' situation at which it both rains and does not rain or, on the contrary, as a 'gappy' situation at which it neither rains nor does not rain. Instead of favoring either of these explanations, von Wright proposes a general domain of values including all these four possibilities.

Let $\mathrm{T}\varphi$ for 'It is true that φ', $\mathrm{T}'\varphi$ for 'It is not false that φ', and $\mathrm{F}\varphi$ for 'It is false that φ'. This results in the following four-valued truth-table:

φ	$\mathrm{T}\varphi$	$\mathrm{T}'\varphi$	$\mathrm{F}\varphi$
B	T	F	T
T	T	T	F
F	F	F	T
N	F	T	F

Roughly speaking, the above truth-table assigns truth to any sentence that includes the truth-value X expressed by the unary operator X, so that $v(\mathrm{X}\varphi) = T$ whenever φ includes the value X and $v(\mathrm{X}\varphi) = F$ otherwise. For example, $v(\mathrm{F}\varphi) = T$ whenever $v(\varphi) = B$ because the glutty value B includes falsity, so that it is true to say that φ is true whenever φ is both-*true*-and-false; at the same time, $v(\mathrm{F}\varphi) = F$ whenever $v(\varphi) = N$ since the gappy value N does not include falsity, so that it is false to say that φ is false whenever φ is neither-true-*nor false*.

This binary treatment of logical values as including or not including truth or falsity is a way to restore 2-valuedness from a 4-valued domain, accordingly. Is von Wright's above logic an adequate semantics for explaining the concepts of truth and falsity, however? We want to develop it into a broader framework, so that the meaning of truth can be weakened or strengthened according to the context of discourse.

The following sections want to develop and extend the case of von Wright's logic of truth in the light of a *bilateralist* approach, according to which truth and falsity are independent values: truth need not entail the failure of falsity, and conversely. In a nutshell, the basic independence of falsity and untruth appears to be instructive and insightful for logical purposes. We can summarize our guideline through the following four statements: truth is made on a par with a number of different affirmations, whilst falsity *and untruth* refer to a number of different negations; the truth-value of falsity F is not equated with untruth $\overline{T}$; truth receives different interpretations, according to the speaker's accepted criteria of justification; different logics are to be explained in terms of different conditions of *correctness*, and these directly interfere with the ensuing domain of truth-values.

3 Structured values

A bias has been mentioned in the preceding section, leading to the rejection of non-classical truth-values. This bias can be caught under what Suszko called the 'Fregean Axiom' [22] and includes two clauses, namely: that the reference of a proposition is a truth-value; that a truth-value is either the True or the False. The latter clause seems to assume a third one, viz. that truth-values are simple objects. In other words, any truth-value is to be viewed as a simple item: either the True T or its failure $\overline{T}$, which is taken to mean the same as the False F. Thus only T and F are properly *logical* values in this view, whereas B or N would be superficial items corresponding to the range of *algebraic* values. In the following, we depart from this pair of statements by endorsing the use of algebraic or *structured* values.

Let us see how truth-values can be constructed and lead to structured items. The first constructive step is the set including a unique element, the singleton $\{T\}$, such that this set $\mathcal{V}_1$ includes $x = 1$ element that is the True. Thus, $\mathcal{V}_1 = \{T\}$. The second constructive step is the set $\mathcal{V}_2$ such that it includes $x = 2$ elements: T, and $\overline{T}$. In $\mathcal{V}_2$, $\overline{T}$ is a failure of value corresponding to the the empty set and can be called the False: $\overline{T} = F$, so that every sentence that is said non-true is thereby false in $\mathcal{V}_2$. And conversely, every sentence that is said non-false is thereby true in $\mathcal{V}_2$. Thus, $\mathcal{V}_2 = \{T, \overline{T}\} = \{T, F\}$. The third constructive step is the set of values $\mathcal{V}_4$ that includes $x = 4$ elements: $T, \overline{T} = F, T\overline{T} = B$, and $\overline{T\overline{T}} = N$. In $\mathcal{V}_4$, the empty set $\overline{T\overline{T}}$ is the Neither-True-Nor-False and this means that not every sentence that is not true is thereby false (and conversely): $\overline{T} \neq F$ in $\mathcal{V}_4$.

A generalization of this constructive process consists in augmenting the number of truth-values by relativizing the denotation of the empty set in an arbitrary domain of values $\mathcal{V}_{2^n}$. For one thing, any such set includes a number of 2^n structured objects x that are made of n simple elements $x_1, \ldots, x_n$ from the previous set $\mathcal{V}_{2^{n-1}}$. Then an empty set can be defined as the denial of all the n simple elements: $\overline{T}$ is the empty set of $\mathcal{V}_2$ including $n = 1$ simple element, $\overline{T\overline{T}}$ is the empty set of $\mathcal{V}_4$ including $n = 2$ simple elements, etc. Let us symbolize each such structured object $x \in \mathcal{V}_{2^n}$ by an integer, according to its order of constructive occurrence: 1 for T, 2 for $\overline{T} = F$, 3 for $T\overline{T} = B$, 4 for $\overline{T\overline{T}} = N$, etc. Then the set of *designated* values $\mathcal{D}$ is a set in which every structured object includes the simple element $x_1 = T$: $\mathcal{D} = \{1\}$ in $\mathcal{V}_2$, $\mathcal{D} = \{1, 3\}$ in $\mathcal{V}_4$, etc. Moreover, all the 3-valued systems $\{\mathcal{V}_3, \mathcal{D}\}$ constructed on the above pattern are subsets of $\mathcal{V}_4$. Three cases of 3-valued domains are given by the logical systems of Łukasiewicz's $Ł_3$ [16], Kleene's (strong) K_3^S [13], and Priest's P_3 [18]. These systems somehow differ from each other by their characteristic subsets of designated values: with $Ł_3$ and K_3, $\mathcal{V}_3^- = \{\{1,2,4\},\{1\}\}$ includes one single designated value, whilst the domain $\mathcal{V}_3^+ = \{\{1,2,3\},\{1,2\}\}$ includes two ones with P_3. Whatever the feature of any logical system may be, truth-values always occur hereby as *Boolean structured objects*. This does not mean that every sentence is every true or false but, rather, that every truth-value results from a combination of two simple values T or $\overline{T}$. For sake of simplicity, any truth-value will be symbolized thereafter as an arbitrary combination of basic Boolean values $x_i = 1$ and $\overline{x_i} = 0$. It results in an increasing set of truth-values such that $V_1 = \{1\}$, $V_2 = \{1,0\}$, $V_4 = \{11, 10, 01, 00\}$, etc. The rest of the paper will be entirely devoted to the latter domain V_4, due to its greater expressive power.

4   Bivalence

Another issue can be clarified in the light of the previous constructive view of truth-values, namely: what is to be meant by a 'classical' logic. In the commonsensical sense, a logical system is said 'classical' whenever it includes a 2-valued domain, V_2. But it can be shown that two-valuedness is neither a necessary nor a sufficient condition for a logical system to be classical. On the one hand, a 2-valued system may not be classical as, e.g., da Costa's hierarchy of logical systems C_1-C_ω [8]. On the other hand, a classical system may not be 2-valued as will be shown in the next framework of 4-valued systems $\mathbf{AR}_{4[O_i]}$. A safer way to characterize classicality is to say that it validates a closed set of characteristic theorems, irrespective of the cardinality of its interpretative domain. These theorems include the main following ones:

Law of Non-Contradiction (LNC)	$\models_{CL} \neg(\varphi \wedge \neg\varphi)$
Law of Excluded Middle (LEM)	$\models_{CL} (\varphi \vee \neg\varphi)$
Law of Double Negation (LDN$_1$)	$\models_{CL} \varphi \rightarrow \neg\neg\varphi$
(LDN$_2$)	$\models_{CL} \neg\neg\varphi \rightarrow \varphi$
De Morgan Laws (DML$_1$)	$\models_{CL} \neg(\varphi \wedge \psi) \rightarrow (\neg\varphi \vee \neg\varphi)$
(DML$_2$)	$\models_{CL} (\neg\varphi \vee \neg\varphi) \rightarrow \neg(\varphi \wedge \psi)$
(DML$_3$)	$\models_{CL} \neg(\varphi \vee \psi) \rightarrow (\neg\varphi \wedge \neg\varphi)$
(DML$_4$)	$\models_{CL} (\neg\varphi \wedge \neg\varphi) \rightarrow \neg(\varphi \vee \psi)$
Law of Explosion (LE)	$\varphi, \neg\varphi \models_{CL} \psi$
Modus Tollens (MT)	$\varphi \rightarrow \psi, \neg\psi \models_{CL} \neg\varphi$

Disjunctive Syllogism (DS) $\varphi \vee \psi, \neg\varphi \models_{CL} \psi$

Finally, classical logic is usually associated with the so-called *Principle of Bivalence* (LBV). Albeit basic, this principle is not so obvious as it may appear and can be formulated in various ways. For example, Kubishkina & Zaitsev [14] distinguished a strong and a weak version of bivalence. The strong version equates bivalence with the structured values $\{T,F\}$ in $\mathcal{V}_2$; it states that "each sentence takes as its value precisely one of two truth values: truth or falsehood". At the same time, the weak version equates bivalence with the Boolean basic values $\{1,0\}$ in $\mathcal{V}_{2^n}$; it states that "there are exactly two possible truth values of a sentence: truth and falsehood" [14, p. 501]. The former version is like a *referential* view of bivalence, in the sense that it deals with truth-values as the proper names 'True' and 'False'; the latter version is like a *descriptive* view of bivalence, in the sense that it deals with truth-values as the definitive descriptions 'what is true and not false' and 'what is not true and false'. In order to combine both versions into a common definition, (LBV) can be depicted as a set of four clauses such that, for any sentence φ:

Definition 1.1: Law of Bivalence
(i)   If φ is true, then φ is not false.
(ii)   If φ is false, then φ is not true.
(iii)   If φ is not true, then φ is false.
(iv)   If φ is not false, then φ is true.

The clauses (i)-(ii) refer to the clause of *exclusiveness*: every sentence must refer to at most one simple value among truth and falsehood, whilst (iii)-(iv) refer to the clause of *exhaustiveness*: every sentence must refer to at least one simple value x_i among truth and falsehood. This shows that (LBV) assumes that $\mathcal{V}_2$ is both the minimal and maximal domain of values to interpret any formal language, insofar as it must include at least and at most the two simple elements T and $\overline{T} = F$. The clause of exclusiveness also entails that no further domain of values can be accepted once any truth-value must remain a simple element. In $\mathcal{V}_2$, (i)-(ii) obviously hold by virtue of the ensuing equation $\overline{T} = F$; but it is not so anymore from the next constructive set $\mathcal{V}_4$ onwards, where $T \neq F$. And nevertheless, (LBV) can be applied outside $\mathcal{V}_2$ by restricting the admitted ways of *affirming* truth and falsity, with the help of the coming unary operators $[A_i]$. More generally, exclusiveness refers to any case in which the occurrence of one value x_i entails the failure of the other one x_j; and exhaustiveness refers to any case in which the failure of any value x_i entails the occurrence of the other one x_j.

Another way to characterize classical logic proceeds by satisfying another pair of metalogical properties that characterize *normal* logics, namely: *consistency* (Cons), and *completeness* (Comp). Thus, for any formal language $\mathcal{L}$ and any formula $\varphi \in \mathcal{L}$:

Definition 2: Consistency

$$\models \mathcal{L}\varphi \Rightarrow \not\models \mathcal{L}\neg\varphi$$

Definition 3: Completeness

$$\not\models \mathcal{L}\varphi \Rightarrow \models \mathcal{L}\neg\varphi$$

Any logic is to be said *normal* whenever is satisfies both (Cons) and (Comp), accordingly. *Paraconsistent* logics PcL do not satisfy (Cons) whilst satisfying (Comp), whereas *paracomplete* logics PmL satisfy (Cons) whilst not satisfying (Comp). Logics that do not satisfy either (Cons) or (Comp) are *paranormal* logics PnL. Finally, *bivalent* logics are those satisfying the above conditions (*i*)-(*iv*).

5   Logics of affirmation and negation

One aim of the present paper is to show that bivalence is neither a necessary nor a sufficient condition for a logical system to be said 'classical', for a logical system may be bivalent without being classical and conversely. Another aim is to present a number of unary operators that modify the conditions of affirmation in various ways and are able to restore the properties of consistency, completeness, or both (normality). Such a challenge has been already faced, with another set of formal tools at hand [7]. To account for the same point, the common semantic framework $\mathbf{AR}_{4[O_i]}$ is going to encompass a variety of logical systems and to show how these partly subscribe to (LBV). The formal language of $\mathbf{AR}_{4[O_i]}$ can be depicted in the usual Backus-Naur form:

$$\varphi ::= \quad [O_i]p \mid [O_i](\varphi \bullet \psi) \mid [O_i]\varphi \bullet [O_i]\psi \mid \neg_1[O_i]\varphi \mid \neg_2[O_i]\varphi$$

where φ is any complex formula of the form $p \bullet q$, $\bullet \in \{\wedge, \vee, \rightarrow\}$ being the set of the usual binary connectives and $[O_i] \in \{[N_i], [A_i]\}$ being a set of two unary connectives, viz. negative and affirmative operators. The above well-formed formulas show that unary operators $[O_i]$ must always be prefixed to sentences: no formula occurs without being affirmed or denied, that is, taken to be true or false or none, so that the expressions 'p' and 'φ' are ill-formed formulas in $\mathbf{AR}_{4[O_i]}$. Furthermore, two kinds of negation must be prefixed to formulas: the first negation $\neg_1$ applies to complex formulas, whereas the second negation $\neg_2$ applies to simple formulas. As $\mathbf{AR}_{4[O_i]}$ is a set of 4-valued logical systems, its domain of interpretation is the constructive set of truth-values $\mathcal{V}_4 = \{11, 10, 01, 00\}$ such that, for any formula φ of $\mathbf{AR}_{4[O_i]}$, $v_4(\varphi) \in \mathcal{V}_4$. Two of these values are *designated*: $\mathcal{D} = \{11, 10\}$, so that any formula of $\mathbf{AR}_{4[O_i]}$ is designated whenever it is told true. Consequence is defined hereby as a relation of truth-preservation such that, for any sets of formulas Γ and a formula Δ,

Definition 4: Logical consequence
For every $\varphi \in \Gamma$, $\Gamma \models_{\mathbf{AR}_{4[O_i]}} \psi \Leftrightarrow v_4(\psi) \in \mathcal{D}$ if $v_4(\varphi) \in \mathcal{D}$.

Let $v_4(\varphi) = \langle a, b \rangle$ and $v_4(\psi) = \langle c, d \rangle$ be the valuation functions of $\mathcal{V}_4$ assigning structured values $x = \langle x_1, x_2 \rangle$ to arbitrary formulas φ, ψ in $\mathbf{AR}_{4[O_i]}$. Then the logical constants can be defined accordingly, assuming the Boolean ordering relation $1 > 0$ between the simple elements x_1, x_2 of x:

Definition 5: Logical connectives
Conjunction
$v_4(\varphi \wedge \psi) = \langle min(a, c), max(b, d) \rangle$
Disjunction
$v_4(\varphi \vee \psi) = \langle max(a, c), min(b, d) \rangle$
Conditional
$v_4(\varphi \rightarrow \psi) = \langle max(b, c), min(a, d) \rangle$

$\wedge$	11	10	01	00
11	11	11	01	01
10	11	10	01	00
01	01	01	01	01
00	01	00	01	00

$\vee$	11	10	01	00
11	11	10	11	10
10	10	10	10	10
01	11	10	01	00
00	10	10	00	00

$\rightarrow$	11	10	01	00
11	11	10	11	10
10	11	10	01	00
01	10	10	10	10
00	10	10	00	00

These connectives match with the standard equivalence of $\varphi \rightarrow \psi$ and $\neg \varphi \vee \psi$.[3] The real innovative part of $\mathbf{AR}_{4[O_i]}$ concerns its unary operators $[O_i]$. Unlike most of the formalized systems, affirmation does not collapse with its sentential content and refers hereby to the attitude of agents with respect to it. In other words: every agent 'affirms' or 'negates' arbitrary sentences, insofar as these are not given a truth-value *per se* but are taken to be true or false by agents.

Affirmation and negation can be defined *intensionally*, as partial operators mapping on structured values $x = \langle x_i, x_j \rangle$. Thus, $T = 10$ means that a sentence is true

---

[3]Unlike the 4-valued logical system characterized in [20] by a non-standard 'defective' or strong conditional, $\mathbf{AR}_{4[O_i]}$ includes the standard conditional or 'material' implication. This is due to our central task of translating logical systems that always assume the paradoxes of material implication in their formal languages.

$(x_i = 1)$ and not false $(x_j = 0)$; $F = 01$ means that a sentence is not true $(x_i = 0)$ and false $(x_j = 1)$; $B = 11$ means that a sentence is true $(x_i = 1)$ and false $(x_j = 1)$; and $N = 00$ means that a sentence is not true $(x_i = 0)$ and not false $(x_j = 0)$.

It may seem queer to talk about sentences being both true and false, due to the usually ontological interpretation of truth-values as correlating with facts or not. However, our point is to use a common domain of values interpreted in terms of available evidence. Thus we will take truth and falsity to mean the occurrence of evidence for or against a sentence, respectively. From this evidential reading of truth-values, the unary operators of affirmation will essentially occur as affording various criteria of justification for sentences.

On the one hand, the operator of *affirmation* means that some given agent accepts or assigns some truth-value to a sentence φ according to its proper criteria for truth, in the sense that affirming a sentence is on a par with taking it *to be* so-and-so. The logical syntax of an affirmation is $[A_i]\varphi$, which can be read 'It is the case that φ is x_i'. This means that, following a deflationist reading, any occurrence of an arbitrary sentence φ means the same as 'φ is x_i'. The lower case subscript i refers to the plurality of affirmative operators; for there may be as much affirmative operators as different criteria of truth assumed by the agents, depending upon whether truth excludes falsity or whether failure of truth entails falsity according to them. A general definition of affirmation may be given for any single values x_i, x_j:

Definition 6.1: Affirmation
$[A]\varphi : x_i \mapsto \overline{x_j}$.

That is, affirming that φ is true (or false) entails that φ is not false (or true) in $\mathcal{V}_4$. It may seem paradoxical to define affirmation in that way, i.e., by introducing a metalinguistic negation $\overline{x_j}$ into its characterizing mapping. Now the point is that any meaningful operator does not admit every value and must exclude at least one of these in order to make sense. And unlike negative operators, affirmative operators are those which do not exclude the value on which they apply but another one. Thus, affirming that φ is true (or false) entails that φ is not false (or true) in $\mathcal{V}_4$.

On the other hand, the operator of *negation* consists in stating that some agent *rejects* or denies some truth-value to φ according to its proper criteria of truth; so by negating a sentence, the agent takes it *not to be* so-and-so. A general definition of negation may be given for any single values x_i, x_j:

Definition 6.2: Negation
$[N]\varphi : x_i \mapsto \overline{x_i}$.

That is, negating that φ is true (or false) entails that φ is not true (or false) in $\mathcal{V}_4$.

A first example of such functions relates to affirmation. Let $[A_1] \in [A_i]$ be an example of partial affirmation mapping from truth to unfalsity, and nothing else. That is:

$[A_1]\varphi : T \mapsto \overline{F}$
which a case of affirmation where $x_i = T$ and $x_j = F$.

Its characteristic truth-table makes a difference between two kinds of output, namely: those which are partly altered by the operators (in boldface), whereas all the outputs are altered by total operators.

φ	$[A_1]\varphi$
11	**10**
10	**11**
01	01
00	00

A second example has to do with negation. Let $[N_1] \in [N_i]$ be an example of partial negation mapping from truth to untruth, and nothing else. That is:

$[N_1]\varphi : T \mapsto \overline{T}$
which a case of negation where $x_i = T$.

Again, not all the outputs are altered in its characteristic truth-table since $[N_1]$ only maps from the unique simple value T onwards.

φ	$[N_1]\varphi$
11	**01**
10	**00**
01	01
00	00

As it has been said earlier, affirmation and negation are intensional operators: these are rules of mapping between domain of truth-values, and there may be several exemplifications or extensions of these mappings in $\mathbf{AR}_{4[O_i]}$. Such an intensional characterization is made possible by our central distinction between simple and complex truth-values, insofar as the mappings of unary operators $[O_i]$ do not range over the compounds like x, but, rather, on either of their components x_i, x_j. This brings more expressive power to the formal language and leads to a further distinction between two kinds of unary operators: *partial*, and *total*.

Definition 7.1: Partial operators
Partial operators map onto simple values of a domain $\mathcal{V}_{2^n} : x_{i/j} \mapsto x_{i/j}$.

Definition 7.2: Total operators
Total operators map onto structured values of a domain $\mathcal{V}_{2^n} : x \mapsto x$.

A third category of unary operator is a *combination* of the preceding two ones. This consists in a special mapping involving $[A]$ or $[N]$.

Definition 8: Combination
If $[O_i]\varphi : x_i \mapsto x_j$, then $[OO_i]\varphi : x_i \mapsto \overline{x_j}$.

Assuming that, for every simple value $x_i : \overline{\overline{x_i}} = x_i$, this means that
$$[AN_i]\varphi = [NA_i]\varphi : x_i \mapsto \overline{\overline{x_j}} = x_i \mapsto x_j$$
$$[AA_i]\varphi = [NN_i]\varphi : x_i \mapsto \overline{\overline{x_i}} = x_i \mapsto x_i$$

Turning back to bivalence, the four clauses (i)-(iv) of (LBV) can be rephrased as a product of affirmative operators. Thus

Definition 1.2: Affirmative Bivalence
$[A_1]\varphi : T \mapsto \overline{F}$
$[A_2]\varphi : F \mapsto \overline{T}$
$[A_3]\varphi : \overline{T} \mapsto F$
$[A_4]\varphi : \overline{F} \mapsto T$

Two main operations can be performed upon such partial operators, whose application is essential to the coming translation of several logical systems. A first operation is *product*, which helps to construct the general principle of bivalence.

Definition 9: Product
The *product* of partial operators $[O_i] \otimes [O_j]$ consists in an aggregation of operators such that, letting $[O_i] : a \mapsto b$ and $[O_j] : c \mapsto d$,

$$([O_i] \otimes [O_j])\varphi = a \mapsto b \otimes c \mapsto d.$$

To give an example, a product of the affirmative operators $[A_3]$ and $[A_4]$ consists in fulfilling the last two clauses of (PBV).

$$([A_3] \otimes [A_4])\varphi = \overline{T} \mapsto F \otimes \overline{F} \mapsto T$$

φ	$[A_3]\varphi$	$[A_4]\varphi$	$([A_3] \otimes [A_4])\varphi$
11	11	11	11
10	10	10	10
01	01	01	01
00	01	10	11

Product can be characterized by a number of algebraic properties such that, for any partial operators $[O_x], [O_y], [O_z]$:

Idempotence	$[O_i]\varphi \otimes [O_i]\varphi = [O_i]\varphi$
Commutativity	$[O_i]\varphi \otimes [O_j]\varphi = [O_j]\varphi \otimes [O_i]\varphi$
Associativity	$[O_i]\varphi \otimes ([O_j] \otimes [O_k])\varphi = ([O_i] \otimes ([O_j])\varphi \otimes [O_k]\varphi$
Distributivity	$[O_i]\varphi \otimes ([O_j] \otimes [O_k])\varphi = ([O_i] \otimes ([O_j])\varphi \otimes ([O_i] \otimes [O_k])\varphi$

Affirmative and negative operators can be constructed following the same pattern. For any domain of structured truth-values $\mathcal{V}_{2^n}$ including n basic elements, there is an amount of $(2^n)^2 - 1$ kinds of unary operators $[O_i]$. Thus, there is $(2^2)^2 - 1 = 15$ cases of operators $[A_i]$ and $[N_i]$ in $\mathcal{V}_4$. Here is a list of the corresponding 15 affirmative operators, where single mappings may be combined variously until the ultimate, total operator $[A_{15}]$ that proceeds as an exhaustive product of partial operators:

$$[A_1]\varphi : T \mapsto \overline{F}$$
$$[A_2]\varphi : F \mapsto \overline{T}$$
$$[A_3]\varphi : \overline{T} \mapsto F$$
$$[A_4]\varphi : \overline{F} \mapsto T$$

$$[A_5]\varphi = ([A_1] \otimes [A_2])\varphi = T \mapsto \overline{F} \otimes F \mapsto \overline{T}$$
$$[A_6]\varphi = ([A_1] \otimes [A_3])\varphi = T \mapsto \overline{F} \otimes \overline{T} \mapsto F$$
$$[A_7]\varphi = ([A_1] \otimes [A_4])\varphi = T \mapsto \overline{F} \otimes \overline{F} \mapsto T$$
$$[A_8]\varphi = ([A_2] \otimes [A_3])\varphi = F \mapsto \overline{T} \otimes \overline{T} \mapsto F$$
$$[A_9]\varphi = ([A_2] \otimes [A_4])\varphi = F \mapsto \overline{T} \otimes \overline{F} \mapsto T$$
$$[A_{10}]\varphi = ([A_3] \otimes [A_4])\varphi = \overline{T} \mapsto F \otimes \overline{F} \mapsto T$$

$$[A_{11}]\varphi = ([A_1] \otimes [A_2] \otimes [A_3])\varphi = T \mapsto \overline{F} \otimes F \mapsto \overline{T} \otimes \overline{T} \mapsto F$$

$$[A_{12}]\varphi = ([A_1] \otimes [A_2] \otimes [A_4])\varphi = T \mapsto \overline{F} \otimes F \mapsto \overline{T} \otimes \overline{F} \mapsto T$$
$$[A_{13}]\varphi = ([A_1] \otimes [A_3] \otimes [A_4])\varphi = T \mapsto \overline{F} \otimes \overline{T} \mapsto F \otimes \overline{F} \mapsto T$$
$$[A_{14}]\varphi = ([A_2] \otimes [A_3] \otimes [A_4])\varphi = F \mapsto \overline{T} \otimes \overline{T} \mapsto F \otimes \overline{F} \mapsto T$$

$$[A_{15}]\varphi = ([A_1] \otimes [A_2] \otimes [A_3] \otimes [A_4])\varphi = T \mapsto \overline{F} \otimes F \mapsto \overline{T} \otimes \overline{T} \mapsto F \otimes \overline{F} \mapsto T$$

The same ordering of partial operators can be obtained with negation $[N_i]$, turning the mapping pattern $x_i \mapsto \overline{x_j}$ into $x_i \mapsto \overline{x_i}$. This results in two sorts of 'degenerate' negations in $\mathbf{AR}_{4[O_i]}$, including more than the two single instances of degenerate operators $\top$ and $\bot$ in CL.

On the one hand, some of these are the *tautology*-forming operators such that, for any truth-value of $\mathcal{V}_4$,

$$v_4(\varphi) \notin \mathcal{D} \Rightarrow v_4([N_i]\varphi) \in \mathcal{D}$$

This is so with any operator including $[N_3]$ by mapping onto truth:

$$[N_3](\varphi) : \overline{T} \mapsto T$$

and such that no other valuation is undesignated. Therefore, any such operator of $\mathbf{AR}_{4[O_i]}$ is a tautology-forming operator: $[N_3], [N_8], [N_{10}], [N_{14}]$.

On the other hand, some of these are the *antilogy*-forming operators such that, for any truth-value of $\mathcal{V}_4$,

$$v_4(\varphi) \in \mathcal{D} \Rightarrow v_4([N_i]\varphi) \notin \mathcal{D}$$

This is so with any operator including $[N_1]$ by mapping onto untruth:

$$[N_1](\varphi) : T \mapsto \overline{T}$$

and such that no other valuation is designated. Therefore, any such operator of $\mathbf{AR}_{4[O_i]}$ is a antilogy-forming operator: $[N_1], [N_5], [N_7], [N_{12}]$.

All the remaining negative operators include both designated and undesignated valuations, accordingly: $[N_2], [N_4], [N_6], [N_9], [N_{11}], [N_{13}], [N_{15}]$.

The following wants to show that only two kinds of negation $[N_i]$ are relevant to translate a number of logical systems in $\mathcal{V}_4$. Unlike the main translation processes, we argue that the various logical systems do not differ from each other by their characteristic negations; these are the same in each of these systems. Rather, the difference between these relies on their characteristic affirmative operators in $\mathcal{V}_4$.

These two operators result from a combination of single operators $[O_i]$, and only one of these can be considered as a 'pure' negation in the sense of being one of the 15 single operators $[N_i]$.

The first main negation is *Boolean* negation $\neg_1$, such that the corresponding formula $\neg_1[A_i]\varphi$ can be read 'It is *not* the case, according to the agent's criteria for truth, that φ is x_i'. Following our deconstruction of unary operators into partial ones, it turns out that Boolean negation can be reconstructed as an operator that turns any designated value into an undesignated one and conversely. Interestingly, the variety of negative operators $[N_i]$ makes more than only one possible candidate for Boolean negation. This means that, given a translation function τ mapping from any formula into a formula of $\mathbf{AR}_{4[O_i]}$, we obtain the following counterparts of Boolean negation into V_4 and their characteristic truth-tables:

$$\tau(\neg_1[A_i]\varphi) = \{[N_{11}][A_i]\varphi, [N_{13}][A_i]\varphi, [N_{15}][A_i]\varphi\}$$

given that
$$[N_{11}]\varphi : ([N_1] \otimes [N_2] \otimes [N_3] = T \mapsto \overline{T} \otimes F \mapsto \overline{F} \otimes \overline{T} \mapsto T$$
$$[N_{13}]\varphi : ([N_1] \otimes [N_3] \otimes [N_4])\varphi = T \mapsto \overline{T} \otimes \overline{T} \mapsto T \otimes \overline{F} \mapsto F$$
$$[N_{15}]\varphi : ([N_1] \otimes [N_2] \otimes [N_3] \otimes [N_4])\varphi = T \mapsto \overline{T} \otimes F \mapsto \overline{F} \otimes \overline{T} \mapsto T \otimes \overline{F} \mapsto F$$

$[A_i]\varphi$	$[N_{11}][A_i]\varphi$	$[N_{13}][A_i]\varphi$	$[N_{15}][A_i]\varphi$
11	**00**	**01**	**00**
10	**00**	**01**	**01**
01	**10**	**11**	**10**
00	**10**	**11**	**11**

For sake of simplicity, however, such a main candidate for 'pure negation' will be equated with $[N_{15}]$ hereafter and may be applied to any affirmation of the sentential content φ.

The second main negation is *Morganian* negation $\neg_2$, whose logical form $\neg_2[A_i]\varphi$ may be read 'It is the case that, according to the agent's criteria for truth, φ is x_j'. This means that, by negating in that way, the agent affirms the second ordered value x_j when negating the first ordered value x_i. Unlike Boolean negation, Morganian negation is not a pure negation of the form $[N_i]$ insofar as it amounts to a kind of combination $[OO_i]$:

$$\tau(\neg_2[A_i]\varphi) = [AN_{15}][A_i]\varphi,$$

given that
$$[AN_{15}][A_i]\varphi : ([AN_1] \otimes [AN_2] \otimes [AN_3 \otimes [AN_4])\varphi = T \mapsto F \otimes F \mapsto T \otimes \overline{T} \mapsto \overline{F} \otimes \overline{F} \mapsto \overline{T}$$

$[A_i]\varphi$	$\neg_2[A_i]\varphi$
11	**11**
10	**01**
01	**10**
00	**00**

Once again, it is important to note that any sentence is affirmed in either way before being negated: agents always assume one's own criteria of truth by introducing any sentence, so that negation does not apply to the sentential content φ but, rather, to the basic formula $[A_i]\varphi$. In other words, agents negate according to one's own commitments into what truth and falsity mean. Thus, the usual syntactic distinction between 'external' and 'internal' negation does not make sense in $\mathbf{AR}_{4[O_i]}$ since formulas like $[A_i]\neg_i\varphi$ are ill-formed formulas in it. The scope of Morganian negation is important to modify the value of its argument $[A_i]\varphi$ in a specific way, as is shown in the following truth-table where the well-formed formula $\neg_2[A_6]\varphi$ does not yield the same valuations as the ill-formed formula $[A_6]\neg_2\varphi$.

φ	$[A_6]\varphi$	$\neg_2[A_6]\varphi$	$[A_6]\neg_2\varphi$
11	10	**01**	10
10	10	**01**	01
01	01	**10**	10
00	10	**01**	10

Note also that the failure of internal negation does not entail any lack of expressive power for $\mathbf{AR}_{4[O_i]}$; for as it will appear in the following (see **Theorem 8** and **Theorem 9**), the usual notion of *contrary negation* can be rephrased hereafter as a composition of Boolean negation and duality.

About the latter expression, the peculiar behavior of both negations is such that *duality* needs to be explained in another way than the usual relation between $[A_i]\varphi$ and $\neg[A_i]\neg\varphi$ in $\mathbf{AR}_{4[O_i]}$. Alternatively, duality can be redefined in terms of mappings between an arbitrary formula $[O_i]\varphi$ and its dual $d([O_i]\varphi)$ by *reverting* the value of the former's inputs and outputs. That is: for any simple values x_i, x_j,

Definition 10: Duality
If $[O_i]\varphi : x_i \mapsto x_j$ then $d([O_i]\varphi) : \overline{x_i} \mapsto \overline{x_j}$.

The above reformulation of duality accounts for its following usual properties, together with a list of duals and self-duals among the affirmative operators of $\mathbf{AR}_{4[O_i]}$:

$d(0) = 1$
$d(1) = 0$
$d(\neg_1[A_i]\varphi) = \neg_1(d([A_i]\varphi))$
$d([A_i]\varphi \wedge [A_i]\psi) = (d(A_i]\varphi) \vee d([A_i]\psi))$
$d([A_i]\varphi \vee [A_i]\psi) = (d(A_i]\varphi) \wedge d([A_i]\psi))$

$d([A_1]\varphi) = [A_3]\varphi, d([A_2]\varphi) = [A_4]\varphi, d([A_5]\varphi) = [A_{10}]\varphi, d([A_7]\varphi) = [A_8]\varphi$
$d([A_6]\varphi) = [A_6]\varphi, d([A_9]\varphi) = [A_9]\varphi, d([A_{15}]\varphi) = [A_{15}]\varphi$

An alternative way of defining duality [19] is by applying Boolean negation to the *reversed* ordered elements of a characteristic *matrix* for a given formula, $\mathfrak{m}([A_i]\varphi) = (a_1, b_1)(a_2, b_2)(a_3, b_3)(a_4, b_4)$, so that $d(\mathfrak{m}([A_i]\varphi)) = \neg_1(a_4, b_4)\neg_1(a_3, b_3)\neg_1(a_2, b_2)\neg_1(a_1, b_1)$.
Taking $[A_1]\varphi$ as an example, its characteristic truth-table $\mathfrak{m}$ is such that

$$(a_1, b_1) = 10, (a_2, b_2) = 10, (a_3, b_3) = 01, (a_4, b_4) = 00.$$

Hence

$$d(\mathfrak{m}([A_1]\varphi)) = \neg_1(00)\neg_1(01)\neg_1(10)\neg_1(10) = (11)(10)(01)(01) = \mathfrak{m}([A_3]\varphi).$$

A proof theory can also be devised for the set of logical systems $\mathbf{AR}_{4[O_i]}$, including a set of axioms (A1)-(A9) that hold for every interpretation of the affirmative operators $[A_i]$.

(A1) $[A_i]\varphi \rightarrow ([A_i]\psi \rightarrow [A_i]\varphi)$
(A2) $([A_i]\varphi \rightarrow [A_i]\psi) \rightarrow (([A_i]\psi \rightarrow ([A_i]\gamma) \rightarrow ([A_i]\psi \rightarrow [A_i]\gamma))$
(A3) $[A_i]\varphi \rightarrow ([A_i]\psi) \rightarrow ([A_i]\varphi \wedge [A_i]\psi))$
(A4.1) $([A_i]\varphi \wedge [A_i]\psi) \rightarrow [A_i]\varphi$
(A4.2) $([A_i]\varphi \wedge [A_i]\psi) \rightarrow [A_i]\psi$
(A5.1) $[A_i]\varphi \rightarrow ([A_i]\varphi \vee [A_i]\psi)$
(A5.2) $[A_i]\psi \rightarrow ([A_i]\varphi \vee [A_i]\psi)$
(A6) $([A_i]\varphi \rightarrow [A_i]\gamma) \rightarrow (([A_i]\psi \rightarrow [A_i]\gamma) \rightarrow (([A_i]\varphi \vee [A_i]\psi) \rightarrow [A_i]\gamma))$
(A7) $([A_i]\varphi \rightarrow [A_i]\psi) \rightarrow (([A_i]\varphi \rightarrow \neg_1[[A_i]\psi) \rightarrow \neg_1[A_i]\varphi))$
(A8) $[A_i]\varphi \rightarrow (\neg_1[A_i]\varphi \rightarrow [A_i]\psi)$

Once a semantics and a general proof theory are set up, a general proof of soundness and completeness should be available. Instead of affording such an important result, however, the main issue that remains is rather the semantic ambiguity of

609

what affirmation and negation mean in various logical systems. For this purpose, it is possible to tackle the main issue of translating the formulas of various logical systems into one and the same semantic background.

6 Logics in translation

We saw in Section 1 that there is a mainstream formalization for such usual statements as affirmation and negation, and that process matches with the deflationist theory of truth. Roughly speaking, any occurrence of a formula like φ is expected to mean 'φ is true'; and any occurrence of a formula like '$\neg\varphi$' is expected to mean 'φ is false'. Let FL be this usual formal language; then we want to show two main things. Firstly, there is a common logical form for every formula in $\mathbf{AR}_{4[O_i]}$. Secondly, the difference between various logical systems in $\mathbf{AR}_{4[O_i]}$ depends on the various interpretations of $[A_i]$ whilst negations $[N_i]$ remain the same.

A general translation of the aforementioned formulas (see section 4) will be used as a test for characterizing logical systems in which these formulas are valid or not. The main properties of τ can be displayed by a set of clauses for translating logical systems into $\mathbf{AR}_{4[O_i]}$. Thus for any translation of a given formula from a source logic FL onto $\mathbf{AR}_{4[O_i]}$:

$\tau(\varphi)_{FL} = [A_i]\varphi$
$\tau(\varphi \bullet \psi)_{FL} = [A_i]\varphi \bullet [A_i]\psi$, for any binary connective $\bullet = \{\wedge, \vee, \rightarrow\}$
$\tau(\neg\varphi)_{FL} = \neg_1[A_i]\varphi$ if φ is a compound formula,
$\qquad = \neg_2[A_i]\varphi$ if φ is a single formula.

The above ambiguity of sentential negation is due to the twofold meaning of it, i.e., as an affirmation of falsity with $\neg_2$ or a mere denial with $\neg_1$. Correspondingly, double negation is to be disentangled as a denial of affirming falsity. This can be implemented by the characteristic formulas of 'classical' logic, especially when these include more than one negation. These can be translated as a set of uniform formulas into $\mathbf{AR}_{4[O_i]}$:

$\tau(\text{LNC})_{FL} = \neg_1([A_i]\varphi \wedge \neg_2[A_i]\varphi)$
$\tau(\text{LEM})_{FL} = [A_i]\varphi \vee \neg_2[A_i]\varphi$
$\tau(\text{LDN}_1)_{FL} = [A_i]\varphi \rightarrow \neg_1\neg_2[A_i]\varphi$
$\tau(\text{LDN}_2)_{FL} = \neg_1\neg_2[A_i]\varphi \rightarrow [A_i]\varphi$
$\tau(\text{DML}_1)_{FL} = \neg_1([A_i\varphi \wedge [A_i]\psi) \rightarrow (\neg_2[A_i]\varphi \vee \neg_2[A_i]\varphi)$
$\tau(\text{DML}_2)_{FL} = (\neg_2[A_i]\varphi \vee \neg_2[A_i]\varphi) \rightarrow \neg_1([A_i]\varphi \wedge [A_i]\psi)$

$\tau(\text{DML}_3)_{FL} = \neg_1([A_i]\varphi \vee [A_i]\psi) \rightarrow (\neg_2[A_i]\varphi \wedge \neg_2[A_i]\varphi)$

$\tau(\text{DML}_4)_{FL} = (\neg_2[A_i]\varphi \wedge \neg_2[A_i]\varphi) \rightarrow \neg_1([A_i]\varphi \vee [A_i]\psi)$

$\tau(\text{LE})_{FL} = [A_i]\varphi, \neg_2[A_i]\varphi \models_{CL} [A_i]\psi$

$\tau(\text{MT})_{FL} = [A_i]\varphi \rightarrow [A_i]\psi, \neg_2[A_i]\psi \models_{CL} \neg_2[A_i]\varphi$

$\tau(\text{DS})_{FL} = [A_i]\varphi \vee [A_i]\psi, \neg_2[A_i]\varphi \models_{CL} [A_i]\psi$

Accordingly, a variety of such interpretations help to translate different logical systems and result in the following theorems.

Theorem 1. A translation of CL into $\mathbf{AR}_{4[O_i]}$ is such that
$\tau(v_4(\varphi))_{CL} = \{10, 01\}$, that is:
$\tau([A_i])_{CL} = \{[A_6], [A_7], [A_8], [A_9]\}$

φ	$[A_6]\varphi$	$[A_7]\varphi$	$[A_8]\varphi$	$[A_9]\varphi$
11	10	10	01	01
10	10	10	10	10
01	01	01	01	01
00	01	10	01	10

Proof. Classical logic CL is such that it validates all the formulas (LNC)-(DS). The reader may observe that all of these are validated in $\mathcal{V}_2 = \{10, 01\}$. Hence every operator $[A_i]$ mapping from $\mathcal{V}_4$ onto $\mathcal{V}_2$ is a translation of CL into $\mathbf{AR}_{4[O_i]}$ by proceeding as a *normality* operator [7], i.e., as a consistency- and completeness-forming operator. $\square$

'Classical' logic CL may be viewed in $\mathbf{AR}_{4[O_i]}$ as a normal and *semi-bivalent* system satisfying both (Cons) and (Comp). 'Semi-bivalence' means hereby that the above affirmative operators $[A_6], [A_7], [A_8]$ and $[A_9]$ include only one half of the four properties (i)-(iv) of (LBV). To be more precise, any proper version of CL is a logical system of $\mathbf{AR}_{4[O_i]}$ such that its characteristic affirmative operator $[A_i]$ includes either one property of consistency (i)-(ii) or one property of completeness (iii)-(iv). Such is the case with the above four affirmative operators, whose valuation processes have the common form $[A_i]\varphi : x_i \mapsto \overline{x_j} \otimes \overline{x_i} \mapsto x_j$.

Theorem 2. A translation of PmL into $\mathbf{AR}_{4[O_i]}$ is such that
$\tau(v_4(\varphi))_{PmL} = \{10, 00, 01\}$, that is:
$\tau([A_i])_{PmL} = \{[A_1], [A_2], [A_5], [A_{11}], [A_{12}]\}$

φ	$[A_1]\varphi$	$[A_2]\varphi$	$[A_5]\varphi$	$[A_{11}]\varphi$	$[A_{12}]\varphi$
11	**10**	**01**	**00**	**00**	**00**
10	**10**	**10**	**10**	**10**	**10**
01	**01**	**01**	**01**	**01**	**01**
00	**00**	**01**	**00**	**01**	**10**

Proof. Paracomplete logics PmL always invalidate the following formulas: (LEM), (LDN$_1$), (DML$_3$). The reader may observe that all of these are respectively validated and invalidated with $V_3^- = \{10, 01, 00\}$, in addition with (DML$_1$). Hence every operator $[A_i]$ mapping from V_4 onto V_3^- is a translation of PmL into $\mathbf{AR}_{4[O_i]}$ by proceeding as a consistency- and incompleteness-forming operator. $\qquad\square$

Paracompleteness in $\mathbf{AR}_{4[O_i]}$ includes either none of the two properties of completeness *(iii)*-*(iv)* or both properties of consistency *(i)*-*(ii)*. That is, each case of PmL includes an affirmative operator $[A_i]$ such that neither $[A_i]\varphi : \overline{x_i} \mapsto x_j$ nor $[A_i]\varphi : \overline{x_j} \mapsto x_i$, or $[A_i]\varphi : x_i \mapsto \overline{x_j} \otimes x_j \mapsto \overline{x_i}$. In other words, each logical system including both properties of consistency entails a domain with undetermined values. The above systems match with the truth-tables of PmL, since their characteristic affirmative operators yield only the three output values $T = 10$, $F = 01$, and $N = 00$. Thus, there cannot be 'glutty' situations therein. Besides that, their characteristic mappings also match with the paracomplete import of PmL by requiring that any false formula be non-true and any true formula be non-false, but not conversely: it does not follow from a formula being non-false or non-true that it should be respectively true or false, in accordance to the intuitionistic interpretation of truth as constructive proof [21].

Theorem 3. A translation of paraconsistent logic PcL into $\mathbf{AR}_{4[O_i]}$ is such that $\tau(v_4(\varphi))_{PcL} = \{11, 10, 01\}$, that is:
$\tau([A_i])_{PcL} = \{[A_3], [A_4], [A_{10}], [A_{13}], [A_{14}]\}$

φ	$[A_3]\varphi$	$[A_4]\varphi$	$[A_{10}]\varphi$	$[A_{13}]\varphi$	$[A_{14}]\varphi$
11	11	11	11	10	01
10	10	10	**10**	10	10
01	01	01	01	01	01
00	01	10	11	11	11

Proof. Paraconsistent logics PcL always invalidate the following formulas: (LNC), (LE), (RAA), (DS). The reader may observe that all of these are invalidated with $V_3^+ = \{11, 10, 01\}$, in addition with (LDN$_2$) and (DML$_4$). Hence every operator $[A_i]$

mapping from $\mathcal{V}_4$ onto $\mathcal{V}_3^+$ is a translation of PcL into $\mathbf{AR}_{4[O_i]}$ by proceeding as a completeness- and inconsistency-forming operator. $\square$

Paraconsistency in $\mathbf{AR}_{4[O_i]}$ includes either none of the two properties of consistency (i)-(ii) or both properties of completeness (iii)-(iv). That is, each case of PcL includes an affirmative operator $[A_i]$ such that neither $[A_i]\varphi : x_i \mapsto \overline{x_j}$ nor $[A_i]\varphi : x_j \mapsto \overline{x_i}$, or $[A_i]\varphi : \overline{x_i} \mapsto x_j \otimes \overline{x_j} \mapsto x_i$. In other words, each logical system including both properties of completeness entails a domain with overdetermined values. The above paraconsistent systems match with the truth-tables of PcL, insofar as their characteristic affirmative operator yield only the three truth-values $B = 11$, $T = 10$, and $F = 01$. This means that there are no 'gappy' situations therein.

Theorem 4. Each paracomplete logic PmL has its proper dual paraconsistent logic PcL in $\mathbf{AR}_{4[O_i]}$.

Proof. The affirmative operators of PmL include (a) neither of the properties of completeness, i.e., neither $[A_i]\varphi : \overline{x_i} \mapsto x_j$ nor $[A_i]\varphi : \overline{x_j} \mapsto x_i$, or (b) both properties of consistency: $[A_i]\varphi : x_i \mapsto \overline{x_j} \otimes x_j \mapsto \overline{x_i}$. The affirmative operators of PcL include (a') neither of the properties of consistency, i.e., neither $[A_i]\varphi : x_i \mapsto \overline{x_j}$ nor $[A_i]\varphi : x_j \mapsto \overline{x_i}$, or (b') both properties of completeness: $[A_i]\varphi : \overline{x_i} \mapsto x_j \otimes \overline{x_j} \mapsto x_i$. Hence every mapping $a \mapsto b$ of PmL has its proper dual $\overline{a} \mapsto \overline{b}$ of PcL whenever $(a) = d(a')$ and $(b) = d(b')$. $\square$

The corresponding duals of PmL and PcL are the following, respectively: $[A_1] = d([A_3]), [A_2] = d([A_4]), [A_5] = d([A_{10}])$.

The above dual relation means that the truth of one of its relata always entails the truth of the other one, by reference to the relation between the modal operators of necessity $\square$ and possibility $\diamond$. Hence some of the affirmative operators leading to PmL and PcL are duals whenever any case of truth of the former is also a case of truth in the latter –but the converse does not hold. PcL being a paraconsistent logic, then PmL can be said a *dual paraconsistent* logic [21]. Their partial mappings help to understand why PmL and PcL are duals by affording two opposed readings of the truth-values: as a conclusive proof, in the former; as an available evidence in the latter, such that the assignment of truth or falsity need not prevent from assigning falsity or truth as well.

Theorem 5. A translation of paranormal logic PnL into $\mathbf{AR}_{4[O_i]}$ is such that $\tau(v_4(\varphi))_{PnL} = \{11, 10, 01, 00\}$, that is:

$$\tau([A_i])_{PnL} = \{[A_{15}]\}$$

φ	$[A_{15}]\varphi$
11	**00**
10	**10**
01	**01**
00	**11**

Proof. The paranormal logic $\mathbf{AR}_{4[A_{15}]}$ is such that it invalidates all the formulas (PNC)-(DS), thus equating with Belnap's logical system FDE [1]. The reader may observe that all of these are invalidated with $\mathcal{V}_4 = \{11, 10, 01, 00\}$. Hence every operator $[A_i]$ mapping from $\mathcal{V}_4$ onto $\mathcal{V}_4$ is a translation of PnL into $\mathbf{AR}_{4[O_i]}$ by proceeding as an incompleteness- and inconsistency-forming operator. $\square$

It follows from these features that $\mathbf{AR}_{4[A_{15}]}$ is both a bivalent and paranormal system: it satisfies the four properties (i)-(iv) of (LBV) without satisfying any of (Comp) and (Cons). Therefore, bivalence and normality are opposed to each other into $\mathbf{AR}_{4[O_i]}$ whereas normality usually calls for bivalence in most logical systems. Furthermore, $\mathbf{AR}_{4[A_{15}]}$ can be depicted as an *anti-classical* logic, following the terminology adopted in Bensusan & Costa-Leite & De Souza [2]: for any formula φ, if $\models_{CL} \varphi$ then $\not\models_{PnL} \varphi$.

Theorem 6. A translation of TL into $\mathbf{AR}_{4[O_i]}$ is such that
$\tau(\mathrm{T}\varphi) = [A_6]\varphi$, that is:
$[A_6] = ([A_1] \otimes [A_3])\varphi : T \mapsto \overline{F} \otimes \overline{T} \mapsto F$
$\tau(\mathrm{T}'\varphi) = [A_9]\varphi$, that is:
$[A_9]\varphi = ([A_2] \otimes [A_4])\varphi : F \mapsto \overline{T} \otimes \overline{F} \mapsto T$
$\tau(\mathrm{F}\varphi) = \neg_2[A_9]\varphi.$

Proof. The above truth-table is equivalent to that of von Wright's logic TL [23], according to the following symbolic translation of truth-values from TL to $\mathbf{AR}_{4[O_i]}$ (see Section 1): $B = 11$; $T = 10$; $F = 01$; $N = 00$, where $\mathcal{D} = \{B, T\}$.

φ	$[A_6]\varphi$	$[A_9]\varphi$	$\neg_2[A_9]\varphi$
11	10	01	10
10	10	10	01
01	01	01	10
00	01	10	01

It follows from this that every theorem of TL is also a theorem of $\mathbf{AR}_{4[A_6]}$.

The translations of the above truth-operators T and T′ as $[A_6]$ and $[A_9]$, respectively, belong to the class of normalization operators in CL, as shown in **Theorem 1**. This means that there is more than one way to express truth in $\mathbf{AR}_{4[O_i]}$, including two further operators that may translate the normal modal logic S5 (see **Theorem 8**).

Theorem 7. The operators T and T′ of TL are not duals.

Proof. We know that, according to **Theorem 6**:
$\tau(\mathrm{T}\varphi) = [A_6]\varphi$
$\tau(\mathrm{T}'\varphi) = [A_9]\varphi$
$[A_6] = d[A_6]$.
Therefore $[A_6]\varphi \neq d([A_9]\varphi)$, and $[A_6]\varphi$ is self-dual since $[A_6]\varphi = d[A_6]\varphi$.

This result shows that von Wright's operators T and T′ are no counterparts of the modalities of necessity and possibility, contrary to the formal appearances. This is due to differences in the structural relations between the four truth-values of TL and the two modalities of modal logic. However, the following is going to show that modal logic can also be translated into $\mathbf{AR}_{4[O_i]}$.

Theorem 8. $\mathbf{AR}_{4[A_8]}$ is a translation of S5, such that:
$\tau(\varphi) = [NN_{15}]\varphi : T \mapsto T \otimes F \mapsto F \otimes \overline{T} \mapsto \overline{T} \otimes \overline{F} \mapsto \overline{F}$
$\tau(\Box\varphi) = [A_8]\varphi = ([A_2] \otimes [A_3])\varphi : (F \mapsto \overline{T}) \otimes (\overline{T} \mapsto F)$
$\tau(\Diamond\varphi) = [A_7]\varphi = ([A_1] \otimes [A_4])\varphi : (T \mapsto \overline{F}) \otimes (\overline{F} \mapsto T)$
$\tau(\Box\neg\varphi) = \tau(\neg\Diamond\varphi) = \neg_2[A_7]\varphi$
$\tau(\Diamond\neg\varphi) = \tau(\neg\Box\varphi) = \neg_2[A_8]\varphi$

φ	$[NN_{15}]\varphi$	$[A_7]\varphi$	$[A_8]\varphi$	$\neg_2[A_7]\varphi$	$\neg_2[A_8]\varphi$
11	11	10	01	01	10
10	10	10	10	01	01
01	01	01	01	10	10
00	00	10	01	01	10

Proof. Modal logic S5 can be characterized by the following logical relations:
(1) $\models_{S5} \Box\varphi \leftrightarrow \neg\Diamond\neg\varphi$
(2) $\models_{S5} \Box(\varphi \wedge \psi) \leftrightarrow (\Box\varphi \wedge \Box\psi)$
(3) $\models_{S5} \Box(\varphi \rightarrow \psi) \leftrightarrow (\Box\varphi \rightarrow \Box\psi)$
(4.1) $\Box(\varphi \vee \psi) \not\models_{S5} (\Box\varphi \vee \Box\psi)$ (4.2) $(\Box\varphi \vee \Box\psi) \models_{S5} \Box(\varphi \vee \psi)$

(5.1) $\Diamond(\varphi \wedge \psi) \models_{S5} (\Diamond\varphi \wedge \Diamond\psi)$ (5.2) $(\Diamond\varphi \wedge \Diamond\psi) \not\models \Diamond(\varphi \wedge \psi)$

(6.1) $\Box\varphi \models_{S5} \varphi$ (6.2) $\varphi \not\models_{S5} \Box\varphi$

(7.1) $\varphi \models_{S5} \Diamond\varphi$ (7.2) $\Diamond\varphi \not\models_{S5} \varphi$

(8) $\models_{S5} \Box\varphi \leftrightarrow \Box\Box\varphi$

(9) $\models_{S5} \Diamond\varphi \leftrightarrow \Diamond\Diamond\varphi$

(10) $\models_{S5} \Diamond\varphi \leftrightarrow \Box\Diamond\varphi$

Then it can be checked that each of the above valid and invalid formulas of S5 are so in $\mathbf{AR}_{4[A_8]}$, provided that the non-modal formulas φ be also translated by the redundant combined operator $[NN_{15}]$. Therefore, $\mathbf{AR}_{4[A_8]}$ is an appropriate translation of S5 and its dual operators $\{\Box, \Diamond\}$, given the dual relation between their affirmative counterparts $\{[A_8], [A_7]\}$. $\Box$

Two related notes are in order about the above theorem.

First, such a translation overcomes what Béziau [3] called 'Łukasiewicz's Paradox', according to which the formula (5.2) is valid in Łukasiewicz's modal system [15]. Thus, Łukasiewicz's failure did not entail that no many-valued system would be unable to characterize modal logic after all.

Nevertheless, Dugundji [10] has shown that no finitely many-valued matrix was able to characterize any modal system between Lewis' systems S1 and S5. Actually, some other theorems that lie outside the Lewis systems –including non-normal modal systems, might be also characterized by $\mathbf{AR}_{4[A_8]}$. Indeed, it can be shown that any formal language including at least k atomic formulas $\{p_1, \ldots, p_k\}$ results in an equivalence formula (D_k),

$$p_1 \leftrightarrow p_2 \vee \cdots \vee p_{k-1} \leftrightarrow p_k.$$

Such a formula can be easily validated in any 4-valued logical system of $\mathbf{AR}_{4[O_i]}$, thereby including the above system $\mathbf{AR}_{4[A_8]}$. Now given that such an equivalence formula does not belong to the characteristic theorems of S5, this means that $\mathbf{AR}_{4[A_8]}$ includes at least one more formula than the S5-theorems and cannot characterize it strictly speaking. At any rate, it can be said again that S5 is a semi-bivalent logic through this translation of $\Box$ and $\Diamond$ resorting to two dual properties of (LBV).

Theorem 9. The logical relations of *opposition* can be translated in $\mathbf{AR}_{4[O_i]}$ such that, for any affirmative operator $[A_i]$: its dual is its *subaltern* or *superaltern*; the Boolean negation of its dual is its *contrary* or *subcontrary*; its Boolean negation is its *contradictory*.

Proof. $[A_i]\varphi$ is the superaltern of $[A_j]\varphi$ iff the truth of $[A_i]\varphi$ entails the truth of $[A_j]\varphi$, that is, $v([A_i]\varphi) \in \mathcal{D} \Rightarrow v([A_j]\varphi) \in \mathcal{D}$; and $[A_i]\varphi$ is the subaltern of $[A_j]\varphi$ iff the untruth of $[A_j]\varphi$ entails the untruth of $[A_i]\varphi$, that is, $v([A_j]\varphi) \notin \mathcal{D} \Rightarrow v([A_i]\varphi) \notin \mathcal{D}$. Now duality is to be viewed into $\mathbf{AR}_{4[O]}$ as inverting mappings, so that $[A_i]\varphi : x_i \mapsto \overline{x_j}$ entails $d([A_i]\varphi) : \overline{x_i} \mapsto x_j$. Hence $v(d([A_i]\varphi)) \in \mathcal{D}$ whenever $v([A_i]\varphi) \in \mathcal{D}$, and $v(d([A_i]\varphi)) \notin \mathcal{D}$ whenever $v([A_j]\varphi) \notin \mathcal{D}$. Therefore, $[A_i]\varphi = d([A_i]\varphi)$ whenever $[A_i]\varphi$ is the superaltern of $[A_j]\psi$ or the subaltern of $[A_j]\varphi$.

$[A_i]\varphi$ is the contrary or subcontrary of $[A_j]\varphi$ iff both formulas cannot be true at once or false at once, that is, $v([A_i]\varphi) \in \mathcal{D}$ entails $v([A_j]\varphi) \notin \mathcal{D}$ or $v([A_i]\varphi) \notin \mathcal{D}$ entails $v([A_j]\varphi) \in \mathcal{D}$, respectively. Now it has been proved here above that duality is such that $v([A_i]\varphi) \in \mathcal{D} \Rightarrow v(d([A_i]\varphi)) \in \mathcal{D}$ and $v([A_i]\varphi) \notin \mathcal{D} \Rightarrow v(d([A_i]\varphi)) \notin \mathcal{D}$, and Boolean negation $\neg_1$ is such that $v([A_i]\varphi) \in \mathcal{D}$ iff $v([A_j]\varphi) \notin \mathcal{D}$. Hence $v([A_i]\varphi) \in \mathcal{D}$ entails $v(\neg_1(d([A_i]\varphi))) \notin \mathcal{D}$ and $v([A_i]\varphi) \notin \mathcal{D}$ entails $v(\neg_1(d([A_i]\varphi))) \in \mathcal{D}$. Therefore, $[A_i]\varphi = \neg_1(d([A_i]\varphi))$ whenever $[A_i]\varphi$ is the contrary or subcontrary of $[A_j]\varphi$.

$[A_i]\varphi$ and $[A_j]\varphi$ are contradictories iff $v([A_i]\varphi) = x \Leftrightarrow v([A_j]\varphi) \neq x$. Now Boolean negation $\neg_1$ is such that $v([A_i]\varphi) \in \mathcal{D}$ iff $v([A_j]\varphi) \notin \mathcal{D}$. Therefore $[A_i]\varphi$ and $[A_j]\varphi$ are contradictories iff $[A_i]\varphi = \neg_1[A_j]\varphi$. $\square$

Note that the central notion of duality restricts to only one case the values of contraries and subcontraries, although formulas may have more than one contrary or subcontrary. Likewise, the commutative behavior of duality and Boolean negation explains why contrariety and subcontrary are defined in the same way, and the same holds with subalternation and superalternation. Indeed, the Boolean negation of duality equates with the duality of Boolean negation in $\mathbf{AR}_{4[O_i]}$:

$$\neg_1(d([A_i]\varphi)) = d(\neg_1[A_i]\varphi)$$

The above kinds of logical relations between affirmative operators may be summarized into the usual square of opposition.

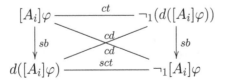

More precisely, let $\mathfrak{m}_{/2}([A_i]\varphi) = (a_1)(b_1)(c_1)(d_1)$ be *semi-matrices* including only the first ordered single values x_1 depicting the truth $x_1 = T$ or untruth $x_1 = \overline{T}$ of any formulas $[A_i]\varphi$ in a truth-table of $\mathcal{V}_4$. This yields the following set of common truth-valuations:

$$\mathfrak{m}_{/2}([A_2]\varphi) = \mathfrak{m}_{/2}([A_5]\varphi) = \mathfrak{m}_{/2}([A_8]\varphi) = \mathfrak{m}_{/2}([A_{11}]\varphi) = (0)(1)(0)(0);$$
$$\mathfrak{m}_{/2}([A_9]\varphi) = \mathfrak{m}_{/2}([A_{12}]\varphi) = \mathfrak{m}_{/2}([A_{14}]\varphi) = \mathfrak{m}_{/2}([A_{15}]\varphi) = (0)(1)(0)(1);$$
$$\mathfrak{m}_{/2}([A_1]\varphi) = \mathfrak{m}_{/2}([A_3]\varphi) = \mathfrak{m}_{/2}([A_6]\varphi) = (1)(1)(0)(0);$$
$$\mathfrak{m}_{/2}([A_4]\varphi) = \mathfrak{m}_{/2}([A_7]\varphi) = \mathfrak{m}_{/2}([A_{10}]\varphi) = \mathfrak{m}_{/2}([A_{13}]\varphi) = (1)(1)(0)(1).$$

Then the above relations of opposition obtain between the 15 affirmative operators of $\mathbf{AR}_{4[O_i]}$, where the arrows express the up-bottom relation of superalternation.

$$(0)(1)(0)(0)$$
$$[A_2]\varphi, [A_5]\varphi, [A_8]\varphi, [A_{11}]\varphi$$

$$\swarrow \qquad\qquad\qquad \searrow$$

$$(1)(1)(0)(0) \qquad\qquad\qquad (0)(1)(0)(1)$$
$$[A_1]\varphi, [A_3]\varphi, [A_6]\varphi \qquad\qquad \downarrow\ [A_9]\varphi, [A_{12}]\varphi, [A_{14}]\varphi, [A_{15}]\varphi$$

$$\searrow \qquad\qquad\qquad \swarrow$$

$$(1)(1)(0)(1)$$
$$[A_4]\varphi, [A_7]\varphi, [A_{10}]\varphi, [A_{13}]\varphi$$

The bottom-up relation of subalternation between affirmative operators $[A_i]$ and $[A_j]$ does not entail that $\mathbf{AR}_{4[A_i]}$ is a *sublogic* of $\mathbf{AR}_{4[A_j]}$ in the sense given by Bensusan & Costa-Leite & De Souza [2], correspondingly. Indeed, the relation of subalternation between such affirmative operators does not mean that every theorem of the former corresponding logical system of $\mathbf{AR}_{4[O_i]}$ is also a theorem of the latter, insofar as neither paracomplete nor paraconsistent systems are sublogics of either of them. Classical logic is a sublogic of both, but such a metalogical property differs from duality.

7   Conclusion: Semi-bivalent logics

Only some properties of (LBV) are required to translate logical systems into the general semantic framework $\mathbf{AR}_{4[O_i]}$. Not only classical logic CL, but also the modal logic S5 can be 'grasped' alongside the 'non-classical' logics, namely: paracomplete logics PmL, paraconsistent logics PcL, and paranormal logics PnL. It has been shown that the main difference between these systems relies upon their affirmative operators, whilst sentential negations are the same throughout. An open question is about whether how many other logical systems could be characterized in $\mathbf{AR}_{4[O_i]}$ in a similar way.

A connection has also made between this work and the well-known theory of oppositions. We have seen that the operators of truth and falsity are not contraries in von Wright's logic of truth TL, whereas the Boolean and Morganian negations are both main negations in any system of $\mathbf{AR}_{4[O_i]}$. Duality has also been shown to be a property connected with to the relation of subalternation.

Other similar features will be investigated in a later work in order to reconstruct the theory of opposition as a semantic set of (partial or total) mappings, in addition to non-Boolean semantics in which the constitutive elements $x_1, \ldots, x_n$ of structured values x do not merely map into $\{1,0\}$.[4] Such a kind of extension beyond (PBV) would also consist in going from $\mathbf{AR}_{2^2[O_i]}$ to $\mathbf{AR}_{m^n[O_i]}$, thus including more than $2^n = 4$ structured 'truth-values'. The whole result leads to a so-called 'Question-Answer Semantics', wherein 4-valuedness occurs as a particular case of truth-values based on $n = 2$ questions (whether φ is *true*, or whether it is *false*) and $m = 2$ answers (*yes*, and *no*).[5] Exploring new logical systems with more than $m = 2$ answers should constitute the real novelty of our semantic program, pending an account of their philosophical relevance.

References

[1] Belnap, N. A Useful Four-Valued Logic. In J. M. Dunn & G. Epstein (eds.), *Modern Uses of Multiple-Valued Logic.* Dordrecht Reidel Publishing Company, Dordrecht, 1977: 8-37.

[2] H. Bensusan & A. Costa-Leite & E. De Souza. Logic and their Galaxies. In: *The Road to Universal Logic*, Vol. II, Springer: 243-252.

[3] J.-Y. Béziau. A New Four-Valued Approach to Modal Logic. *Logique et Analyse.* 54(2011): 109-121.

[4] J.-Y. Béziau. S5 is a Paraconsistent Logic and so is First-Order Logic. *Logical Investigations*, 9(2002): 301-309.

[5] W. Carnielli. Possible-Translations Semantics for Paraconsistent Logics. In: *Frontiers in Paraconsistent Logic: Proceedings of the I World Congress on Paraconsistency*, Ghent, 1998, pp. 159-72, edited by D. Batens et al., Kings College Publications, 2000.

[6] A. Costa-Leite & F. Schang. Une sémantique générale des croyances justifiées. *CLE e-prints*, 16(3), 2016: 1-24.

---

[4]See F. Schang, "Quasi-concepts of logic", in *Abstract Consequence and Logics Ɖ Essays in Honor of Edelcio G. de Souza*, A. Costa-Leite (ed.), London: College Publications, 2020: 245–266.

[5]See F. Schang, "Question-Answer Semantics", *Revista de Filosofia Moderna e Contemporânea*, 8(1), 2020: 73–102.

[7] R. Cuini & M. Carrara. Normality Operators and Classical Collapse. In T. Arazim & P. Lavicka (ed.), *The Logica Yearbook 2017*. London, United Kingdom (2018): 2-20.

[8] N. C. da Costa, On the Theory of Inconsistent Formal Systems. *Notre Dame Journal of Formal Logic*, 15(1974): 497-510.

[9] S. Jaśkowski. On the Discussive Conjunction in the Propositional Calculus for Inconsistent Deductive Systems. *Logical and Logical Philosophy*, 7(1999): 57-59.

[10] J. Dugundji. Note on a Property of Matrices for Lewis and Langford's Calculi of Propositions. *The Journal of Symbolic Logic*, 5(1940): 150-151.

[11] K. Gödel. Eine Interpretation des intuitionistischen Aussagenkalküls. *Ergebnisse Math. Colloq.*, 4, 1933): 39-40.

[12] A. Kapsner. *Logics and Falsifications: A New Perspective on Constructivist Semantics*, Springer, Cham (2014).

[13] S. K. Kleene. On a Notation for Ordinal Numbers. *The Journal of Symbolic Logic*, 3(1938): 150-155.

[14] E. Kubishkina & D. Z. Zaitsev. Rational Agency from a Truth-Functional Perspective. *Logic and Logical Philosophy*, Vol. 25(2016): 499-520.

[15] J. Łukasiewicz. A System of Modal Logic. *The Journal of Computing Systems*, 1(1953): 111-149.

[16] J. Łukasiewicz. On Three-Valued Logic. In S. McCall (ed.). *Polish Logic 1920-1939*, Oxford: Clarendon, 1967: 16-18.

[17] H. J. Ohlbach. Semantics-Based Translation Methods for Modal Logics. *Journal of Logic and Computation*, 1(1991): 691-746.

[18] G. Priest. The Logic of Paradox. *Journal of Philosophical Logic*, 8(1979): 219-241.

[19] F. Schang. Epistemic Pluralism. *Logique et Analyse*, 60(2017) 337-353.

[20] F. Schang. A Four-Valued Logic of Strong Conditional". *South American Journal of Logic*, 3(2017): 59-86

[21] Y. Shramko. Dual Intuitionistic Logic and a Variety of Negations: The Logic of Scientific Research. *Studia Logica*, 80(2005): 347-367.

[22] R. Suszko. The Fregean Axiom and Polish Mathematical Logic in the 1920's. *Studia Logica*, 36(1977): 373-380.

[23] G. H. von Wright. Truth logics. *Logique et Analyse*, 30(1987): 311-334.

Received 22 November 2019